HVACR 301

AIR CONDITIONING CONTRACTORS OF AMERICA
PHCC EDUCATIONAL FOUNDATION
REFRIGERATION SERVICE ENGINEERS SOCIETY

 DELMAR
CENGAGE Learning

Australia • Brazil • Japan • Korea • Mexico • Singapore • Spain • United Kingdom • United States

DELMAR
CENGAGE Learning

HVACR 301
John Hohman

Vice President, Technology and Trades
Professional Business Unit: Gregory L.
Clayton

Director of Building Trades: Taryn Zlatin
McKenzie

Product Development Manager: Ed Francis

Product Manager: Vanessa L. Myers

Development: Kelly Henthorne

Editorial Assistant: Nobina Chakraborti

Director of Marketing: Beth A. Lutz

Marketing Manager: Marissa Maiella

Production Director: Carolyn Miller

Production Manager: Andrew Crouth

Senior Content Project Manager: Kara A.
DiCaterino

Art Director: Benjamin Gleeksman

For product information and technology assistance, contact us at
**Professional Group Cengage Learning Customer & Sales Support,
1-800-354-9706.**

For permission to use material from this text or product,
submit all requests online at **cengage.com/permissions.**

Further permissions questions can be e-mailed to
permissionrequest@cengage.com.

Library of Congress Control Number: 2009938195

ISBN-13: 978-1-4180-6666-6
ISBN-10: 1-4180-6666-4

Delmar
5 Maxwell Drive
Clifton Park, NY 12065-2919
USA

Cengage Learning is a leading provider of customized learning solutions with office locations
around the globe, including Singapore, the United Kingdom, Australia, Mexico, Brazil and Japan.
Locate your local office at **international.cengage.com/region.**

Cengage Learning products are represented in Canada by Nelson Education, Ltd.

Visit us at **www.InformationDestination.com.**

For more learning solutions, please visit our corporate website at **www.cengage.com.**

Notice to the Reader
Publisher does not warrant or guarantee any of the products described herein or perform any
independent analysis in connection with any of the product information contained herein. Publisher does not assume, and expressly disclaims, any obligation to obtain and include information
other than that provided to it by the manufacturer. The reader is expressly warned to consider
and adopt all safety precautions that might be indicated by the activities described herein and to
avoid all potential hazards. By following the instructions contained herein, the reader willingly
assumes all risks in connection with such instructions. The publisher makes no representations
or warranties of any kind, including but not limited to, the warranties of fitness for particular
purpose or merchantability, nor are any such representations implied with respect to the material
set forth herein, and the publisher takes no responsibility with respect to such material. The publisher shall not be liable for any special, consequential, or exemplary damages resulting, in whole
or part, from the readers' use of, or reliance upon, this material.

Printed at CLDPC, USA, 10-20

Brief Contents

Contents

Preface

This is the third book in a four-volume series of HVACR training textbooks, and it is intended for technicians in the third year of a four-year formal training program. It can also be used for continuing education classes or as a reference for professionals preparing for certification exams. The book is formatted in a way that allows for use both in the classroom and in training programs that involve a combination of field work and formal education. Apprentice or internship training programs can use this book to address the important topics and concepts required for real-world applications. This book emphasizes troubleshooting and problem solving for HVACR systems while introducing the trainee to new systems and concepts such as cooling towers, testing and balancing, water treatment, and indoor air quality.

Throughout the series, topics are revisited in more depth and with additional applications to coincide with students' increasing work experience in the field. The aim of *HVACR 301* is to build on the learning objectives students have already covered in *HVACR 201*. For example, Chapter 11 of *HVACR 201* introduces electrical troubleshooting. In *HVACR 301*, Chapter 5, "Applied Electrical Problem Solving," revisits troubleshooting methods and skills with more troubleshooting application. This type of relationship in the HVACR series helps students build knowledge while they are developing skill. All of the concepts are presented in a spiraling and connective way to reinforce previously learned information.

This book was developed in partnership with the following organizations: Air Conditioning Contractors of America (ACCA), Plumbing-Heating-Cooling-Contractors Educational Foundation (PHCC Educational Foundation), and Refrigeration Service Engineers Society (RSES). These organizations appointed a team of subject matter experts composed of outstanding HVACR professionals and educators, who reviewed the material during development to ensure the curriculum reflects industry needs and takes an approach to which today's technicians can relate. The subject matter experts guided the contents by developing the outline and approving each chapter. The writer was asked to follow and improve upon the suggestions of the subject matter expert group. Every effort was made to provide a text that is germane to the requirements of the HVACR industry and meets the needs of contractors, manufacturers, instructors, and trainees.

Acknowledgments

The publisher wishes to thank the following companies and individuals who provided illustrations and permission to reproduce them:

Aerotech Laboratories, Inc.
Air Conditioning Contractors of America
Bill Johnson
Bryant Heating and Cooling Systems
Emerson Electric Company
Eugene Silberstein
Honeywell, Inc.
Ice-O-Matic
John A. Tomczyk
NIBCO INC.
Nu-Calgon Wholesaler, Inc.
Nutech Energy Systems, Inc. (Life Breath)
Refrigeration Service Engineers Society
UEi

ACCA–PHCC Educational Foundation–RSES Subject Matter Experts

A special thank you is extended to Merry Beth Hall for her time and dedication as the HVACR 301 Project and Subject Matter Expert Committee Coordinator. Merry Beth is the Director of Apprentice & Journeyman Training with the PHCC Educational Foundation.

We would also like to recognize and thank the following individuals for their time, effort, and expertise:

Tim Gioe (RSES), Educational Publications Manager, HVACR 301 Project Coordinator

Merry Beth Hall (PHCC Educational Foundation), Director of Apprentice & Journeyman Training, HVACR 301 Project and Subject Matter Expert Committee Coordinator

Michael Honeycutt (ACCA), Educational Consultant, HVACR 301 Project Coordinator

Renee Tomlinson (RSES), Director, Training and Testing, HVACR 301 Project Coordinator

Hugh Cole (RSES), Certification and Training Services, Lawrenceville, GA

Greg Goater (ACCA), Isaac Heating & Air Conditioning, Inc., Rochester, NY

Terry Miller (ACCA), Energy Management Specialists, Cleveland, OH

Ron Newman (PHCC Educational Foundation), Lakes Plumbing, Heating & Cooling, Inc., Spirit Lake, IA

Donald Prather (ACCA), Technical Services Specialist, ACCA, Arlington, VA

Dick Shaw (ACCA), Technical Education Consultant & Standards Manager, Grand Rapids, MI

Jamie Simpson (PHCC Educational Foundation), Schaal Heating & Cooling, Des Moines, IA

About the Author

Dr. John E. Hohman has 40 years of experience in various education and industry fields. He has been an instructor and administrator at both the university and community college level. He is currently a consultant and has spent the last 10 years developing national certification, employee enhancement, apprenticeship programs, and assessments.

Dr. Hohman has worked in the industry as a technician, engineer, contractor, wholesaler, manufacturer representative, instructor, and educational administrator. He holds a patent in a refrigeration specialty: cascade refrigeration, which is part of environmental test conditioning. He has served in the capacity of consultant for his own business and other nationally known companies, professor of HVACR at Ferris State University and Mid Michigan Community College, Department Head of Technology at Mid Michigan Community College, Consultant and Trainer for Marshall Institute, Professional Development Coordinator at Central Michigan University and Ferris State University, and Consulting Engineer to Sexton ESPEC. Dr. Hohman is the State Director of the Michigan Society for Healthcare Engineers (MiSHE), the National Director of Mechanic Evaluation and Certification for Healthcare (MECH), a past Director of Research and Assessment Validation for HVAC Excellence, and a past Director of the National Occupational Competency Testing Institute (NOCTI). He received his associate's degree in refrigeration, heating, and air conditioning; his bachelor's degree in teacher education; and his master's degree in educational administration from Ferris State University. He received his doctorate in education with emphasis on evaluation systems from Capella University. John Hohman holds many other licenses and certifications in both education and technology.

CHAPTER 1

Servicing and Troubleshooting Systems

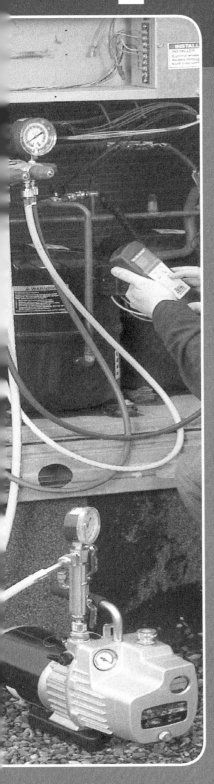

LEARNING OBJECTIVES

The student will:

- Compare the similarities and differences in checklists for maintenance and troubleshooting various types of HVAC equipment
- Describe the conversion of fuel-burning systems to use a different fuel
- Discuss the purpose of combustion testing
- Describe the general steps of a combustion test
- Describe water system checks
- Describe fresh-air intake requirements
- Describe cooling system checks
- Discuss alternative energy systems
- Describe solar water system checks

INTRODUCTION

Servicing and troubleshooting in HVAC requires that the technician be analytical and flexible in his or her thinking. Your servicing skills can be applied to many dissimilar pieces of equipment. Good technicians keep the following basic principles in mind while conducting servicing and troubleshooting.

- All energy needs to be converted to produce heat and motion (heating, cooling, and motor operation).
- All heat exchangers need to be clean and unobstructed to function efficiently and safely.
- Air and water that is used as a heating medium (heat transfer solution) must flow with little (or engineered) resistance through all of the equipment in the system.
- Controls must function properly to maintain comfort levels.
- All safety and limit controls must be properly adjusted and tested to ensure proper function.
- Systems must be tested, determined to operate efficiently, and documented.
- The customer must be satisfied.

These basic principles do not come in any particular order. Sometimes one may supersede all of the others, but all of them are necessary during each and every service call.

This chapter emphasizes checking HVAC systems as a process of service and maintenance. While on a service call or conducting scheduled maintenance, every system will have a set of checks that need to be performed. It is important that technicians follow and adhere to a company- or contractor-developed checklist and service procedures, which may incorporate some or all of the national standards (such as the ACCA 4 Maintenance of Residential HVAC Systems or ASHRAE/ACCA 180 Commercial Maintenance) that are recognized by ANSI. Manufacturers supply specific maintenance information with every piece of equipment, and this information describes the minimum that must be done. Always consult the manual for specific maintenance that is required. Included in this chapter are generic

Field Problem

During a maintenance call for a natural gas furnace, the technician noticed that there was substantial rusting around the area of the draft diverter for this atmospherically vented furnace. Because this system was new to him, he read the documentation from previous service on the unit as well as all the observations that had been recorded. There was no mention of motorized exhaust vent equipment in the home. The system had gone through a complete season of operation since the last maintenance work. Finishing the routine maintenance and checking the vent draft, the technician found nothing that could account for the rust that had appeared over the last heating season—but he had a suspicion.

After completing the work, the technician discussed his findings and the work that was done. Coming back to the rust observation, the technician asked if there were any changes to the home or any construction work performed over the last year. The customer responded that he had not done any remodeling. Probing a little more, the technician specifically asked about bathroom and kitchen exhaust fans. This time, the customer had a funny look on his face. He explained that he had purchased a range hood as a present for his wife and installed it himself. He wanted the best, so he got a commercial unit with a big, quiet fan. His wife liked it so much that she used it all of the time.

After asking the customer to turn on the range hood and then checking the draft, it was discovered that the fan was overcoming the natural draft venting system and so combustion by-products were venting back into the home. Further conversation with the customer indicated that his children had experienced a bad year of colds, flu, and headaches. The tech-

Field Problem (Continued)

nician explained that the range hood created a dangerous condition and that CO (carbon monoxide) poisoning may have occurred. Flue gases were also condensing on the furnace, which caused the rusting. He suggested that a fresh-air ventilation system be installed to prevent the range hood from creating a negative pressure in relation to the outside. The technician called the office to make an appointment for the company engineer to come review the situation and make specific recommendations to the customer.

checklists that can be used to ensure that major system components have been visually or physically reviewed for cleanliness and operation. Although each checklist is specific to a class of equipment, the lists are designed to mix and match depending on the system and equipment attached. For instance, an air source heat pump with electric strip backup could be inspected by using the checklist for heat pumps and the checklist for electric furnaces.

HEATING SYSTEMS: ELECTRIC FURNACE, GAS FURNACE, OIL FURNACE, HEAT PUMP

Every heating system needs service (maintenance) and at some point will probably need troubleshooting. From a service prospective, every system should be inspected and cleaned before being placed into service for the next heating season. At that time, specific checks should be made of each type of system. Documentation should be made of the initial and final operating characteristics. Customers should be apprised of the findings and of any options to maintain or upgrade the system.

We will take a look at the service and maintenance required for specific systems, from the least intensive to the most. Each system analysis will provide an overview of system operation followed by specific service and checks to make.

Electric Furnace

Electric furnaces (Figure 1–1) have electric resistance heat strips that are turned on in stages to meet specific heating needs. Each strip (or groups of strips) can be controlled by a multistage thermostat or, with a single-stage thermostat, be brought on in steps or stages by using delay relays or sequencers until all are turned on. Each strip of heat has an overload safety that monitors temperature and an overcurrent safety that monitors amperage. A blower is operated while heat strips are on in order to transfer heat from the heating strips to the airstream and to provide conditioned air to the living space. Electric furnaces usually can be configured to deliver air in an upflow, downflow, or horizontal pattern.

Because of the electric furnace's simplicity, fewer service calls are required than all other systems. This fact alone may lull owners into thinking that electric furnaces do not need maintenance. It is still important to check the system once per year before placing it into service for the heating season (see Checklist 1–1). *Note:* The checklists presented in this chapter come from the *ACCA Standard: Maintenance of Residential HVAC Systems.* It is suggested that the latest version of these standards be consulted to obtain similar standards for systems other than those used as examples in this chapter.

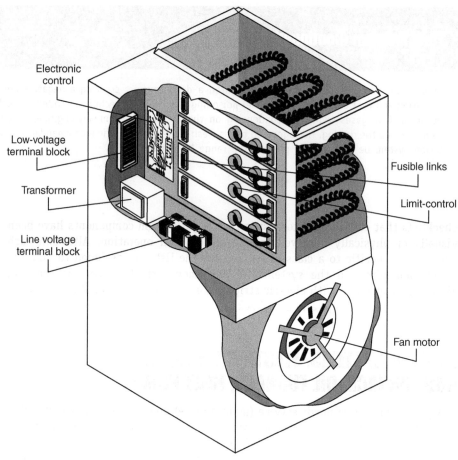

Electronic control

Low-voltage terminal block

Transformer

Line voltage terminal block

Fusible links

Limit-control

Fan motor

Figure 1–1
Electric forced air furnace.

Checklist 1–1 Electric Furnace (Courtesy of ACCA)

Inspection Task	Recommended Corrective Actions
Cabinet	
Inspect cabinet, cabinet fasteners, and cabinet panels.	Repair or replace insulation to ensure proper operation. Replace lost fasteners as needed to ensure integrity and fit/finish of equipment (as applicable). Seal air leaks.
Electrical	
Inspect electrical disconnect box.	Ensure electrical connections are clean and tight. Ensure fused disconnects use the proper fuse size and are not bypassed. Ensure case is intact and complete. Replace as necessary.
Ensure proper equipment grounding.	Tighten, correct, and repair as necessary.
Measure and record line voltage.	Compare to OEM (original equipment manufacturer) specifications or equipment nameplate data. Notify homeowner and/or utility.
Inspect and test contactors and relays.	Look for pitting or other signs of damage. Replace contactors and relays that show evidence of excessive contact arcing and pitting.

Checklist 1-1 Electric Furnace *(Continued)*

Inspection Task	Recommended Corrective Actions
Inspect electrical connections and wire.	Ensure wire size and type match the load conditions. Tighten all loose connections, replace heat-discolored connections, and repair or replace any damaged electrical wiring.
Inspect all stand-alone capacitors.	Replace those that are bulged, split, or incorrectly sized per OEM specifications.
Measure and record amperage draw to motor/nameplate data (FLA) (as available).	If outside OEM rating or specification, inspect for cause and repair as necessary.
Test electric heater's capacity sequence of operation.	If outside OEM rating or sequencer specification, inspect for cause and repair as necessary.
Blower Assembly	
Determine and record airflow across heating elements.	Adjust, clean, replace, and repair as necessary to ensure proper airflow.
Test variable-frequency drive for proper operation.	Replace if necessary to ensure proper operation.
Inspect fan belt tension. Inspect belt and pulleys for wear and tear.	Repair or replace as necessary to ensure proper operation (if applicable).
Confirm that the fan blade or blower wheel has a tight connection to the blower motor shaft. Inspect fan for free rotation and minimal end play. Measure and record amp draw.	Lubricate bearings as needed, but only if recommended by OEM. If amp draw exceeds OEM specifications, adjust motor speed or otherwise remedy the cause. Replace failed blower motor if necessary.

Tech Tip

There is a difference between taking an amperage reading for single-phase or three-phase systems when converting to kilowatts. Single-phase systems are easy: measured amperage multiplied by voltage equals wattage; then divide by 1,000 to obtain kilowatts. For example, 240 volts × 5 amps = 1,200 watts, and 1,200 watts ÷ 1,000 = 1.2 kW. With a three-phase electric heat strip, if each leg measures the same amperage (or is showing "balance"), then the calculation requires a three-phase factor of 1.73 in the calculation. Example: 240 volts × 2.9 amps × 1.73 factor = 1,200 watts (and again 1,200/1,000 = 1.2 kW).

Heat Pump

Heat pumps are normally sized for the cooling load, but they deliver both heating and cooling in one system. In the heating mode, the heat pump converts electrical energy into mechanical energy to concentrate and move heat from an outdoor location to an interior living space. Heat can be pulled from the outside air, from the ground, or from water. Indoor and outdoor heat exchangers switch function through the use of a reversing valve. There are two major categories of heat pumps, air-source and ground-source. Ground-source heat pumps use the ground or a water source to make an "earth" connection. Essentially, the earth connection (loop) absorbs or rejects heat directly or indirectly from the ground. "Directly" means that the coil is in direct contact with the ground, whereas "indirectly" means that water transfers heat from the ground to the coil. This category of heat pump has many ground coil configurations: horizontal loop, vertical loop, coiled loop (slinky), pumped, standing well, and direct coupled. An indoor blower delivers air over the indoor (air coil) heat exchanger, transferring energy from the coil to the living space. Heat pumps also have backup electric strip heat for use during colder weather conditions and emergency heat that looks and operates in the same way as an electric furnace. Many heat pumps can discharge the air as an upflow, downflow, or horizontal unit. Homes and businesses using a heat pump may be able to take advantage of special programs or rates from the electrical supplier.

Heat pumps have a few more components than electric furnaces, including those parts that are considered the same as electric furnaces. Because heat pumps operate in both the heating and cooling season, it is expected that they should be checked at the beginning of both of those seasons. As a standard procedure, follow the guidelines given in Checklist 1-2 (the example is for a geothermal heat pump).

Checklist 1–2 Geothermal (Courtesy of ACCA)

Inspection Task	Recommended Corrective Actions
Cabinet	
Inspect cabinet, cabinet fasteners, and cabinet panels.	Repair or replace insulation to ensure proper operation. Replace lost fasteners as needed to ensure integrity and fit/finish of equipment (as applicable). Seal air leaks.
Electrical	
Inspect electrical disconnect box.	Ensure electrical connections are clean and tight. Ensure fused disconnects use the proper fuse size and are not bypassed. Ensure case is intact and complete. Replace as necessary.
Ensure proper equipment grounding.	Tighten, correct, and repair as necessary.
Measure and record line voltage.	Compare to OEM specifications or equipment nameplate data. Notify homeowner and/or utility.
Inspect and test contactors and relays.	Look for pitting or other signs of damage. Replace contactors and relays that show evidence of excessive contact arcing and pitting.
Inspect electrical connections and wire.	Ensure wire size and type match the load conditions. Tighten all loose connections, replace heat-discolored connections, and repair or replace any damaged electrical wiring.
Inspect motor capacitors.	Replace those that are bulged, split, or incorrectly sized per OEM specifications.
Measure and record amperage draw to motor/ nameplate data (FLA) (as available).	If outside OEM rating or specification, inspect for cause and repair as necessary.

Checklist 1–2 Geothermal *(Continued)*

Inspection Task	Recommended Corrective Actions
Indoor Blower Motor	
Determine and record airflow across heat exchanger/coil.	Verify that all grilles and registers are open and free of obstruction. Adjust, clean, replace, and repair as necessary to ensure proper airflow.
Test variable-frequency drive for proper operation.	Replace if necessary to ensure proper operation.
Inspect fan belt tension. Inspect belt and pulleys for wear and tear.	Repair or replace as necessary to ensure proper operation (if applicable).
Confirm the fan blade or blower wheel has a tight connection to the blower motor shaft. Inspect fan for free rotation and minimal end play. Measure and record amp draw.	Lubricate bearings as needed, but only if recommended by OEM. If amp draw exceeds OEM specifications, adjust motor speed or otherwise remedy the cause. Replace failed blower motor if necessary.
Condensate Removal	
Inspect for condensate blowing from coil into cabinet or ADS.	Adjust fan speed, clean coil fins, ensure OEM-supplied deflectors are in place, or replace coil as necessary to eliminate water carryover.
Inspect condensate drain piping (and traps) for proper operation.	Clean, insulate, repair, or replace as necessary.
Inspect drain pan and accessible drain line for biological growth.	Clean as needed to remove growth and ensure proper operation; add algae tablets or strips as necessary. Ensure algae tablets and cleaning agent are compatible with the fin and tube material.
Air Coil Section	
Inspect coil fins.	Ensure that fins are straight and open. Clean and straighten as required.
Inspect for condensate blowing from coil into cabinet or air duct system.	Adjust fan speed, clean coil fins, or replace coil as necessary to eliminate water carryover.
Measure and record temperature difference (TD) across air coil.	Evaluate this measurement with airflow, refrigerant charge, and operating conditions.
Refrigeration	
Inspect accessible refrigerant connecting lines, joints, and coils for oil leaks.	Test all oil-stained joints for leaks and then clean or repair as necessary.
Inspect refrigerant line insulation.	Repair or replace refrigerant line insulation if needed.
Measure pressure drop across the air coil.	Adjust, clean, replace, or repair as necessary to ensure proper airflow.

Gas Furnace

Gas fuel sources are typically either natural gas or LP (liquefied petroleum). However, a gas furnace can operate with almost any gas as long as the orifices, burners, and controls can be converted for the specific gas. Gaseous fuel is converted to heat (burned). Using a heat exchanger and blower, the heat is delivered to the airstream that provides conditioned air to the living space. Gas furnaces come in several efficiency ratings and, depending on their efficiency, will use venting systems constructed of various materials. Higher-efficiency models (Figure 1–2; see the discussion of direct venting later in this chapter) vent with plastic (PVC), which is also used to bring combustion air into the unit. High-efficiency units also require a drain connection for the condensed water that is removed from the flue gas. The placement of safety controls and the configuration of the heat exchanger dictate the direction of furnace discharge. Units are designed to be used specifically as upflow, counterflow (downflow), horizontal, or a combination of these.

See Checklist 1–3 for guidelines pertaining to gas furnaces.

Tech Tip

Always check the manufacturer's installation instructions. Some units are convertible and can be either upflow or downflow or can be horizontal right or left flow. When these units are installed, condensate drain and collectors may change positions or need to be reconfigured for the installation.

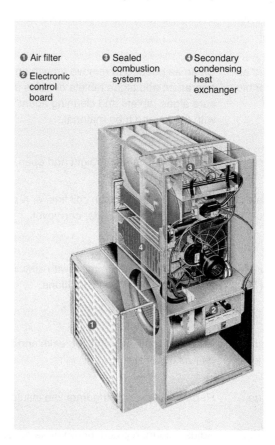

❶ Air filter

❷ Electronic control board

❸ Sealed combustion system

❹ Secondary condensing heat exchanger

Figure 1–2
A high-efficiency condensing gas furnace. (Courtesy of Bryant Heating and Cooling Systems)

Checklist 1–3 Gas Furnace (Courtesy of ACCA)

Inspection Task	Recommended Corrective Actions
Cabinet	
Inspect cabinet, cabinet fasteners, and cabinet panels.	Repair or replace insulation to ensure proper operation. Replace lost fasteners as needed to ensure integrity and fit/finish of equipment (as applicable). Seal air leaks.
Inspect the clearance around cabinet.	Document and report instances where the cabinet does not meet the requirements set by the OEM and applicable codes.
Electrical	
Inspect electrical disconnect box.	Ensure electrical connections are clean and tight. Ensure fused disconnects use the proper fuse size and are not bypassed. Ensure case is intact and complete. Replace as necessary.
Ensure proper equipment grounding.	Tighten, correct, and repair as necessary.
Measure and record line voltage.	Compare to OEM specifications or equipment nameplate data. Notify homeowner and/or utility.
Inspect and test contactors and relays.	Look for pitting or other signs of damage. Replace contactors and relays that show evidence of excessive contact arcing and pitting.
Inspect electrical connections and wire.	Ensure wire size and type match the load conditions. Tighten all loose connections, replace heat-discolored connections, and repair or replace any damaged electrical wiring.
Inspect all motor capacitors.	Replace those that are bulged, split, or incorrectly sized per OEM specifications.
Measure and record amperage draw to motor/nameplate data (FLA) (as available).	If outside OEM rating or specification, inspect for cause and repair as necessary.
Blower Assembly	
Determine and record airflow across heat exchanger.	Verify that all grilles and registers are open and free of obstruction. Adjust, clean, replace, and repair as necessary to ensure proper airflow.
Test variable-frequency drive for proper operation.	Replace if necessary to ensure proper operation.
Inspect fan belt tension. Inspect belt and pulleys for wear and tear.	Repair or replace as necessary to ensure proper operation (if applicable).
Confirm the fan blade or blower wheel has a tight connection to the blower motor shaft. Inspect fan for free rotation and minimal end play. Measure and record amp draw.	Lubricate bearings as needed, but only if recommended by OEM. If amp draw exceeds OEM specifications, adjust motor speed or otherwise remedy the cause. Replace failed blower motor if necessary.
Test inducer fan motor and blower assembly.	Adjust as needed.
Condensate Removal (if applicable)	
Inspect condensate drain piping (and traps) for proper operation.	Clean, insulate, repair, or replace as necessary.

Checklist 1–3 Gas Furnace *(Continued)*

Inspection Task	Recommended Corrective Actions
Gas Combustion	
Inspect combustion chamber, burner, and flue.	Look for signs of water, corrosion, and blockage.
Inspect heat exchanger for signs of corrosion, fouling, cracks, and erratic flame operation during blower operation.	Clean or replace as needed.
Visually inspect burners for signs of contamination.	Clean, repair, or replace as necessary.
Inspect the draft inducer blower wheel.	Clean as needed to ensure proper operation.
Inspect hot surface igniter for cracks (white spots when energized, or check cold with ohmmeter; check for proper supply voltage).	Replace if outside OEM specifications.
Measure and record inlet gas pressure at inlet pressure tap.	If the inlet gas pressure is insufficient for OEM operation specifications, contact the gas supplier.
Measure, record, and adjust manifold pressure as necessary.	Adjust the gas valve to provide a manifold pressure in accordance with OEM installation and operation instructions.
Test main burner ignition.	Clean thermocouple or flame sensor/pilot assembly as needed.
Test burners.	Fire unit and adjust air shutters (if used) for OEM specification compliance.
Test inducer fan motor and blower assembly.	Adjust as needed.
Ensure that combustion air volume is correct.	Check against local code.
Perform combustion analysis test. Measure and record test results.	Adjust as needed per OEM instructions.
Measure and record TD across the heat exchanger.	Clean components or adjust airflow as necessary to meet necessary operating conditions and design parameters.
Venting	
Inspect inside of chimney/flue inlet and exhaust vent for water, signs of condensation, corrosion, cracks, fractures, and blockages.	Clean, remove blockages, repair, or replace as necessary.
Inspect all vent connectors for rust discoloration and signs of condensate.	Ensure that connectors are securely fastened. Repair or replace as necessary.
Inspect inlet and exhaust vent pipe for proper support, slope, and termination.	Repair or replace as necessary.
Inspect for combustible materials placed too close to vent or pipe.	Relocate to safe place. Inform the customer.

Oil Furnace

Oil heating forced air systems convert (burn) oil in a high-temperature combustion chamber designed to reflect heat back to the burning fuel. Operating at much higher temperatures, oil heating systems need heavier heat exchangers and venting systems that are rated for higher temperatures. Oil furnaces come in several efficiency ratings. The highest-efficiency (direct vent) models use plastic pipe to vent and also to bring in combustion air. The placement of safety controls and the configuration of the heat exchanger dictate the direction of furnace discharge. Units are designed to be used specifically as upflow, counterflow (downflow), or horizontal.

Oil furnaces tend to require maintenance more often, and customers expect to have them checked at the beginning of the heating season. Checklist 1–4 supplies guidelines for inspecting an oil unit.

Checklist 1–4 Oil Furnace (Courtesy of ACCA)

Inspection Task	Recommended Corrective Actions
Cabinet	
Inspect cabinet, cabinet fasteners, and cabinet panels.	Repair or replace insulation to ensure proper operation. Replace lost fasteners as needed to ensure integrity and fit/finish of equipment (as applicable). Seal air leaks.
Inspect the clearance around cabinet.	Document and report instances where the cabinet does not meet the requirements set by the OEM and applicable codes.
Electrical	
Inspect electrical disconnect box.	Ensure electrical connections are clean and tight. Ensure fused disconnects use the proper fuse size and are not bypassed. Ensure case is intact and complete. Replace as necessary.
Ensure proper equipment grounding.	Tighten, correct, and repair as necessary.
Measure and record line voltage.	Compare to OEM specifications or equipment nameplate data. Notify homeowner and/or utility.
Inspect and test contactors and relays.	Look for pitting or other signs of damage. Replace contactors and relays that show evidence of excessive contact arcing and pitting.
Inspect electrical connections and wire.	Ensure wire size and type match the load conditions. Tighten all loose connections, replace heat-discolored connections, and repair or replace any damaged electrical wiring.
Inspect all motor capacitors.	Replace those that are bulged, split, or incorrectly sized per OEM specifications.
Measure and record amperage draw to motor/nameplate data (FLA) (as available).	If outside OEM rating or specification, inspect for cause and repair as necessary.
Blower Assembly	
Determine and record airflow across heat exchanger.	Verify that all grilles and registers are open and free of obstruction. Adjust, clean, replace, and repair as necessary to ensure proper airflow.

Checklist 1–4 Oil Furnace *(Continued)*

Inspection Task	Recommended Corrective Actions
Test variable-frequency drive for proper operation.	Replace if necessary to ensure proper operation.
Inspect fan belt tension. Inspect belt and pulleys for wear and tear.	Repair or replace as necessary to ensure proper operation (if applicable).
Confirm the fan blade or blower wheel has a tight connection to the blower motor shaft. Inspect fan for free rotation and minimal end play. Measure and record amp draw.	Lubricate bearings as needed, but only if recommended by OEM. If amp draw exceeds OEM specifications then adjust motor speed or otherwise remedy the cause. Replace failed blower motor if necessary.

Oil Combustion

Inspect combustion chamber, burner, and flue.	Look for signs of water, corrosion, and blockage.
Inspect heat exchanger for signs of corrosion, fouling, cracks, and erratic flame operation during blower operation.	Clean or replace as needed.
Inspect all burner gaskets.	Replace any gaskets that are damaged or that fail to seal adequately.
Inspect retention head, electrodes, and ceramic insulation.	Clean retention head, electrodes, and ceramic insulation of soot and carbon. Change electrodes with ceramic cracks or if tips are rounded.
Inspect electrodes for proper positioning.	Position electrodes as specified by OEM instructions.
Measure and record photocell (cad cell) resistance.	Remove photocell (cad cell), check resistance, and clean as necessary. Ensure resistance is within OEM specifications.
Clean combustion air inlet.	Remove lint or other foreign material around burner combustion air openings that may obstruct airflow.
Verify that burner head or nozzle is of proper type and is correctly positioned.	Make all adjustments as outlined in the burner's OEM specifications.
Replace oil burner nozzle.	Install new (never attempt cleaning) identical flow rated nozzle (verify gallons per hour, spray angle, and pattern) per OEM specifications.
Replace fuel filter.	Replace filter with one of proper type.
Bleed oil line.	With open fuel supply (cap removed) on a one-pipe system, remove any air from oil line.
Measure, adjust, and record oil pressure.	Measure and adjust oil pressure to OEM design pressure.
Inspect oil pump for proper pressure and leaks.	If pump pressure is below OEM specifications or if there are signs of leaks, remove oil pump cover and gasket. Discard gasket. Replace strainer or clean with a fine-bristle brush and solvent. Reassemble with new gasket. Retest pump.

Checklist 1–4 Oil Furnace *(Continued)*

Inspection Task	Recommended Corrective Actions
Test fuel pump for proper operation, pressure, and cutoff. Measure and record line vacuum.	Install a pressure gauge in the nozzle port and run the burner to observe operating pressure and record. Shut the burner off and record cutoff pressure. If the cutoff pressure drops below OEM specifications, replace pump or add check valve.
Measure and record ignition transformer secondary voltage.	Nominal range is 10,000 volts (AC) for iron core transformers. Solid-state igniters cannot be tested with an iron core transformer tester.
Ensure that combustion air volume is correct.	Check against local code.
Perform combustion analysis test. Measure and record test results.	Adjust as needed per OEM instructions.
Measure and record TD across heat exchanger.	Verify with furnace rating plate and adjust airflow until TD is within OEM rating.
Check primary control safety timing.	Disconnect the cad cell; then run the burner and time the lockout. Replace safety control if timing exceeds OEM specifications.
Venting	
Inspect inside of chimney/ flue inlet and exhaust vent for water, signs of condensation, corrosion, cracks, fractures, and blockages.	Clean, remove blockages, repair, or replace as necessary.
Inspect all vent connectors for rust discoloration and signs of condensate.	Ensure that connectors are securely fastened. Repair or replace as necessary.
Inspect inlet and exhaust vent pipe for proper support, slope, and termination.	Repair or replace as necessary.
Inspect for combustible materials placed too close to vent or pipe.	Relocate to safe place.

CONVERTING SYSTEMS OR FUELS

Some fuel-burning systems may be converted from one fuel to another if the manufacturer has conversion kits (see Figure 1–3). You should always follow the manufacturer's instructions for making conversions. Table 1–1 summarizes the elements involved in a typical conversion.

COMBUSTION ANALYSIS

Combustion analysis is the process of measuring: flue gas for the presence and percentage of various gases; the temperature of the flue gas after it leaves the

Cross-section

Figure 1–3
A cross-section of a spud burner orifice, one of several that would need to be replaced as part of a fuel conversion of LP to natural gas or natural gas to LP.

Table 1–1 Elements of Fuel Conversion

Conversion	Necessary Kit Components

Natural gas to LP	• Burner orifice replacements
	• Pilot orifice replacement (units with pilots)
	• Gas control replacement or conversion
	   ○ Pressure regulator conversion parts
	   ○ Complete control replacement
	   ○ Ignition control module

LP to natural gas	Same as above
Oil to gas	This conversion requires the complete replacement of the burner; the oil gun will be replaced with a gas gun, which may be set up for either natural or LP gas.

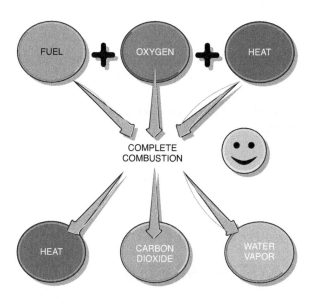

Figure 1–4
Diagram of complete combustion.

heat exchanger; and the amount of draft or venting that is occurring. Combustion characteristics may be changed to favor a more complete burn (see Figure 1–4 and Figure 1–5). For fuel-burning equipment, 100% efficiency cannot be obtained, but it is desirable to maximize the system's design efficiency. Suboptimal combustion efficiencies will result in:

- Lower economic operation
- Shortened equipment life
- More frequent cleanings to remove sooting or smudging

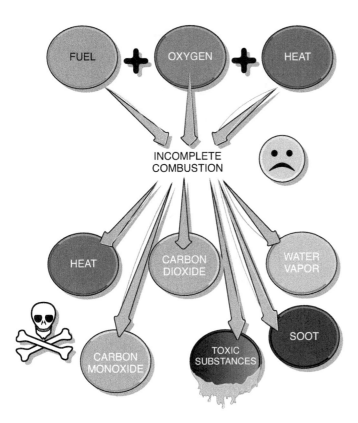

Figure 1–5
Diagram of incomplete combustion.

All combustion analysis equipment (see Figure 1–6 and Figure 1–7) requires some basic training to use. Some equipment is more sophisticated and requires specific training. Setting up for a combustion analysis, conducting the analysis, and understanding the results all require training and experience. Be sure to ask

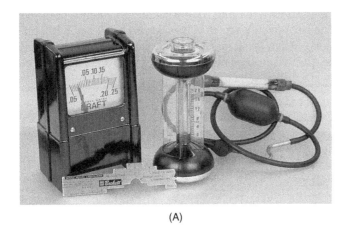

(A)

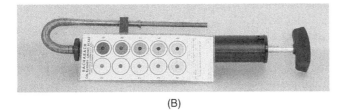

(B)

Figure 1–6
Combustion efficiency testing equipment. (A) A draft gauge, nozzle gauge, and CO_2 tester.
(B) A smoke tester. (Photos by Bill Johnson)

Figure 1–7
An electronic combustion analyzer.

for additional training to conduct combustion testing. Always follow combustion test instructions provided by the manufacturer of test instruments. Table 1–2 lists the steps involved when conducting a combustion analysis.

Carbon Monoxide

Carbon monoxide (CO) is an odorless but deadly gas. A product of incomplete combustion, it must be reduced to minimum levels in flue gas and properly vented. Note that the optimum reading for CO in the delivered air stream is *zero* CO (see Figure 1–8). Any reading higher than zero should be of concern to the technician. The most deadly effect of this gas is that it can build up in the bloodstream over time. When levels get too high, the result is flu-like symptoms that can lead to death. Poisoning can occur slowly in low concentrations. The symptoms include:

- Frequent headache
- Stuffy feeling
- Tiredness
- Dizziness
- Nausea
- Chest pain

In the later stages or in high concentrations, the symptoms increase in severity:

- Confusion
- Shortness of breath
- Impaired judgment
- Hallucinations
- Agitation
- Vomiting
- Abdominal pain
- Fainting
- Seizure
- Death

Table 1–2 Combustion Analysis

Steps	Process
Setup	• Drill access holes (vent pipe and fresh air pipe). • Warm up test equipment (if necessary). • Set up the slide calculator with the correct card for the fuel being burned.
Initial test	• Measure vent draft. • Measure overfire draft (oil). • Measure soot (oil) using smoke test. • Measure stack temperature. • Measure CO_2. • Measure O_2 (vapor). • Measure CO. • Adjust spark ignition electrodes. • Adjust oil or gas pressure.
Steady-state conditions	• Operate the equipment to normal operating temperature. • Visually check for burner operation; clean and adjust as necessary. • Set primary air controls (if installed). • Adjust blower or circulator for temperature rise.
Post-adjustment test	• Measure vent draft (gas diverter/hood or oil barometric draft damper). • Measure overfire draft (oil). • Measure soot (oil) using smoke test. • Measure stack temperature. • Measure CO_2. • Measure O_2. • Measure CO.
Readjustment (until highest efficiency is reached)*	• Readjust vent draft; ensure that draft measurements are within manufacturer specifications or recommendations. • Readjust primary air setting (if able). • Adjust or readjust the blower to meet temperature rise requirements for the system.
Ideal conditions	The ideal conditions for each installed system will be different. Readjust the system and test until OEM efficiencies are obtained or nearly obtained. Generally, equipment cannot be made more efficient than it was designed. Combustion efficiency testing and adjustment aims to obtain the highest possible efficiency for the unit with the lowest amount of CO production. Always test for: • Low CO production • Low stack temperature • Low soot production (smoke test) • High CO_2 production

*Take one set of measurements after every single adjustment; do not adjust several things before retaking measurements.

Symptoms do not always arise in this order and may occur differently for different people and for different ages. Children and the elderly are most susceptible to CO poisoning. The concentrations—in PPM (parts per million) and percentage (by volume) in the air—that can induce symptoms are given in Table 1–3.

Figure 1–8
A carbon monoxide test instrument. (Courtesy of Bacharach, Inc.)

Table 1–3 Symptoms Resulting from Various Concentrations of Carbon Monoxide

Concentration	Exposure Time	Result
0 PPM	8 hours	• No physical problems
35 PPM (0.0035%)	8 hours	• Headache • Dizziness
400 PPM (0.04%)	1–2 hours	• Frontal headache
800 PPM (0.08%)	45 minutes	• Dizziness • Nausea • Convulsions
1,600 PPM (0.16%)	20 minutes	• Headache • Dizziness • Nausea • Death in less than 2 hours

It is becoming more acceptable to take indoor air and flue gas samples before and after service is conducted on a fuel-burning appliance. Vent draft measurements must also be taken, since positive draft is often responsible for CO problems. Additionally, bad operating characteristics and combustion combine to produce CO and related negative furnace conditions. Any data obtained through tests of the systems and observation are recorded on the work report. To be able to report accurate data, a technician should be trained in testing for CO and must understand the difference between *air free* (or *corrected*) and standard measurements in PPM. Training should include the use of CO sampling and measurement instruments by the instrument manufacturer.

CO₂ and Fuel Relationships

Carbon dioxide (CO_2) is a product of combustion that should make up a high percentage of any flue gas sample. High levels of CO_2 mean that more of the fuel has been converted or combusted. The amount of CO_2 necessary to produce an efficient burn is different for each fuel and temperature of the flue gas. Table 1–4 gives some optimal CO_2 percentages at given stack temperatures (net flue temperatures, compensating for the temperature of the combustion air) that might be found for furnaces that are 80% efficient. A combustion test kit or flue gas analyzer (see Figure 1–9) is required to make the measurement.

To get fuel to burn as completely as possible, excess air is introduced to the combustion process to ensure complete combustion. Air is a mixture of mostly nitrogen (N, about 79%) and oxygen (O_2, about 21%). Nitrogen goes through the combustion process without bonding unless it produces NO_2 (nitrous oxide). However, nitrogen needs to be warmed from the entering temperature to the flue gas temperature. As this occurs, some of the heat produced by the flame heats only the nitrogen.

Table 1–4 Optimal CO_2 Concentrations for 80% Efficiency

Fuel Type	CO₂ Concentration and Net Flue Temperature
Natural gas	6.5% at 300°F
LP	6% at 300°F
Oil	11% at 500°F

Figure 1–9
A handheld monitor that tests for carbon monoxide, temperature, humidity, and gas levels.

Tech Tip

Dual- or variable-fired systems are manufactured to be the most efficient when they operate at their maximum firing rates. Combustion at rates other than maximum may not be as efficient. It is difficult to change the input of a system and maintain high efficiency without also reducing the size of the heat exchanger. Combustion testing should be done across the range of firing rates (low, mid, and high) or as instructed by the manufacturer. Be sure to read and understand the manufacturer's suggestions for measuring combustion efficiency.

Tech Tip

Gaseous fuels like natural gas and LP will mix and burn more easily than will liquid fuels like oil. Oil will produce more carbon (soot), which is measured during a combustion analysis. Zero smoke is optimal but seldom attainable. Soot will build up in a gas furnace over time, but large buildups or spillage of sooty deposits are a sign that the system is not burning correctly.

Compensating for Altitude

Different levels of elevation throughout the country may require adjustment or compensation for altitude (see Figure 1–10). At sea level, oxygen concentrations are said to be 21% of the total volume of air. Concentrations of oxygen drop as the altitude increases. Mountain climbers need to carry oxygen when they reach certain altitudes—there is not enough oxygen to breathe.

Just as humans do, fuel-burning equipment needs oxygen. When levels of oxygen drop as the altitude increases, more air is necessary for combustion. Most fuel-burning equipment is designed to operate from sea level to an altitude of 2,000 feet. For altitudes higher than this you may need to consult the manufacturer unless there is specific information in the installation manual. Some OEMs have equipment designed to operate to 5,000 feet and specialized equipment that will operate over 7,000 feet.

Safety Tip

With new and energy-efficient burners and conversion burners, do not assume that a standard w.c. pressure setting is safe. Always review the installation instructions for specific field test conditions.

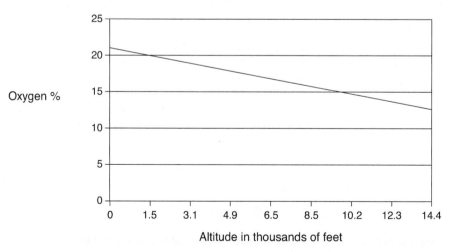

Figure 1–10
The percentage of oxygen decreases as the altitude increases.

Altitude also has an effect on gas regulators. Natural gas regulators set at 3.5" w.c. (water column) may not allow enough gas to come through at higher altitudes. Make sure to obtain specific pressure settings before adjusting and/or lighting a fuel-burning unit at high altitude. For example, a manufacturer may instruct the technician to set a natural gas regulator for 4.5" w.c. as a starting point. Combustion test equipment would permit fine-tuning of pressure to meet efficiency standards.

Compensating for Temperature

The density of air entering the combustion process depends on the temperature. Cold air brought into the combustion process from the outside will be denser than warm air brought from the conditioned space. Just as with altitude, the amount of O_2 is more when the air is denser (colder). Manufacturers design "direct" vent systems to handle colder air. When combustion analysis is done, the temperature of air entering the combustion process is measured and accounted for as part of the calculation.

Tech Tip

Never guess at gas pressure regulator settings for an altitude over 2,000 feet. Check with the local gas company for incoming pressure requirements and contact the OEM for specific gas valve pressure settings.

Combustion Air

Adequate ventilation air is necessary for the occupants of a home or building and for the fuel-burning equipment. If there is not enough ventilation, products of combustion may spill out of venting systems and into living spaces. Powered exhaust equipment is largely responsible for negative pressures in the living space. If negative conditions exist inside the building, outside air will be pulled in through any crack or opening. The opening with the least resistance will supply outside air. The largest openings tend to be chimneys and venting systems.

Natural venting will supply the least resistance when the bathroom vent is turned on. If there is no other opening to allow air back into the building, then air will come from outdoors through the venting system, pulling products of combustion into the interior space (see Figure 1–11). Rust or soot smudging around vent pipe openings, draft hoods, and draft diverters are a telltale sign that flue gas is spilling. The dangers are that CO (carbon monoxide) is entering the living space.

Tech Tip

To ensure adequate combustion air, there is a minimum requirement of 50 cubic feet of space for every 1,000 BTUs of connected load. If there is less than 50 cubic feet per 1,000 BTUs, you must duct in combustion air. Generally, duct is sized at a rate of 1 square inch for every 4,000 BTUs. Consult local fuel gas codes for details.

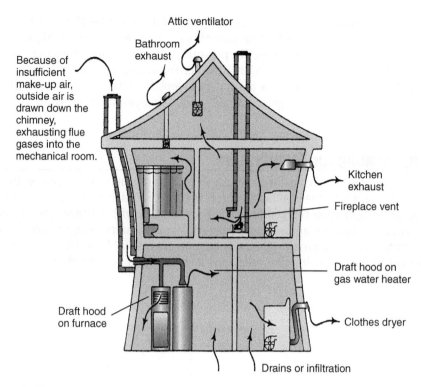

Figure 1–11
Air that has been exhausted needs to be made up by ventilation air. If power exhaust is used, it will easily overcome naturally vented appliances. (Courtesy of Bacharach, Inc.)

To ensure that there is enough ventilation for flue gas venting, turn on all exhaust equipment and then conduct a test of the exhaust venting system. There should be a minimum of −0.02" of draft during initial firing of the fuel burning equipment. Be sure to turn on such exhaust equipment as:

- Bathroom exhaust vent
- Kitchen exhaust vent
- Range hood exhaust vent
- Clothes dryers
- Commercial/industrial process exhaust vents
- Commercial/industrial room or lab exhaust vents

Tech Tip

Some homes have "whole house" ventilation fans. These are typically used during the cooling season when windows are open. These fans should not be turned on when the house is closed up.

HYDRONIC SYSTEMS

Hydronic systems consist of any piece of equipment that heats, distributes, and uses hot water for heating purposes (see Figure 1–12). This category of heating equipment includes boilers, radiant panels, baseboards, circulators, and more. If the boiler produces steam, then this type of system is referred to as a steam system that produces, distributes, or uses steam to heat.

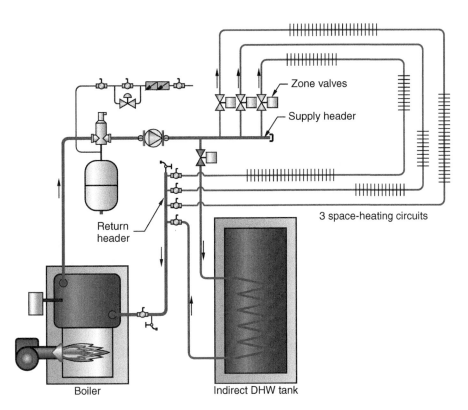

Figure 1–12
Hydronic heating system diagram showing three zones and DHW (domestic hot water).

Tech Tip

Hydronic systems are sometimes referred to as "wet heat" systems. There is a complete separation between the heating medium and the room air. There is no humidification effect obtained from a boiler, unless there is a leak.

In addition to fuel-burning component checks, hydronic systems should generally be inspected as described in Checklist 1–5 (the example is for gas boilers).

Radiant Floors and Panels

Radiant panels can be any building surface, such as floors, ceilings, or walls. Radiant floors are heated from below the finished surface. Recommendations are that the finished surface be any hard surface (usually not carpeting). Radiant heated floors can be heated electrically or with "low-temperature" hot water (see Figure 1–13).

The most popular systems embed PEX (cross-linked polyethylene) in concrete and sandwiched wood floor systems. The flexible PEX tubing can be made to conform to many different configurations to cover the floor area. Some popular piping arrangements in floors are:

- Slab-on-grade
- Thin-slab
- Stapled-up

Checklist 1–5 Gas Boiler (Courtesy of ACCA)

Inspection Task	Recommended Corrective Actions
Cabinet	
Inspect cabinet, cabinet fasteners, and cabinet panels.	Repair or replace insulation to ensure proper operation. Replace lost fasteners as needed to ensure integrity and fit/finish of equipment (as applicable).
Inspect the clearance around cabinet.	Document and report instances where the cabinet does not meet the requirements set by the OEM and applicable codes.
Electrical	
Inspect electrical disconnect box.	Ensure electrical connections are clean and tight. Ensure fused disconnects use the proper fuse size and are not by-passed. Ensure case is intact and complete. Replace as necessary.
Ensure proper equipment grounding.	Tighten, correct, and repair as necessary.
Measure and record line voltage.	Compare to OEM specifications or equipment nameplate data. Notify homeowner and/or utility.
Inspect and test contactors and relays.	Look for pitting or other signs of damage. Replace contactors and relays that show evidence of excessive contact arcing and pitting.
Inspect electrical connections and wire.	Ensure wire size and type match the load conditions. Tighten all loose connections, replace heat-discolored connections, and repair or replace any damaged electrical wiring.
Inspect all stand-alone capacitors.	Replace those that are bulged, split, or incorrectly sized per OEM specifications.
Measure and record amperage draw to motor/nameplate data (FLA) (as available).	If outside OEM rating or specification, inspect for cause and repair as necessary.
Gas Combustion	
Inspect combustion chamber, burner, and flue.	Look for signs of water, corrosion, and blockage.
Inspect heat exchanger for signs of corrosion, fouling, cracks, and erratic flame operation.	Clean or replace as needed.
Visually inspect burners for signs of contamination.	Clean, repair, or replace as necessary.
Inspect the burner blower wheel.	Clean as needed to ensure proper operation.

Checklist 1–5 *Gas Boiler (Continued)*

Inspection Task	Recommended Corrective Actions
Inspect hot surface igniter for cracks (white spots when energized, or check cold with ohmmeter and proper supply voltage).	Replace if outside OEM specifications.
Measure and record inlet gas pressure at inlet pressure tap.	If the inlet gas pressure is insufficient for OEM operation specifications, contact the gas supplier.
Measure, record, and adjust manifold pressure as necessary.	Adjust the gas valve to provide a manifold pressure in accordance with OEM installation and operation instructions.
Test main burner ignition.	Clean thermocouple or flame sensor/pilot assembly.
Test burners.	Fire unit and adjust air shutters (if used) for OEM specification compliance.
Test inducer fan motor and blower assembly.	Adjust as needed.
Ensure combustion air volume is correct.	Check against local code.
Perform combustion analysis test. Measure and record test results.	Adjust as needed per OEM instructions.
Measure and record TD across the heat exchanger.	Clean components or adjust airflow as necessary to meet necessary operating conditions and design parameters.

Hydronic Loop

Inspection Task	Recommended Corrective Actions
Inspect screen on reducing valve, pressure reducing valve, and "Y" strainer if available.	Clean or replace as necessary.
Test bladder/expansion tank for proper air cushion.	Adjust to ensure proper air cushion per manufacturer's specifications.
Inspect water pump.	Clean or clear as needed to reduce cavitation and ensure proper operation.
Measure and record pressure difference of the water loop across the refrigerant water heat exchanger.	Adjust water pump or control valve as necessary.
Measure and record TD between water entering and leaving coil/heat exchanger.	Add or remove refrigerant or adjust firing rate as necessary.

Checklist 1–5 Gas Boiler *(Continued)*

Inspection Task	Recommended Corrective Actions
Venting	
Inspect vent termination for water, signs of condensation, corrosion, cracks, fractures, and blockages.	Clean, remove blockages, repair, or replace as necessary.
Inspect all vent connectors for rust discoloration and signs of condensate.	Ensure that connectors are securely fastened. Repair or replace as necessary.
Inspect inlet and exhaust vent pipe for proper support, slope, and termination.	Repair or replace as necessary.
Inspect for combustible materials placed too close to vent or pipe.	Relocate to safe place. Notify building owner.

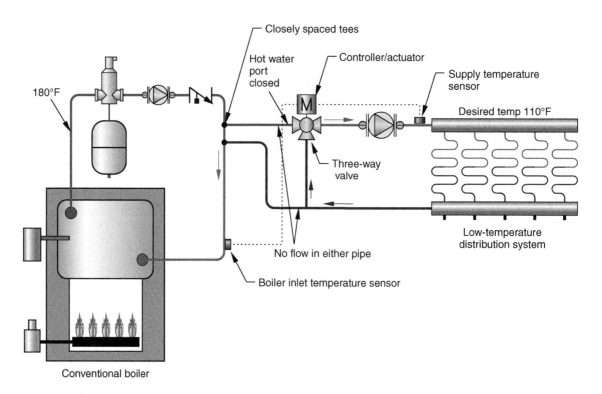

Figure 1–13
A low-temperature hydronic system showing connections to a radiant floor panel. A primary and secondary loop are used to maintain manufacturers' minimum return water temperature limits to prevent thermal stresses to the boiler and to prevent excessive condensation in the combustion chamber.

Table 1-5 Checklist for Radiant Panels

Check	Procedure
Water flow	• Check water conditions (excessive corrosion products and, if treated, chemical levels). • Check circuit flow using temperature or pressure drop. • Check for similar temperature across the panel (maximum difference of 2–5 degrees).
Amperage	• Take pump (circulator) motor amperage (for each circuit if each circuit has a pump) and check against the name-plate rating.

Radiant floors operate at temperatures between 90 and 100 degrees Fahrenheit. Higher temperatures feel uncomfortable to walk on. Manifold arrangements control the slab system, allowing each loop to be individually regulated. Because of the relatively low operating temperature, radiant panels are ideally suited for solar heating applications.

Radiant panel checks are given in Table 1-5.

Pressure Relief Valves

Pressure relief valves are safety valves. The *Boiler and Pressure Vessel Code* of the American Society of Mechanical Engineers requires that all boilers be tested and have a rated pressure relief valve mounted on the boiler. Most low-pressure hydronic systems operate at around 15 psig (see Figure 1–14). Relief valves for these systems are set to open (relieve pressure) at 30 psig. These valves need to be

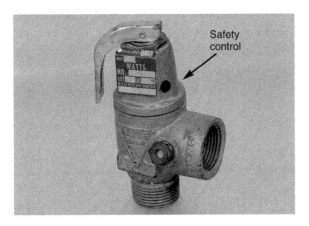

Safety control

Figure 1-14
Pressure relief valve used for hydronic systems as a safety. Relief valve pressure settings are stamped on the valve tag. (Courtesy of Bill Johnson)

Tech Tip

A pressure relief valve will open to relieve pressure when the valve senses pressure above the rating on the valve. The maximum rating on the relief valve must be lower than or equal to the maximum rating on the pressure vessel (boiler) and higher than a combination of the static fill pressure and operating pressure. The pressure setting for the pressure relief valves used on most low-pressure boilers is 30 psig. Note that the input BTUH rating of the pressure relief valve should be greater than the input BTUH rating of the boiler.

checked for operation. They need to open when system pressure rises to 30 psig, which can be caused by overfiring, waterlogging of the expansion tank, or faulty operation of automatic water fill regulators. The valve must open and reseal. Many times the valve will not reseal because of corrosion or valve seal deterioration. Any valve that does not operate and seal must be replaced.

Zone Valves

Zone valves (see Figure 1–15) are used to turn on and off the flow of water in a hydronic loop. Each loop may heat a single room or group of rooms. Zoning allows individual rooms to be controlled with an independent thermostat to any comfort level. Zone valve checks are summarized in Table 1–6.

Pumps

Circulating pumps, also referred to as circulators (see Figure 1–16), are used to create flow in the hydronic loop, supplying heated water to heat exchangers and returning this water to the boiler to be reheated. Circulators are usually centrifugal pumps and create a pressure difference between suction and discharge to create and maintain water flow. These pumps are rated in feet of head, which is a pressure rating indicating the maximum pressure difference that the particular pump can create. This is also known as "shutoff head" or "dead head" pressure.

Figure 1–15
Electrically driven, motor-style zone valves (one with the cover off showing the motor). (Courtesy of Eugene Silberstein)

Table 1–6 Checklist for Zone Valves

Check	Procedure
Water flow	• Check the operation of the zone valve (visual position). • Check the flow through the zone valve by feeling or measuring temperature difference; no temperature difference with water flowing means the valve is open. • Check water flow through loop with a listening device. • Check for leakage around the mechanical seal of the valve; crusty deposits indicate leakage.
Voltage	• Voltage to the zone valve should not drop below 22 volts when all secondary voltage devices are operating.

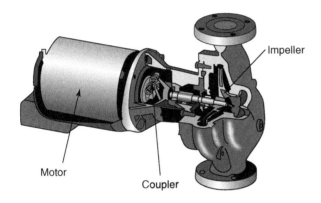

Impeller

Motor

Coupler

Figure 1–16
Cutaway of a centrifugal circulator pump.

Many systems have gauges on either side of the pump that read in feet of head, but some have gauges that indicate psig instead, so you may need to convert from psig to feet of head. A column of water 2.3 feet high will exert a pressure of 1 psig, so:

1 psig = 2.3 feet of head
and
psig = feet of head ÷ 2.3
or
feet of head = psig × 2.3

Knowing the pump pressure difference allows you to plot the operating conditions on a pump curve (see Figure 1–17), available from the pump manufacturer, to find the flow rate in gpm for a given set of conditions.

Note that in Figure 1–17, at 0 gpm, the pressure difference the pump can create is 75 feet of head, which is the maximum rated pump head for this pump. In an open system such as a cooling tower, this is the maximum height that this pump could handle (as illustrated on the right side of Figure 1–17), and the flow rate would be at or near zero. The term "feet of head" was derived from this concept for rating pumps.

Most hydronic systems, however, are not open-loop systems; they are closed-loop systems. In a closed-loop system, the height of the system has no pumping head effect because the resistance to flow of pumping the water up against

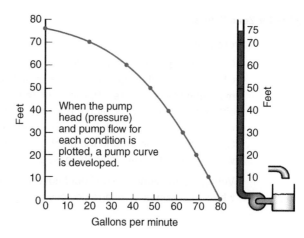

Figure 1–17
Using a pump curve plotted by the manufacturer, the technician can determine the flow by measuring the head pressure in feet of water column.

gravity is balanced out by gravity assisting the water in coming back down to the pump. The only resistance to flow in a closed-loop system is that created by fluid friction through the piping, fittings, valves, and components (heat exchangers).

To find the flow rate through an open or closed system, you must know three things:

1. Operating pump suction pressure
2. Operating pump discharge pressure
3. Pump curve for the pump being checked

Subtract the suction pressure from the discharge pressure to find pumping head, and then plot the value on a pump curve to find the gpm.

For example, a pump is running in a closed-loop system and you need to find out the flow rate through the pump. The suction pressure indicated is 5 psig, and the discharge pressure is 22 psig.

Discharge − Suction = Pumping head
22 psig − 5 psig = 17 psig
17 psig × 2.3 = 39.1 feet of head

For this problem, use the pump curve in Figure 1–17. (Keep in mind that not all pump curves are the same; they are specific to each pump.) Finding 39.1 feet of head on the left side of the graph, draw a straight line across horizontally until it intersects the pump curve. Then draw a line straight down to the bottom axis to find the gpm. In this example, the result is approximately 57 gpm.

Tech Tip

System head can be estimated by multiplying the length of the piping by 1.5 and again by 0.04. This is helpful in the field when you suspect that an incorrectly sized circulator has been installed.

Circulators are sized for a design flow rate against a design pressure drop (system head). Circulator pump checks are given in Table 1–7.

Table 1–7 Checklist for Circulator Pumps

Check	Procedure
Water flow	• Check for water leakage at the mechanical seal and pump housing; crusty deposits are an indication of leaking. • Check pump head and compare to pump curve from manufacturer for correct flow.
Amperage	• Take circulator pump motor amperage (for each circuit if each circuit has a pump) and check against the nameplate rating.
Lubrication	• Check wicking or lubrication points and lubricate per manufacturer's specifications. • Check pump and motor bearings for wear.

Expansion Tanks

Expansion tanks are used to help maintain pressure in a closed-loop hydronic system. They may be as simple as a tank with an air pocket at the top connected to the system piping below the tank. They may be of the bladder type, which can be mounted above or below system piping, usually with an air valve (similar to a tire valve) to allow for checking or adjusting air charge pressure in the bladder. See Figure 1–18. Expansion tanks are needed to accommodate the expansion of system water as it is heated or to maintain pressure in the system as it cools during the off cycle. They are sized based on system operating temperature range and system volume.

The best place for the expansion tank to be connected to the system is near the suction of the circulator, although some system designs vary. This helps to maintain a positive pressure at the suction of the pump, also known as "net positive suction head," or NPSH. This positive pressure helps to ensure that air is not drawn into the system through the pump seals and prevents excessive wear of seal surfaces as a result of entrained contaminants. Air in the system also promotes corrosion and fouling of heat exchanger surfaces and impedes heat transfer in heat exchangers where the air may collect. Air in the system may also result in reduced flow and flow noise in building systems. A handy way to remember this

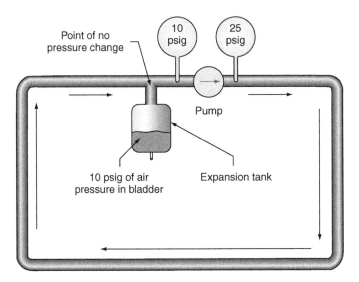

Figure 1–18
The position of the expansion tank to the suction of the circulator creates a point of no pressure change and allows the circulator to exert a positive pressure in the hydronic circuit.

is that NPSH prevents a "not pumping so hot (NPSH)" hydronic system. NPSH also aids in the prevention or minimization of pump cavitation, which may rapidly and permanently damage the pump impeller. Centrifugal pumps should never be run with the suction pressure in a vacuum.

As stated earlier, in a closed-loop system, the height of the system has no effect on the pumping head of the system, but it does have an effect on the static head of the system. Static head is the pressure existing in the piping with the circulator turned off. Most bladder-type expansion tanks come pre-charged to 12 psi. In addition, most pressure reducing valves used in the make-up water lines also come preset to 12 psi. This pressure is adequate for most systems with a total height up to just under 28 feet. This is done to ensure that all parts of the system remain under positive pressure during the run and off cycles. This is important to keep air from getting into the system in case of leaks, but more importantly, it allows for the venting of air from the system, which will tend to collect in the top parts of the system.

Expansion tanks should be checked periodically and at least at the beginning of the heating season to ensure that they are not waterlogged. Waterlogging is a condition in which the air volume in the tank does not exist and therefore there is no room for expansion of system water. For bladder-type tanks, the pressure should be checked and some air should be bled off to check for the presence of water. If water is found at the air fitting, then the bladder is probably leaking and should be replaced. Always check final pressure and replace the cap on the air fitting. Non–bladder-type tanks are checked by looking at the sight glass level. Always verify that the sight glass valves are open, or the level in the tank may not be accurately indicated in the sight glass. Often, these valves are shut to prevent air leaks at the sight glass or valve packing.

VENTILATION AND FRESH-AIR REQUIREMENTS

Ventilation is necessary for occupants and fuel-burning equipment. Enough ventilation air must be coming in from the outside to satisfy indoor air and burner requirements, in addition to making up for any air being exhausted. Residential structures have smaller ventilation needs but are often limited in the amount of fresh air they allow to enter. Commercial buildings usually have ventilation systems that are designed to deliver prescribed amounts of fresh air (see Figure 1–19). For instance, ASHRAE Standard 62.2 requires that a certain amount of fresh air be brought into a residence, based on square footage and number of bedrooms. A house of between 1,501 to 3,000 square feet with two or three bedrooms requires a minimum of 60 cfm of ventilation air. Ventilation air must come from the outside. This can mean that air is either forced into or exhausted from the residence.

In a commercial application, ventilation dampers are installed or part of the equipment (such as economizers). Ventilation dampers open or close to allow more or less fresh air into a building. These dampers can be a combination of fresh air and exhaust that open at the same time to maintain building pressure. The amount of air exhausted always equals the amount of air brought in from the outside. If the system exhausts only, then the interior building pressure will be negative. This may also mean that any outside contaminants (dust, gases, and vapors) are pulled into the building through cracks around doors, windows and other openings. If you enter a building and air pressure moving into the building tends to hold the door open, the building interior may be negative. Positively pressurized buildings bring in more air than is exhausted.

Basic ventilation and fresh-air intake checks are summarized in Table 1–8.

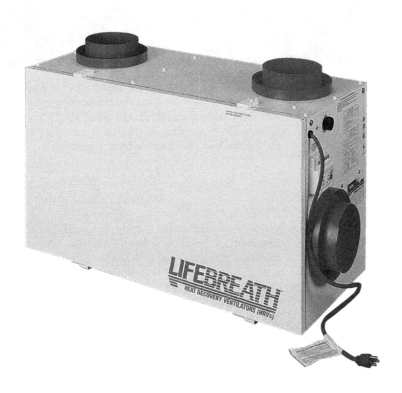

Figure 1–19
An energy recovery ventilator. (Courtesy of Nutech Energy Systems, Inc.)

Table 1–8 Checklist for Ventilation

Check	Procedure
Intake grille/louver	• Check for blockage. • Clean screen/pre-filter (if installed). • Replace filter (if installed).
Damper (intake and exhaust)	• Check the operation of automatic, barometric, or motorized damper. • Check the minimum fresh-air setting (check system requirements or local code for minimum settings). • Check for smooth operation.
Lubrication	• Check and lubricate damper and motor bearing surfaces as needed. • Check and lubricate motor and blower bearings as needed. • Check for bearing wear.
Airflow	• Check for minimum airflow (cfm). • Check for maximum airflow (cfm). • Change filter. • Check for economizer control operation and calibration (if installed). • Check heat recovery temperatures at inlet and outlet (if installed).

Case Study

During a service check, the economizer damper was partially open. Air from the outside was being pulled in and mixed with return air from the building. The mixed air was being conditioned and distributed to occupied spaces. As the technician continued her service checks, the damper moved toward the closed position and stopped. She had been instructed to set the fresh air at a minimum of 20%, but it stopped in a position that looked like more air would be pulled into the building. She had already determined that the unit was moving 1,200 cfm. If the mixed air temperature was right for 20% fresh air, then the damper would be in the right position.

The damper had a spring return motor that would always return the damper to the most closed position. The technician disconnected the damper motor from the control system and watched to see if the damper moved—but it stayed in the same position. Next she measured the air temperatures: return air (70°F), outside air (80°F), and mixed air (72°F). Using the formula for % Outside air, the technician determined that the damper had moved to the correct position.

$$\%_{\text{Outside air}} = \frac{T_{\text{Mixed air}} - T_{\text{Return air}}}{T_{\text{Outside air}} - T_{\text{Return air}}} \times 100$$

For example, suppose the outside air temperature is 80°F and the return air temperature is 70°F; then 80 − 70 = 10. If the mixed air temperature is 72°F and the return air temperature is 70°F, we get 72 − 70 = 2. Now dividing 2 by 10 gives the answer: 0.2, or 20% outside air.

Economizers

The economizer (see Figure 1–20) is part of the ventilation air system. An economizer is a device that adjusts the amount of fresh air delivered to the building when outside air can be used to cool the building instead of running the a/c system. If, for instance, the outside air condition is 50°F and low in humidity, then it would be more economical to use outside air to cool the building than to use electricity to create a cooling effect. In this case, economizer louvers open

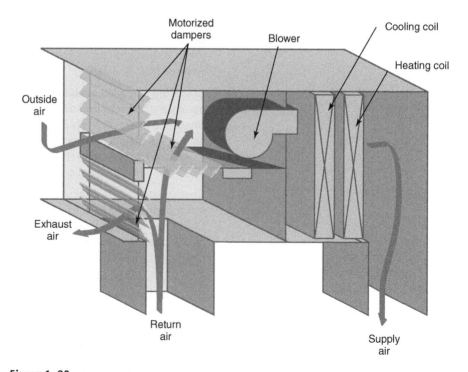

Figure 1–20
An economizer for an air system; the motorized dampers open and close proportionally to allow more or less fresh air into the building.

fully to allow 100% ventilation air into the building and exhaust dampers open to exhaust 100% of the return air from the building. The economizer works between the minimum fresh air setting for the building to full open, 100% outside air.

Building Pressure Control

Relief dampers (also called barometric and static dampers; see Figure 1–21) are used to relieve the pressure caused by pressure difference between the inside and outside of a building. Pressure differences occur because more air is brought into or exhausted from a building. Exhaust systems tend to cause lower or negative pressure, whereas ventilation systems that bring in fresh air tend to cause higher or positive pressures. Commercial HVAC systems are designed to purposefully discharge stale air to the outside or bring fresh air into the building for ventilation. Relief dampers are sized to allow replacement air to move into the delivery system to be conditioned and distributed to the entire building. Relief dampers are also sized to include other building exhaust systems and the air needed for fuel burning systems (furnaces and boilers).

In addition to barometric dampers that use no power, building pressure control can be accomplished by using the system (as in the case of an economizer). Fans can also be independent, providing building pressure control by exhausting or bringing in ventilation air. Ventilation systems that transfer heat between exhausting air and ventilation air (heat recovery systems) can also control building pressure. Positively pressurized buildings are able to control the infiltration of outside contaminants. Positive pressure means that air leaks out through cracks in doors and windows, and outside contaminants cannot be pulled into the building. This type of pressurization is better able to control indoor air quality than are negative pressure systems.

Direct and Indirect Venting

The terms *direct venting* and *indirect venting* are terms used to describe how fuel-burning appliances vent and receive fresh air. Direct vent systems (see Figure 1–22) use one pipe to exhaust combustion air and a second pipe to bring fresh air to the burner. Indirect venting systems use one pipe to exhaust the products of combustion and require that combustion air be drawn from the living space. Draft hoods, draft diverters, and barometric dampers all allow room air to mix with the flue gas being exhausted and allow a chimney effect.

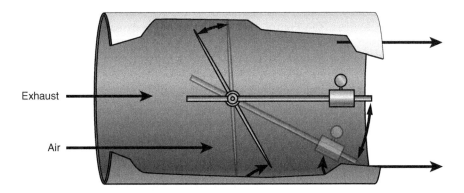

Exhaust

Air

Figure 1–21
The barometric damper opens when exhaust air pressure is higher than the balance weight pressure.

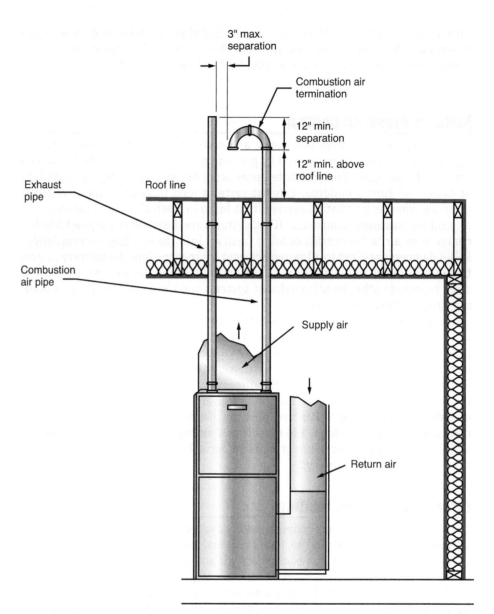

Figure 1–22
A direct vent high-efficiency gas furnace.

Heat Recovery

Heat recovery devices (see Figure 1–23) capture the heat leaving a building as stale air and use that heat to warm incoming fresh air. This saves on the amount of energy needed to heat cold air to room temperature and beyond. The amount of heat saved can be substantial. In addition, some recovery devices are capable of recovering humidity lost with the stale air.

Also called "cooling recovery" devices, they act by cooling incoming air using stale but cool interior air. Because interior air is also dryer, this stale, drier air is used to reduce humidity (in those systems that are capable of recovering humidity).

Recovering moisture (humidity) requires that the device be able to move moisture from one air stream to another. In most cases this requires the condensation of moisture or the absorption of moisture within the device. Some devices present the capturing material to the stale air and then move the material into the incoming fresh air stream. In this way, the moisture is transferred from one air stream to the other.

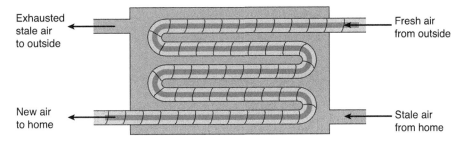

Figure 1–23
A heat recovery unit.

COOLING SYSTEMS

Specific procedures and check sheets are usually provided to the technician by the employing contractor and are usually available for all standard cooling systems. Each manufacturer provides installation (see Figure 1–24), technician, and owner manuals that list all of the specific maintenance required for the system. Always consult the manufacturer's manuals. General checks for cooling systems are presented from the ACCA standards in two sections: evaporator coil (Checklist 1–6) and condensing unit (Checklist 1–7).

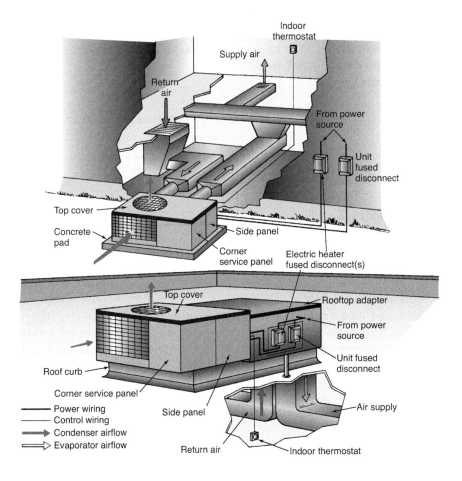

Figure 1–24
Alternative installations for a packaged a/c system.

Checklist 1–6 Evaporator Coil (Courtesy of ACCA)

Inspection Task	Recommended Corrective Actions
Cabinet	
Inspect cabinet, cabinet fasteners, and cabinet panels.	Repair or replace insulation to ensure proper operation. Replace lost fasteners as needed to ensure integrity and fit/finish of equipment (as applicable). Seal air leaks.
Condensate Removal	
Inspect condensate drain piping (and traps) for proper operation.	Clean, insulate, repair, or replace as necessary.
Inspect for condensate blowing from coil into cabinet or ADS.	Adjust fan speed, clean coil fins, ensure OEM-supplied deflectors are in place, or replace coil as necessary to eliminate water carryover.
Inspect drain pan and accessible drain line for biological growth.	Clean as needed to remove growth and ensure proper operation; add algae tablets or strips as necessary. Ensure algae tablets and cleaning agent are compatible with the fin and tube material.
Refrigeration	
Measure and record TD across evaporator coil.	Evaluate this measurement with airflow, refrigerant charge, and operating conditions.
Inspect coil fins.	Ensure that fins are visibly clean, straight, and open. Clean and straighten as required.
Inspect accessible refrigerant lines, joints, components, and coils for oil leaks.	Test all oil-stained joints for leaks and then clean or repair as necessary.
Inspect refrigerant line insulation.	Repair or replace refrigerant line insulation if needed.
Measure pressure drop across the coil.	Adjust, clean, replace, or repair as necessary to ensure proper airflow.

Heat Exchangers

Used in a number of HVAC heating and cooling applications, heat exchangers transfer heat media (air, water, or refrigerant) of higher or lower temperature. Air coils are fin and tube devices made of copper tubing and aluminum fins (see Figure 1–25). Fin spacing and tube size combine with the number of passes to determine heat transfer rates.

Water-based heat exchangers are configured in two general styles: tube-in-shell and tube-in-tube. Tube-in-shell exchangers use a tube bundle filled with water that is surrounded by refrigerant. Tube-in-tube exchangers use a water-filled tube placed inside a larger tube that carries refrigerant. (The reverse can also be true, where water surrounds the refrigerant; it depends on the manufacturer's design and the application.) The tube-in-tube is then formed into a bundle to take up less space. In both configurations, water is generally flowing in the opposite direction of the refrigerant. Coolest water is exchanging with the coolest refrigerant in an effort to maintain the greatest temperature difference between one end of the heat exchanger and the other.

Checklist 1–7 Condensing Unit (Courtesy of ACCA)

Inspection Task	Recommended Corrective Actions
Cabinet	
Inspect cabinet, cabinet fasteners, and cabinet panels.	Repair or replace insulation to ensure proper operation. Replace lost fasteners as needed to ensure integrity and fit/finish of equipment (as applicable). Seal air leaks.
Inspect the clearance around cabinet.	Document and report instances where the cabinet does not meet the requirements set by the OEM and applicable codes.
Electrical	
Inspect electrical disconnect box.	Ensure electrical connections are clean and tight. Ensure fused disconnects use the proper fuse size and are not bypassed. Ensure the case is intact and complete. Replace as necessary.
Ensure proper equipment grounding.	Tighten, correct, and repair as necessary.
Measure and record line voltage.	Compare to OEM specifications or equipment nameplate data. Notify homeowner and/or utility.
Inspect and test contactors and relays.	Look for pitting or other signs of damage. Replace contactors and relays that show evidence of excessive contact arcing and pitting.
Inspect electrical connections and wire.	Ensure wire size and type match the load conditions. Tighten all loose connections, replace heat-discolored connections, and repair or replace any damaged electrical wiring.
Inspect all motor capacitors.	Replace those that are bulged, split, or incorrectly sized per OEM specifications.
Measure and record amperage draw to motor/nameplate data (FLA) (as available).	If outside OEM rating or specification, inspect for cause and repair as necessary.
Refrigeration	
Inspect accessible refrigerant lines, joints, components, and coils for oil leaks.	Test all oil-stained joints for leaks and then clean or repair as necessary.
If indoor airflow is within OEM specifications but TD is not: measure and record system refrigeration charge, in "cooling" mode, as prescribed by OEM.	Add or remove refrigerant as necessary.
Inspect refrigerant line insulation.	Repair or replace refrigerant line insulation if needed.

Checklist 1–7 Condensing Unit *(Continued)*

Inspection Task	Recommended Corrective Actions
Condenser Fan Motor	
Confirm the fan blade or blower wheel has a tight connection to the blower motor shaft. Inspect fan for free rotation and minimal end play. Measure and record amp draw.	Lubricate bearings as needed, but only if recommended by OEM. If amp draw exceeds OEM specifications then adjust motor speed or otherwise remedy the cause. Replace failed blower motor if necessary.
Condenser Coil	
Inspect coil fins.	Ensure that fins are clean, straight, and open. Clean, straighten, and repair as required.

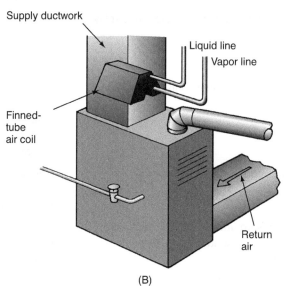

Supply ductwork

Liquid line

Vapor line

Finned-tube air coil

Return air

(A)   (B)

Figure 1–25
(A) A heat exchanger made of aluminum finned copper tubing is used for an air conditioning evaporator. (B) The unit is located in the plenum of a forced air furnace.

End bells of some tube-in-shell exchangers can be removed so that the tubing can be cleaned. Rotary brushes are used to scrape off buildup that would cause an insulating effect (see Figure 1–26). In situations where the exchanger cannot be opened, an acid-type cleaning can be done (see Figure 1–27). Acid-based fluid in introduced to the heat exchanger's water-carrying tubes and flushes out debris. (Please refer to additional information on heat exchangers in *HVACR 101* and *HVACR 201*.)

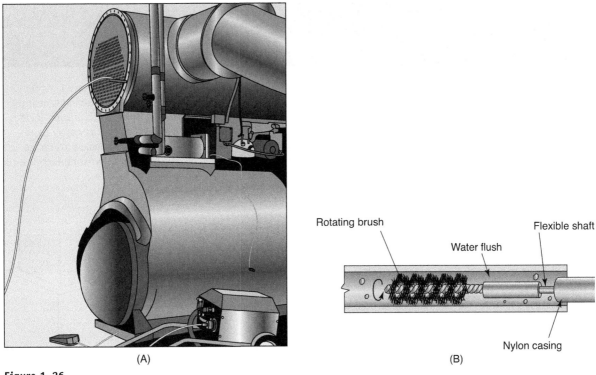

Figure 1–26
(A) Rotary brush being pushed through the water tube of a tube-in-shell condenser. (B) Cutaway of the brush inside the tubing.

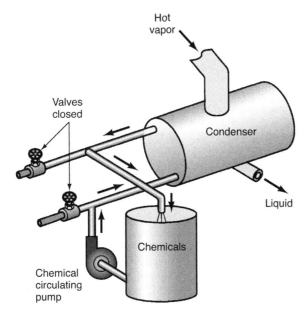

Figure 1–27
Chemical cleaning of a water-based condenser.

Air-source Heat Pumps

In the heating mode, air-source heat pumps (see Figure 1–28) capture the heat in outside air, concentrate it, and pump it to the inside. Outdoor units and indoor coils look similar to split-system air conditioning. The main component that changes the function of indoor and outdoor heat exchangers is the reversing

Case Study

The product cooler of a convenience store is too warm. The system is a reach-in with a remote, water-cooled condensing unit. As the technician asks questions of the customer, it is discovered that the store manager is also experiencing higher water bills. The water-cooled condenser is suspect. Compressor discharge temperatures are high. Checking the inlet and outlet water temperatures, the technician finds a difference of only 2 degrees, indicating that not enough heat is being transferred. Disconnecting the inlet and outlet water connections, the technician discovers that the water tubing is nearly blocked. The condenser is a tube-in-tube configuration. Acid cleaning would take too long and be nearly as costly as replacing the small condenser. The technician makes his recommendation to the store owner, who chooses to replace the condenser.

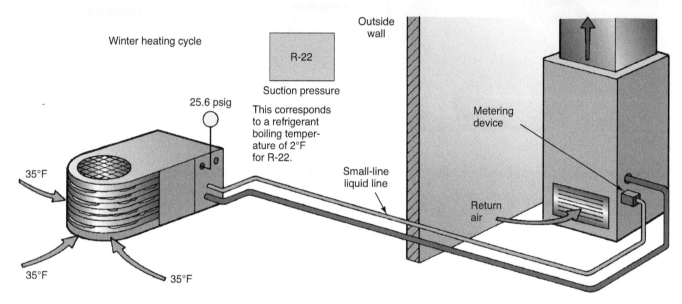

Figure 1–28
An air-source heat pump connection diagram that shows operating characteristics.

Case Study

A customer complained that her (air-to-air) heat pump—which had operated satisfactorily at the end of the summer—did not perform well at the beginning of the winter. But when questioned by the technician about humidity levels in the summer, the customer agreed that the heat pump did not perform well when removing late summer humidity, either. Until this heating season, the system had been operating well for the past two summers.

Some of the first checks were the routine amperage and temperature checks of the refrigerant system, which seemed to point to higher than normal indoor coil temperatures. Checking for airflow, the technician found that the volume of air across the indoor coil was lower than it should be and that the filter had just been replaced (verified by the customer). The technician also noticed an unusual noise when the indoor fan started. Checking the fan, he found that the set screw that held the blower wheel to the motor shaft had become loose. The blower wheel had been spinning on the shaft. The motor shaft was not scored, so the set screw was the only repair needed. Airflow then returned to the proper level.

valve. This valve is responsible for reversing the discharge from the compressor to the indoor coil (heating mode) to the outdoor coil (cooling mode) and back. (Please refer to additional information on air source heat pumps in *HVACR 101* and *HVACR 201*.)

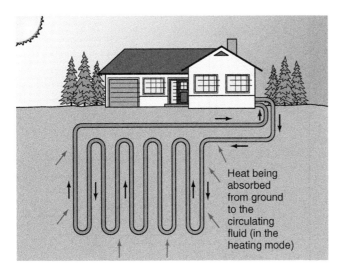

Figure 1–29
A ground-loop geothermal heat pump.

Geothermal Heat Pumps

Geothermal or ground-source heat pumps (see Figure 1–29) connect the outdoor coil to the ground or to a water source. At about five feet of depth (depending on location), ground temperatures are constant regardless of outdoor conditions. The ground-source heat pump takes advantage of this constant temperature and abundant heat for heating structures. Because of the constant temperature, ground-source heat pumps are more efficient than air-source systems. (Please refer to additional information on geothermal heat pumps in *HVACR 101, HVACR 201,* and *HVACR 401.*)

ALTERNATIVE ENERGY SYSTEMS

Alternative energy generally means energy sources other than conventional ones. Conventional sources are considered to be "fossil fuels." Energy that does not use a fossil fuel to generate light, heat, or mechanical motion is considered to be an alternative energy source.

As more alternative systems are installed, there will be an increase in the amount of service needed. Most alternative energy systems are specialized and require specific training for technicians.

Solar

Power from the sun is considered to be direct solar energy. Solar energy is responsible for the creation of fossil fuels, plant growth (biomatter), wind, and tidal action. But direct solar energy can be used to heat and to create electrical energy. Although direct solar energy can be used to produce mechanical motion, the result is so slight that it can only be demonstrated by a device called a radiometer (sometimes called a light-mill), which moves when sunlight is presented to a vane in a vacuum and held on a nearly frictionless bearing surface.

The most popular solar heating system is solar hot water. Water (or a propylene glycol mix) is heated by the sun in solar collectors, which are connected into an array (many panels). Water is circulated in much the same way as a boiler system, with all of the same controls and boiler system components: expansion tank, circulator, mixing valves, temperature differential controls, and so forth (see Figure 1–30). If the solar system is being used to heat water, there is a double-wall heat exchanger that separates the glycol and domestic hot-water systems (see Figure 1–31).

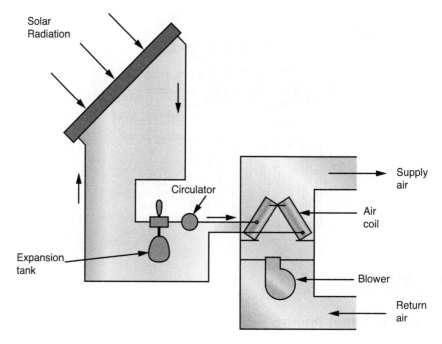

Figure 1–30
A diagram of a solar hot-water heating system connected to an air delivery system.

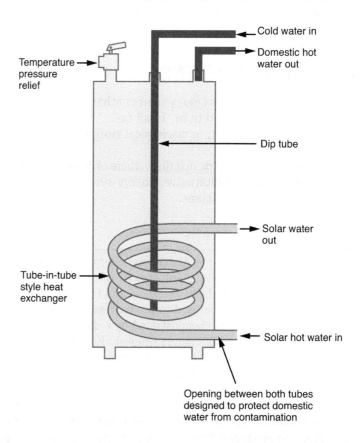

Figure 1–31
Cutaway of a solar domestic hot-water heater showing the solar double-walled heat exchanger.

Table 1–9 Checklist for Solar Hot-water Systems

Check	Procedure
Cleanliness	• Check water chemistry and/or glycol solution. • Check solar panel pressure drop. • Check and clean filters (if any).
Amperage	• Take circulator motor amperage and check against the name-plate rating (there may be more than one circulator).
General condition	• Check all electrical connection points for discoloration and looseness. • Check all wiring for discoloration and intact insulation. • Check condition of the solar panels (bolts and screws). • Check pipe insulation for gaps and missing insulation. • Check pipe hangers. • Check boots around pipe penetrations for leaks to the structure.
Water flow	• Ensure that the proper water level or pressure is achieved. • Measure the pressure drop to calculate flow. • Check flow for each solar panel and terminal device.
Safety controls	• Check the operation of high-limit controls by blocking the water flow and monitoring the temperature; compare against ratings on the control or manufacturer's specification. • Check relief valves. • Check expansion tank pressure.

Solar panels can also be used to heat radiant panels. Radiant floors operate at temperatures between 90 and 100 degrees Fahrenheit. Solar panels can take advantage of this operating temperature and deliver warm water to the radiant panel. Return water is only 10 degrees cooler, and the cycle continues.

Solar hot-water systems have similar checks as hydronic boilers; see Table 1–9.

Photovoltaic

Direct solar radiation can be transformed into electricity when photons strike semiconductor material (solar cell) and dislodge electrons. Electrons flow only in one direction through the semiconductor material. More photons striking the solar cell creates more free electrons. Solar cells are grouped into an array (many cells electrically connected) to produce greater voltages. Like batteries, a single

Case Study

The technician arrived at the job site, where the solar domestic hot-water system was not producing hot water. The system was a glycol solar loop with an intermediary heat exchanger to an auxiliary hot-water storage tank. The storage tank acted as a pre-heater during low solar times and as a storage tank during peak solar times. Inspecting the system, the technician noticed that the solar loop was not hot, even though it was a sunny day. Both the supply and return lines to the solar array were only warm. The solar TD controller was calling for solar storage. The pump was being signaled to operate and it was running. From this point, the technician knew that there were two things to check: the mechanical connection to the pump impeller and that the solar fluid system was not air-locked. The pressure gauge showed pressure of about 15 psig, indicating that the solar system had not leaked fluid. The only other check was the pump impellor. The pump was a magnetic drive system, so the motor could be removed from the pump body without losing any solar fluid. The pump and magnet drive seemed to work fine; therefore, the technician concluded that the impellor was seized. A new pump assembly was ordered, and the repair was rescheduled to replace the pump impellor assembly.

solar cell will produce only a limited voltage; in this case, approximately 0.5 volts. Solar cells are wired in series to produce a designed voltage, and then the series-wired gang of cells is wired in parallel with other gangs to produce larger amperages. This series-parallel group of cells is known as an *array*.

Photovoltaic electricity is DC (direct current). An inverter is used to convert DC to AC (alternating current) before being used with standard 120-V equipment. In some applications, DC can be used directly to operate such components as lights and heaters. The increasing prevalence of photovoltaic electricity should lead to more equipment being produced that will operate on DC (i.e., without the conversion losses associated with transforming DC to AC). All checks of the system require voltage and amperage testing.

Biofuels

Biofuel is a term that is used to describe a number of different energy sources. It is defined as a fuel (which can be solid, liquid, or gaseous) that come from recently dead biological matter. More recently, the *bio* prefix has been added to a fuel designation to indicate that it is derived from biomatter. For example, ethanol (which is an alcohol derived from fermenting wood products) has been renamed as bioethanol. A short list of biofuels includes:

- *Biogas*—sewer gas; methane; wood gas
- *Bioethanol*—ethanol alcohol
- *Biodiesel*—vegetable oil products
- *Biomass*—wood chips; wood products; plants; grasses

Some applications of biofuel use in HVAC are:

- *Heating with biomass*—integrated wood heating systems with conventional fuel backup
- *Biodiesel*—oil furnaces converted to operate on this oil rather than #2 diesel
- *Biogas*—converting conventional gas furnaces to operate on methane

No matter what the fuel, the same system and burner checks are conducted. Wood systems may use augers and additional sensors for control in the case of pellet stoves, but they all respond in the same way as conventional systems. A technician needs additional training and certification for these biofuel systems.

SUMMARY

This chapter focused on the differences and similarities in service and maintenance requirements for HVAC systems. System checklists were presented that
could be used as a general guide while on a service or maintenance call if there is
no other information available. Always consult the manufacturer's information.

REVIEW QUESTIONS

1. Describe three ways in which gas and oil furnace checks are similar.
2. Describe what is necessary to make the conversion from natural gas to LP.
3. Why is combustion testing so important?
4. What needs to be done to conduct a complete combustion analysis?
5. Why is it necessary to bring in fresh air?
6. What are the differences in checks made for air-source and ground-source
cooling systems?
7. What are alternative energy systems?
8. What are the similarities between solar hot-water systems and hydronic
systems?

CHAPTER

2

TAB (Testing, Adjusting, and Balancing)

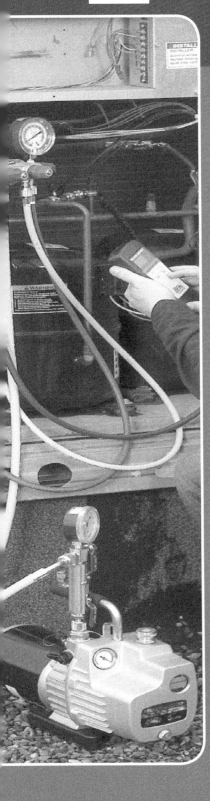

LEARNING OBJECTIVES

The student will:

- Describe terminal devices
- List meters and instrumentation that are part of TAB
- Describe the differences in manometers
- Describe the relationship between velocity and volume
- Describe the use of flow control devices
- Explain duct traversing
- List TAB methods
- Describe how TAB data is stored

INTRODUCTION

It has only been within the last 30 years that systems have been tested, adjusted, and balanced after they were installed. System design has become more complicated, and systems are now being designed, installed, and maintained to conserve energy while providing optimum comfort levels. For both of these reasons, TAB (testing, adjusting, and balancing) has become more important. System engineers and building owners want to know that the system is operating as specified with efficiency and dependability.

During this time period, the Associated Air Balance Council (AABC) as well as the National Environmental Balancing Bureau (NEBB), Testing, Adjusting, and Balancing Bureau (TABB), and Air Conditioning Contractors of America (ACCA) have worked to develop programs that increase the standards for installation, TAB, and maintenance. These organizations have members that conduct TAB work on a daily basis, and the organizations certify individuals and businesses to conduct TAB work for the general public. These organizations are good sources of information about the TAB field and on specialized TAB requirements for residential and commercial applications. Persons interested in this part of the HVAC industry should investigate the resources supplied by associations like these.

SYSTEM COMPONENTS

Testing, adjusting, and balancing relies on the use of existing valves, dampers, gauges, and terminal devices. Each component plays an important role in the TAB of a system.

Dampers

Dampers are air-devices used to slow, modify, or regulate the flow of air (see Figure 2–1). Dampers come in many different styles and configurations. The two main classifications are single- and multiple-blade dampers. A single-blade

Field Problem

All of the components of the heating system seemed to be working, but the customer was complaining that the last room on the duct run was continually cold on cold days. All of the specifications from the TAB work were available and showed all of the test and balance data.

Taking the data printout of the ducting and distribution system, the technician selected the meters and test instruments needed to verify the recorded readings. Starting with the last room on the run, she measured the speed (velocity in feet per minute) of the air with an anemometer and used the Ak factor of the diffuser to calculate the cfm, which was much lower than the original TAB data indicated (see Figure 2–2). Something must have changed since the TAB work was conducted. The equation for calculating the cfm is:

cfm = velocity (feet per minute) × area (square feet) × Ak factor

A test of the next two rooms on the run also showed lower than specified volume. Without testing any more of the rooms,

the technician conducted an inspection of the air handling system. She already had a suspicion and wanted to confirm or rule it out. Looking for the balance damper on the first run, she looked for the position of the handle. The balance damper was located in a maintenance closet close to the system plenum. The ceiling of the closet was not finished, so it was easy to find the damper and see the position of the handle. As expected, she found the damper handle in the closed position.

Using the information supplied by the TAB report, she measured airflow and adjusted the damper so that the first room was receiving the right volume of air. Going to the last room on the run and conducting another test, she measured a volume of air that was close to the value shown on the TAB report. She put some duct tape on the handle of the first run damper to keep it from moving and then reported her findings and adjustment to the building owner.

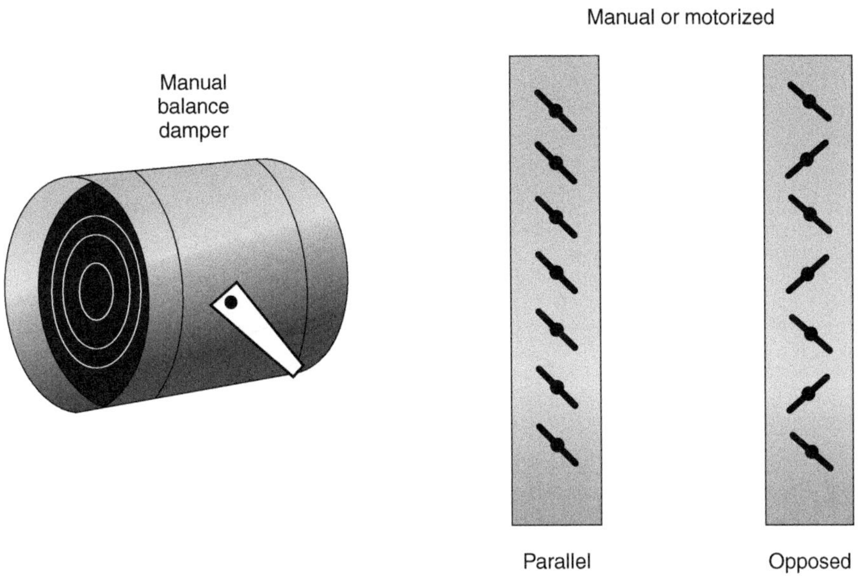

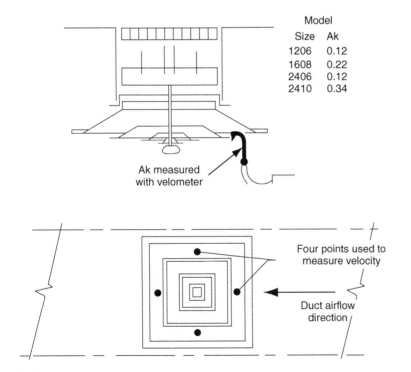

Manual
balance
damper

Manual or motorized

Parallel

Opposed

Figure 2–1
Dampers used to control airflow. Parallel dampers are better suited for on–off control, and opposed dampers are better suited for mixing and modulation.

Model

Size	Ak
1206	0.12
1608	0.22
2406	0.12
2410	0.34

Ak measured
with velometer

Four points used to
measure velocity

Duct airflow
direction

Figure 2–2
Ak factors (manufacturer data).

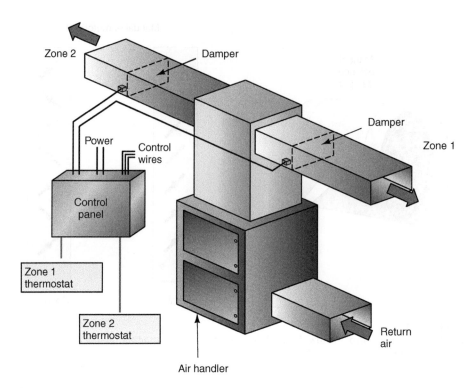

Figure 2–3
Motorized dampers connected to a control panel in a two-zone system.

damper can be as simple as a manual damper found in a round duct branch as it leaves the main supply duct. More complicated are the multi-blade dampers found in air handlers to control total air supply. Multi-blade dampers can be further categorized as opposed (blades closing toward each other) or parallel (blades closing with each other).

Dampers can be manual or automatic. Manual dampers use handles or screw mechanisms to manually open or close the damper. Manual dampers are only periodically adjusted as needed to affect air flow. Automatic dampers are designed to operate to meet demands required for the occupied space. A motorized damper (see Figure 2–3) is an example of an automatic damper that is designed for zoned forced air systems. This damper is directly controlled by the room or space controller, and it opens or closes in response to the room thermostat. Each room could have its own damper and a thermostat that could be used to meet the needs of the occupant, regardless of the heating or cooling needs of other rooms in the building.

Valves

Valves are generally considered to be fluid flow controls. Valves come in various styles and are designed to be used for specific purposes. Some valves are designed for variable flow control, and others are designed simply to stop the flow.

Valves used for flow control have two classifications: manual and automatic. Manual valves are used to adjust flow semipermanently. Automatic valves allow controls to adjust the valve to meet changing conditions.

Common valves that are adjusted and operated during TAB include:

- Flow control valves (Figure 2–4)
- Balance valves (circuit setter) (Figure 2–5)
- Bypass valves (Figure 2–6)
- Mixing valves (Figure 2–7)
- Three-way valves (Figure 2–8)

Figure 2–4
Flow control valve used to prevent the zone from overheating because of gravity circulation
(commonly referred to as "ghost flow").

Figure 2–5
Balance valve (also called a circuit setter) showing percentage scale and two access fittings
used to connect a manometer for measuring pressure drop across the valve.

Figure 2–6
Bypass valve used to allow access water to bypass during times when there is reduced heat demand.

Figure 2–7
Mixing valve used to mix water to maintain a constant output water temperature.

Figure 2–8
Three-way valve used to redirect water flow from one heating loop to another or from a bypass loop to a heating loop.

Gauges

Standard and specialty gauges are used to determine temperature, pressure, and/ or flow of fluids (air, water, and other heat transfer fluids). Measurements of temperature in Fahrenheit or Celsius can be read when thermometers are permanently attached to the system. When it is necessary to take external measurements, handheld temperature measuring equipment is used. The same is true with other gauges that are permanently attached to the system: conditions such as velocity, static pressure, and volume may be read directly. Otherwise, handheld equipment must be used.

Many installations today are attached to digital monitoring and control systems. Sensors placed throughout the system allow an operator to check, from a central location, the condition of the system or of a particular occupied space. The operator can also adjust the system to meet specific temperatures, pressures, and volume.

Common gauges used for testing and balance are:

- Thermometer (liquid, dial, contact, infrared, and electronic)
- Bourdon tube gauge
- Electric (digital) manometer (Figure 2–9)
- Inclined manometer
- U-tube manometer (Figure 2–10)
- Magnehelic® gauge (Figure 2–11)
- Digital and analog sensors
- Hot-wire anemometer
- Water pressure differential gauge
- Vacuum gauges

Terminal Devices: Diffusers (Air) and Radiators (Water)

Terminal devices are considered to be the last point of control in an HVAC system. In an air delivery system, the terminal device is the diffuser. Air moving through the diffuser is controlled for things like volume, pressure drop, velocity, and

Figure 2–9
Digital manometer. (Courtesy of UEi)

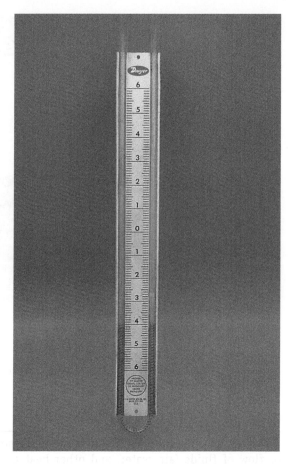

Figure 2–10
U-tube manometer. (Photo by Bill Johnson)

Figure 2–11
Magnehelic® gauge. (Photo by Bill Johnson)

directional pattern. After the air moves outside of the directional pattern, the air is outside of the control of the air delivery system.

Common terminal devices are:

- Diffusers, grilles, or registers (Figure 2–12)
- Baseboard convectors (Figure 2–13)
- Fan-coil units (Figure 2–14)
- Radiators and radiant panels (Figure 2–15 and Figure 2–16)

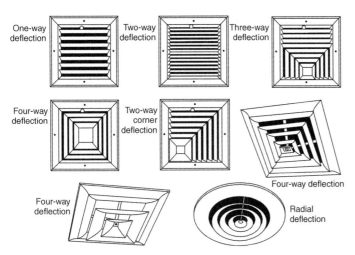

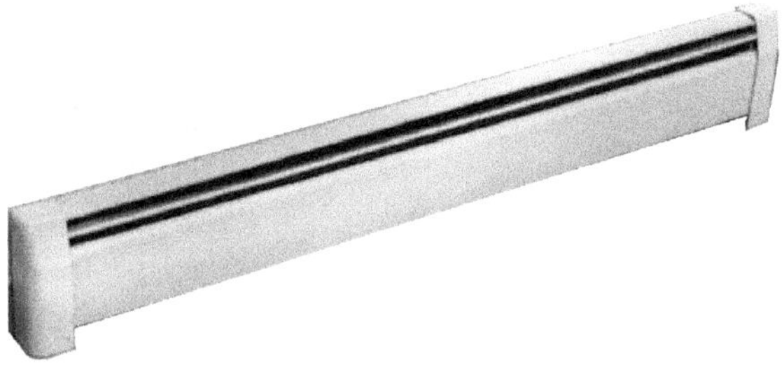

Figure 2–13
Baseboard terminal unit. (Courtesy of RSES)

Figure 2–12
An assortment of air terminal devices. (Courtesy of RSES)

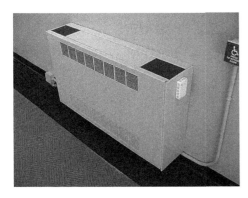

Figure 2–14
The fan coil unit is similar to baseboard units but has a blower that circulates air over the hot surface of the terminal unit. (Courtesy of Ferris State University. Photo by John Tomczyk.)

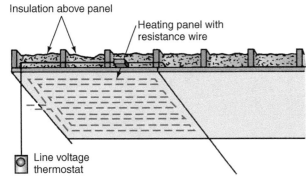

Figure 2–16
A radiant ceiling heating panel.

Figure 2–15
Water and steam in the radiator radiate heat out into the room. (Photo by Bill Johnson)

METERS AND INSTRUMENTATION

In addition to the typical or installed measuring equipment, the following specialized tools, meters, and instrumentation are used for testing and balancing operations. All of the instruments, tools, and meters that measure temperature, pressure, and air flow must be calibrated. *Calibration* means that the meter has been certified to be accurate within a specific tolerance before it can be used in the testing and balancing procedure. Tools, meters, and instrumentation must also be recorded in the TAB forms as having been calibrated within a certain period of time along with their certified tolerances.

Manometers

There are several types of manometers. Each type is used to measure specific pressures.

- *U-tube manometer* (see Figure 2–10)—measures in inches (and fractions of inches) of w.c. (water column). The U-tube manometer is typically used for checking natural gas or LP fuel gas pressure.
- *Inclined manometer* (see Figure 2–17)—measures w.c. in tenths of an inch. Used to measure static pressure in air ducts or pressure drop across various components in the HVAC system as well as draft in chimneys. The manometer needs to be filled and leveled before it can be used.
- *Digital manometer*—measures pressure in inches of w.c. as well as other selectable pressure scales (Pa, kPa, MPa, mbar, bar, mmHg, and psi). The digital manometer does not require a liquid column and so is easier to carry and use in the field.

Temperature Measurement and Recorders

Temperature measurements can be made by reading thermometers, gauges, and sensors. There are also some specialty thermometers that are used for testing and balancing.

Common temperature measurement instruments include:

- Liquid thermometers—standard fluids are alcohol and mercury
- Contact thermometers
- Infrared thermometers

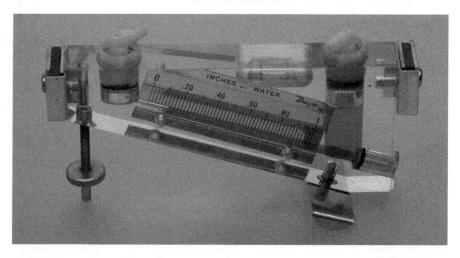

Figure 2–17
Inclined manometer. (Photo by Bill Johnson)

- Bimetallic/dial indicating thermometers
- Thermocouple/digital thermometers
- Psychrometer (sling or aspirating; compares wet- and dry-bulb measurements)
- Thermal hygrometer

Temperature recording can be done using digital thermometer probes connected to a time recording device. The temperature is recorded either on paper or digitally to be retrieved by wire, infrared, or wireless signal to a computer.

Anemometer

The anemometer (also known as a velometer) is an instrument that measures the velocity of air. The instrument comes in two styles.

- *Hot-wire anemometer*—this device utilizes the temperature change of a hot wire. As air moves over the hot wire, the temperature of the wire and the resistance of the wire changes. Changes in the resistance are calibrated to the gauge that measures air speed in fpm (feet per minute).
- *Rotating vane anemometer*—this device uses a propeller that spins as air moves through the blades (see Figure 2–18). The speed of the propeller is calibrated to the gauge, which reads in fpm. The rotating anemometer gauge can be either mechanical (like a stop-watch) or digital with an LCD (liquid crystal display; see Figure 2–19 and Figure 2–20).

Air Measuring Device (Capture Hood)

The air measuring device is placed over a diffuser or grille to direct all of the air through a test port in the capture hood (see Figure 2–21). Because it captures all of the air, the hood can measure the volume of air in velocity and cfm. Direct measurement of the volume of air can thus be made quickly and easily. The alternative to using the hood is to measure the velocity and calculate the volume using an Ak (average constant) factor supplied by the manufacturer of the diffuser.

Figure 2–18
Vane anemometer. (Photo by Bill Johnson)

Figure 2–19
Digital anemometer. (Photo by Eugene Silberstein)

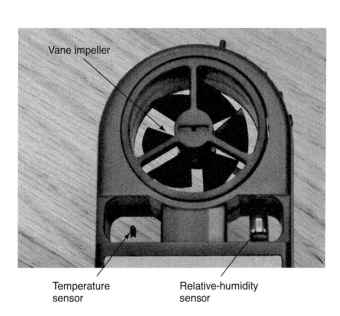

Vane impeller

Temperature
sensor

Relative-humidity
sensor

Figure 2–20
Close-up of digital anemometer showing temperature and
humidity sensors. (Photo by Eugene Silberstein)

Figure 2–21
Capture hood for measuring cfm at a diffuser.

Pitot Tube

Pronounced *PEA-tow* tube, this device is used to probe air ducts to determine air velocity and air pressure. Both of these measurements can be used to determine the volume of air moving through a section of the duct. Often the pitot tube is used to "traverse" a duct or duct opening. Traversing is the process of taking many readings by dividing the opening into many imaginary squares. There are specific points where you take the air velocity readings based on the dimensions and shape of the duct (see Figure 2–22). The more readings that are averaged, the more accurate is the final measurement calculation.

Tachometer

The tachometer is used to measure the speed of rotating shafts of blowers, blower motors, and other rotating equipment. The tachometer measures in rpm (revolutions per minute) and can be used to verify the operational speed of rotating equipment. There are two basic types of tachometers.

- *Contact tachometers*—the tip of the tachometer is placed on the end of a rotating shaft, and the measurement of speed is read directly on the gauge.
- *Noncontact tachometers*—light is used to measure the rotational speed in one of two ways:
    - *Strobe-tach*—a strobe light is used to match the speed of the rotational equipment. As the strobe light is increased or decreased to the same speed of the equipment, the rotating component will appear to be stationary.
    - *Photo-tach*—counts the number of times a "chalked" mark passes the sensor. A piece of chalk or paint stick is used to mark one spot on a rotating shaft. When the shaft is rotating, the sensor of the photo-tach is placed where light from the photo-tack can bounce off of the white mark. The photo-tach counts the number of passes per minute (i.e., the rpm).

For example, you can use rpm and the fan laws to determine cfm, static pressure, and motor horsepower with the aid of a small paper calculator (see Figure 2–23). The beginning values of cfm, static pressure, horsepower, and rpm are aligned under the movable cursor. Holding the disk in place, the cursor can be moved to a new rpm value; then, under the cursor will be found the corresponding values of cfm, static pressure, and horsepower at the new rpm value.

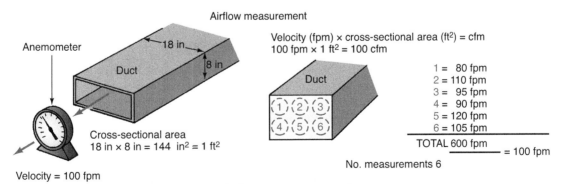

Figure 2–22
To measure rectangular duct, divide the duct into six to twelve areas and then measure the velocity of the air in each area. Add all the measurements together and then divide by the number of measurements to obtain the average velocity.

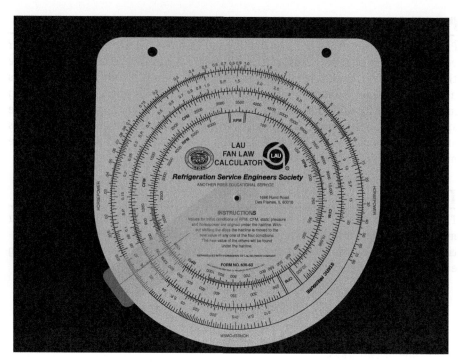

Figure 2–23
Fan law paper calculator provided by RSES and LAU.

Flow Meter (Air Measuring Device)

Flow meters read directly in cfm (cubic feet per minute) or gpm (gallons per minute). All or a calibrated portion of the flow—either air or water—must go through the flow meter.

- *Air flow meter*—this device measures the amount of air moving through the meter in cfm. It can be attached to a capture hood to gather air moving from a terminal device. The instrument can also be connected directly to a section of ductwork.
- *Water flow meter*—this device is physically placed in the piping system. All or a calibrated portion of fluid moves through the device, which is sized for the flow and pipe diameter at the point of measurement. The meter must also be calibrated to the fluid type, though most are designed for water. If a fluid other than water is used, the meter will need to be calibrated for that particular fluid.

Water Differential Pressure Gauge

This gauge (which is also known as a flow meter) is installed upstream and downstream across a circuit sensor (a calibrated orifice plate; see Figure 2–24). The water flow creates a pressure difference (drop) that the meter can read (see Figure 2–25). Pressure drop is directly related to volume in gpm.

VELOCITY AND VOLUME RELATIONSHIPS

Velocity and volume have direct relationships when they are under a particular pressure. An example is an air system that moves conditioned air through a diffuser. The air is under a specific static pressure as it moves through a diffuser. Both the velocity (speed of air) and the volume (quantity of air) are directly

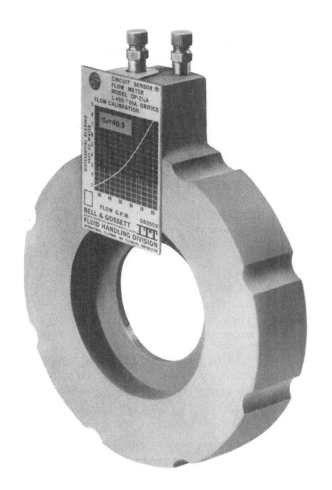

Figure 2–24
Circuit sensor. (Courtesy of RSES)

Figure 2–25
Circuit sensor placed in the piping and connected to the water differential pressure gauge.
(Courtesy of RSES)

related. More or less pressure will change both the air velocity and volume. And, as the speed of air slows through the diffuser, the volume of air reduces. With a change in velocity at the terminal unit, the distribution pattern of the diffused air can change significantly.

Pumps (Water)

Pumps used to produce water flow are called *circulators* or *centrifugal pumps* (see Figure 2–26). Circulators are not positive-displacement pumps (like piston-style pumps). Circulators are able to respond to variable fluid flow and will continue to spin even if the fluid flow stops (sometimes referred to as dead-heading the pump). The centrifugal action of the water as it is slung away from the center of the pump impeller and captured in the scroll explains why these are also called centrifugal pumps (see Figure 2–27).

Pumps can be modified in several ways to meet test and balance specifications:

- Impellers can be cut down
- Different impellers can be swapped out
- Motor size can be changed
- The speed of the motor can be changed with motor controls

Each change to a circulating pump means changes in the pump curve. Only knowledgeable design people should make modifications to pumps. If a modification is needed, application engineers should be consulted before it is actually made.

Pressure Curve Charts (Water)

Pressure curves for heat exchangers and pumps show the relationship between pressure and volume (see Figure 2–28). As the pressure changes, the amount of

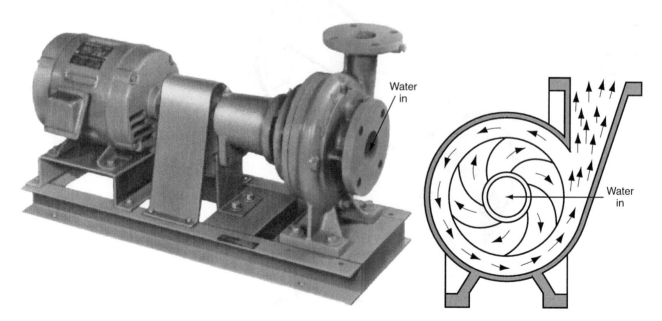

Figure 2–26
Horizontal base-mounted centrifugal pump. (Courtesy of RSES)

Figure 2–27
Water flow through a centrifugal pump. (Courtesy of RSES)

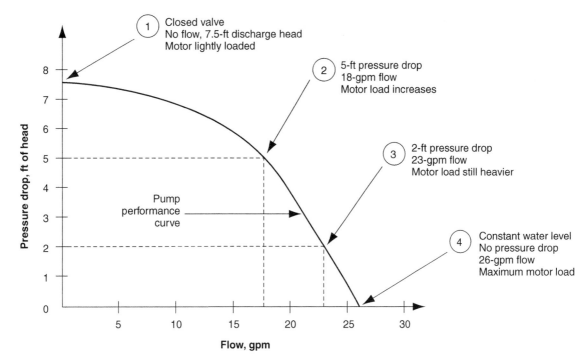

Figure 2–28
Generic pump performance curve. (Courtesy of RSES)

volume (in gallons per minute) changes. If there were no piping or other components, the pump would pump at the maximum volume. As the flow of water in a system encounters resistance (head pressure), the flow of water reduces in a relationship that can be plotted on a graph (see Figure 2–29).

Tech Tip

Pressure differential simply means pressure difference. The pressure differential of a pump is determined by measuring the suction pressure and subtracting it from the discharge pressure (see Figure 2–30). Maximum pump differential is the same as maximum head pressure. The higher the head pressure (pressure differential), the lower the flow volume (gpm). In heat exchangers, pressure differential works in the opposite way: the higher the pressure differential (difference in pressure between the inlet and outlet), the greater the fluid flow (gpm). In a closed system, a pump approaches the maximum flow rate to an extent that depends on the resistance of the piping and heat exchangers. The pump head pressure will be maximum head pressure—or less, because of the frictional resistance of fluid flow in the system.

The system resistance can be plotted in graph form. Where the system resistance (pressure) intersects the pump curve is called the *point of operation* (see Figure 2–31). At these conditions of pressure and volume is when the system will operate best.

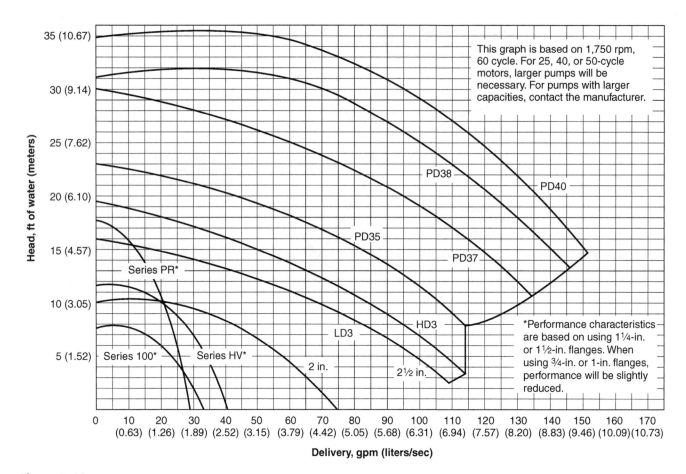

Figure 2–29
Pump curves for several different line-mounted pumps. (Courtesy of RSES)

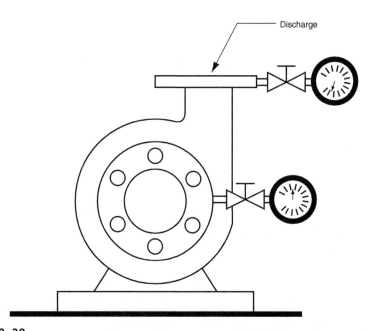

Figure 2–30
Gauge readings on the suction and discharge port of a pump can be subtracted, one from the other, to calculate the pump's operational head pressure. (Courtesy of RSES)

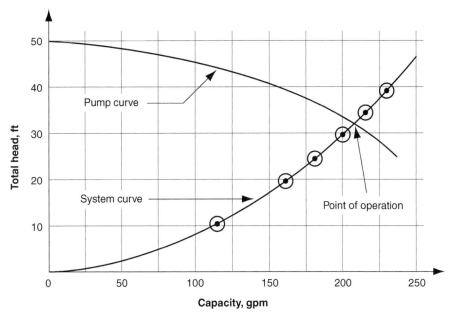

Gallons per minute	Feet of head
115	10
165	20
185	25
200	30
215	35
230	40

Figure 2–31
The system curve can be plotted on the same chart with the pump curve. These two curves intersect at the point of operation (volume and pressure). (Courtesy of RSES)

Fans (Air)

Fans are similar in many ways to pumps. Both devices are built and operate in similar ways. Both circulate a fluid rather than positively pump the fluid. Fans use centrifugal force to move air from the center of the device and into the scroll. Two types of fans are used in HVAC, axial and radial. Axial fans (see Figure 2–32) are propeller-style fans, and radial fans (Figure 2–33) are often called squirrel cages. There are three styles of blade configurations for squirrel cage fans: forward curved, backward curved, and flat. Each blade configuration gives the blower a different fan curve characteristic.

Squirrel cage fans can modify the pressure and volume of air by:

- Changing squirrel cage blade configurations
- Changing the diameter of the squirrel cage
- Changing the speed of the blower wheel

As with pumps, fan changes result in changes to the fan curve. Only knowledgeable people should make modifications to fans. If a modification is needed, application engineers should be consulted before it is made.

Pressure Curve Charts (Air)

All of the things that can be said for the water pressure curves can be said also for the air curves (see, for example, Figure 2–34). Both water and air are

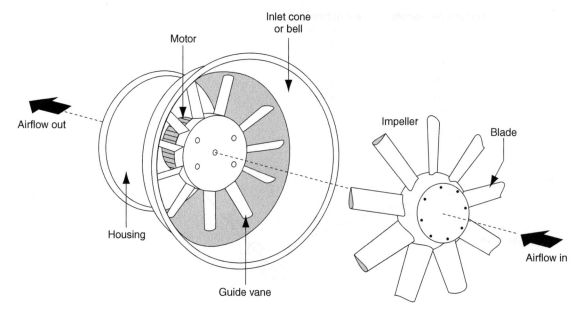

Figure 2–32
Axial fan and components. (Courtesy of RSES)

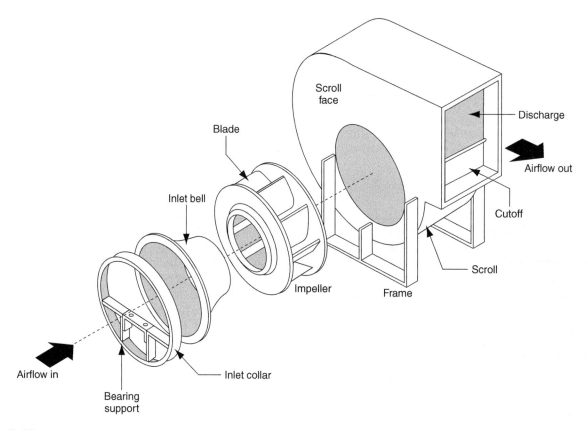

Figure 2–33
Radial fan and components. (Courtesy of RSES)

fluids—albeit with different viscosities. Water and air operate in much the same fashion. One difference that is not seen in the operation of these two fluids is that air has a compressive ability (air can be compressed); water cannot.

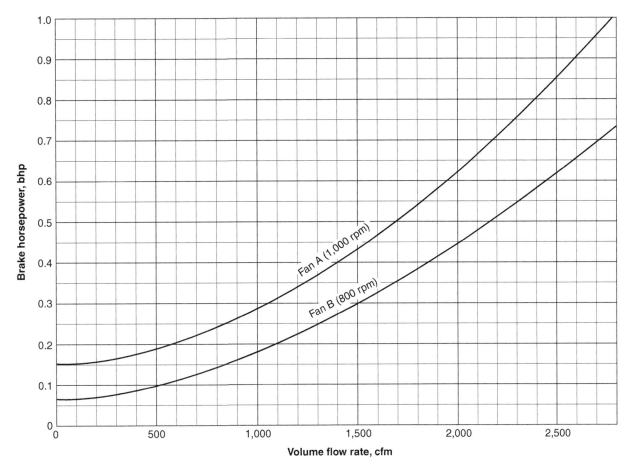

Figure 2–34
Fan curve for the same fan operating at two different rpm values. (Courtesy of RSES)

The system has a resistance that can be plotted. All of the system components in an air system offer some form of resistance (static) to air flow. Where the system static and the fan static intersect on the graph is where you will find the greatest volume of air the fan can deliver to the system (see Figure 2–35).

FLOW CONTROL (WATER)

Water-based cooling systems are part of hydronic heating and cooling systems. Hydronic systems are used to move large quantities of heat over long distances. There are many applications where water-based systems are used in commercial and industrial HVAC.

Application

Volume control is applied to many different heating and cooling configurations. It can be applied to heating runs for an entire floor and ceiling of radiant panels, yet it can also be applied to a single heat exchanger. Typical applications are:

- Turning fluid flow on or off (Figure 2–36)
- Mixing fluids of two different temperatures
- Bypassing components or heat exchangers

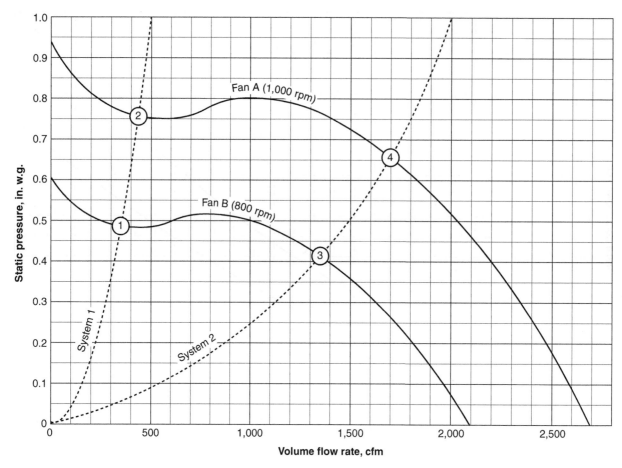

Figure 2–35
System curves plotted with the fan curve for the same fan operating at two different rpm values. (Courtesy of RSES)

- Limiting and/or setting the flow for a circuit
- Regulating flow on a variable-demand component

Types and Operation

The most common types of volume control may be listed as follows.

- *Circuit setter*—a regulating valve with a scale that shows the percentage of valve opening (Figure 2–37).
- *Three-way and four-way valves*—mixing valves that combine two or more fluid streams and produce a mixed volume (Figure 2–38, Figure 2–39).
- *Bypass valves* (also known as *diverting valves*)—valves that allow fluid flow around other devices in the system (Figure 2–40). A heat exchanger could be fitted with a bypass valve to allow a portion of the water to flow around the heat exchanger, modifying the heat exchanger's capacity.
- *Globe valves*—a valve configuration, hand operated or motorized, in which the fluid flow must move at right angles to the flow in the connecting pipe (see Figure 2–41). This configuration allows for better control of the fluid through the valve seat.
- *Regulating valves*—this valve could be set for a particular pressure to allow or disallow flow to system components. It could also be a motorized valve that regulates the amount of fluid passing through the valve.

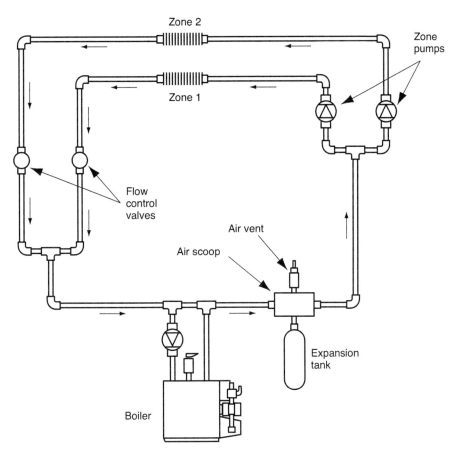

Figure 2–36
Flow control valves in a multizone pump system. (Courtesy of RSES)

Figure 2–37
Circuit setter (or pressure differential balance) valve. (Courtesy of RSES)

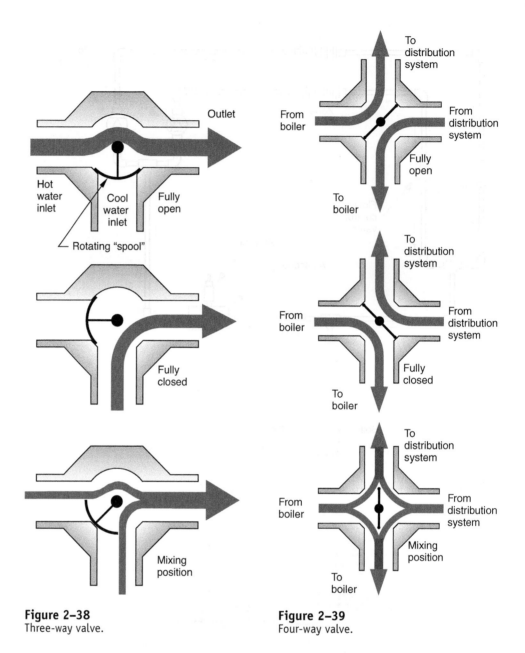

Figure 2–38
Three-way valve.

Figure 2–39
Four-way valve.

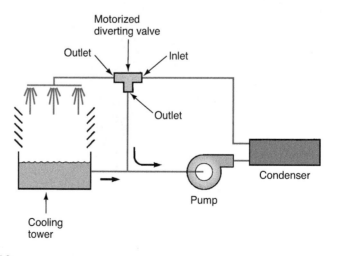

Figure 2–40
A bypass (diverting) valve used to bypass the cooling tower.

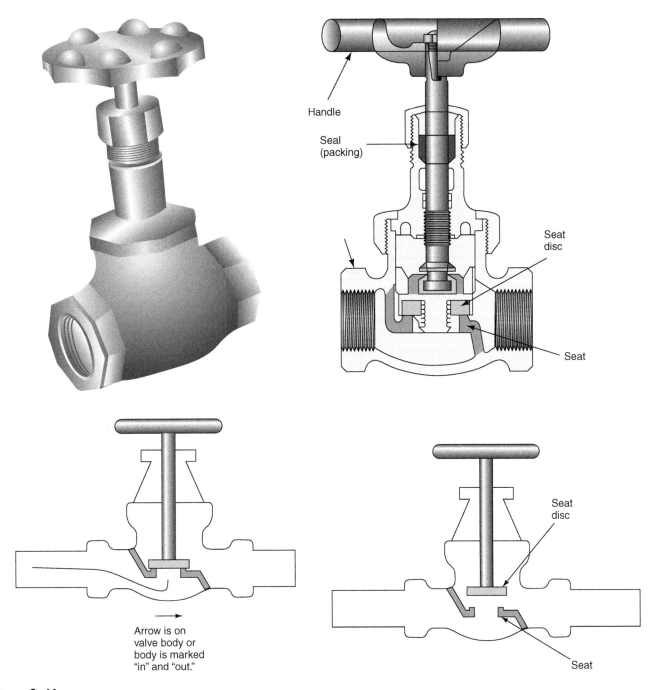

Figure 2–41
Globe valve operation; four manual valves are shown, but motorized valves operate the same way.

Testing and Setting

Balancing valves can be tested, evaluated, and set for a particular HVAC application. Typically, they come from the manufacturer with test specifications. Placed in a system, the valve can exhibit some different operating characteristics. One of the tests is pressure drop: valves can be tested for pressure drop at various percentages of operation (more open or more closed).

Testing and setting the balancing valve assures that fluid flow is correct at that point. Before you can test and set a valve, there must be test fittings on the valve or installed ahead of and behind the valve. The valve test steps are as follows:

1. Connect a water pressure differential gauge to the upstream and downstream fittings.
2. Read the pressure difference.
3. Adjust the valve to the new pressure difference that corresponds to the new fluid flow.
4. Disconnect the water pressure differential gauge.

Troubleshooting

Balancing valves are often checked for operation and setting. Because these valves can move, they are the first thing suspected if the system is not working. Table 2–1 provides a short troubleshooting checklist.

TEST AND BALANCE METHODS

Traversing

In *traversing*, many measurements are made in a regular pattern and then averaged to determine airflow. The traverse is made using instruments that will not appreciably disturb the airflow. Traversing is done for coils, filters, supply ducts, and return ducts. Either an anemometer or a pitot tube connected to a manometer is used to determine the air velocity at predetermined points.

Air does not uniformly flow in a duct. If it did then we could simply take a single reading. Instead, many readings are necessary, after which the readings are averaged to determine the overall flow and pressure in the duct. Some type of grid pattern is necessary to keep all of the readings in order. Technicians can make up their own grids. The more grid spaces and readings, the more accurate the resulting calculation will be. Figure 2–42 depicts a recommended pattern and describes how the readings are recorded. If twelve velocity readings are made, then the readings should be added together and divided by twelve to determine the average velocity. Here's an example:

Point 1 = 80 fpm

Point 2 = 150 fpm

Point 3 = 140 fpm

Point 4 = 60 fpm

Point 5 = 100 fpm

Point 6 = 300 fpm

Point 7 = 300 fpm

Point 8 = 90 fpm

Point 9 = 40 fpm

Point 10 = 125 fpm

Point 11 = 130 fpm

Point 12 = 70 fpm

Total 1,585 fpm ÷ 12 readings = 132 fpm

Pressure and Volume

There are two ways to look at pressure and volume. In the first, pressure and volume are directly related and increase or decrease together through a defined opening. The key to remember is that the defined opening is fixed (if it becomes

Table 2–1 Troubleshooting Checklist for Balancing Valves

Problem	Check
No flow	Check for power at the valve or valve position.
Flow is too high	Check pressure drop across the valve and valve position (close valve); also check for high pressure to the valve.
Flow is too low	Check pressure drop across the valve and valve position (open valve); also check for low pressure to the valve.
Not mixing	Check temperatures on all pipe connections to the valve. Check for valve position.
Too much heat at the heat exchanger	Check bypass valve position (open valve).
Too little heat at the heat exchanger	Check bypass valve position (close valve).
System flow is too high	Check pump operation and adjust.
Circuit flow is too high	Adjust circuit setter; check for high system pressure.

larger or smaller then this relationship no longer holds). When pressure increases there is more force on the fluid, whether it is air or water. That force pushes more fluid through the defined opening, which results in an increase of volume. Likewise, when the fluid volume goes down there is a corresponding drop in pressure. An example where this effect can be measured is that of air moving through a diffuser.

The second way of looking at pressure and volume incorporates valves. If a valve is used to regulate fluid flow then pressure and volume again work together, but now in reverse. Starting with a valve in the open position, fluid volume and pressure are stabilized (or have maximized flow for the opening). As the valve closes, pressure drop across the valve begins to change. Increasing pressure across the valve, in this case, starts to reduce the flow because the valve is acting as a resistive force to fluid flow. Continuing to close the valve increases the pressure drop across the valve and further reduces the flow of fluid.

The relationships between (static) pressure, (air) volume, and fan speed are summarized by the following three *fan laws* (also known as *affinity laws*).

1. Air volume is directly proportional to fan speed. This means that if the fan rpm doubles, then so does the cfm: $cfm_1 \div cfm_2 = rpm_1 \div rpm_2$.
2. Fan static is proportional to the *square* of the fan speed. This means that if the fan speed doubles, then fan static increases by a factor of 4 ($= 2^2$): $sp_1 \div sp_2 = (rpm_1 \div rpm_2)^2$.
3. Power is proportional to the *cube* of the fan speed. This means that if the fan speed doubles, then the amount of power consumed increases by a factor of 8 ($= 2^3$): $hp_1 \div hp_2 = (rpm_1 \div rpm_2)^3$.

TAB Procedures

During a testing, adjusting, and balancing procedure, the controls specialist will often need to operate the system. Representatives from the OEM (original equipment manufacturer) may also be on-site. If the building has a facilities manager, then that individual will be one of the most important participants and will work

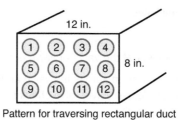

Pattern for traversing rectangular duct

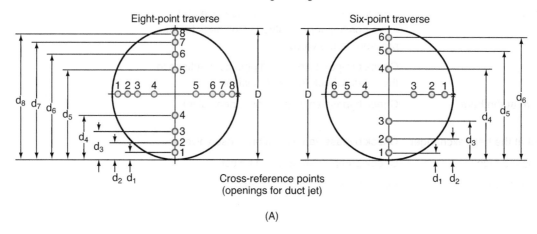

Cross-reference points
(openings for duct jet)

(A)

TRAVERSE METHOD ⬇	PROBE IMMERSION IN DUCT DIAMETERS									
	d_1	d_2	d_3	d_4	d_5	d_6	d_7	d_8	d_9	d_{10}
6 POINT	0.043	0.147	0.296	0.704	0.853	0.957	–	–	–	–
8 POINT	0.032	0.105	0.194	0.323	0.677	0.806	0.895	0.968	–	–
10 POINT	0.025	0.082	0.146	0.226	0.342	0.658	0.774	0.854	0.918	0.975

(B)

DUCT. DIA. (IN.)	PROBE IMMERSION FOR 6-PT. TRAVERSE					
	d_1	d_2	d_3	d_4	d_5	d_6
10	3/3	1-1/2	3	7	8-1/2	9-5/8
12	1/2	1-3/4	3-1/2	8-1/2	10-1/4	11-1/2
14	5/8	2	4-1/8	9-7/8	12	13-3/8
16	3/4	2-3/8	4-3/4	11-1/4	13-5/8	15-1/4
18	3/4	2-5/8	5-3/8	12-5/8	15-3/8	17-1/4
20	7/8	3	6	14	17	19-1/8
22	1	3-1/4	6-1/2	15-1/2	18-3/4	21
24	1	3-1/2	7-1/8	16-7/8	20-1/2	23

(C)

DUCT. DIA. (IN.)	PROBE IMMERSION FOR 8-PT. TRAVERSE							
	d_1	d_2	d_3	d_4	d_5	d_6	d_7	d_8
10	5/16	1	2	3-1/4	6-3/4	8	9	9-5/8
12	3/8	1-1/4	2-3/8	3-7/8	8-1/8	9-5/8	10-3/4	11-1/2
14	7/16	1-1/2	2-3/4	4-1/2	9-1/2	11-1/4	12-1/2	13-1/2
16	1/2	1-5/8	3-1/8	5-1/8	10-7/8	12-7/8	14-3/8	15-1/2
18	9/16	1-7/8	3-1/2	5-7/8	12-1/4	13-1/2	16-1/8	17-1/2
20	5/8	2-1/8	3-7/8	6-1/2	18-1/2	16-1/8	17-7/8	19-3/8
22	11/16	2-3/8	4-1/4	7-1/8	14-7/8	17-3/4	19-3/4	21-1/4
24	3/4	2-1/2	4-5/8	7-3/4	16-1/4	19-1/2	21-1/2	23-1/4

(D)

Figure 2–42
To determine the average velocity of a round duct, measure the velocity for either the eight-point or six-point patterns shown in (A). Use the tables in (B)–(D) to determine the velocity probe measurement points. Add all velocity measurements and divide by the number of measurements to determine the average velocity.

directly with the TAB technician during the testing, adjusting, and balancing. Listed below, by category, are the general procedures for testing, adjusting, and balancing operations.

General Preliminary Procedures
1. Review all documents and plans for all HVAC system components.
2. Review "as-builds" (if any)—these are shop drawings or changes to the prints.
3. Review system schematics of piping and duct runs.
4. Record preliminary data.
5. Review electrical requirements of system components.
6. Review balance control and locations.
7. Confirm that all test equipment and measurement sources are calibrated.
8. Confirm that all system components are flushed, clean, and have been tested for operation.
9. Confirm that all system components are sealed and in place (ducts sealed, access doors are in place, etc.).

Air System Preliminary Procedures
1. Set all dampers in the full open position.
2. Set outside air dampers in minimum position.
3. Verify fan rotation speed.
4. Check and adjust fans to design volume.
5. Verify blower motor amperage and voltage.

Hydronic (Water) System Preliminary Procedures
1. Set all balancing devices to full open position.
2. Set mixing valves to full flow.
3. Close any bypass valves.
4. Verify pump rotation and drive alignment.
5. Verify pump motor amperage and voltage.
6. Confirm that the fluid system is full or at correct fluid levels.
7. Confirm automatic air purge is working.
8. Confirm operation of systems (boiler, expansion tank, chiller, cooling tower, etc.).

Air System TAB
1. Ensure that air exhaust systems are functioning.
2. Make sure all air flow dampers are in the full open position.
3. Traverse the supply duct and branch runs.
4. Adjust balance dampers to meet specifications, starting at the supply plenum.
5. Record data for each trunk and branch.
6. Check for duct leakage and seal as required.
7. Make at least two repetitions of the entire checklist when adjusting dampers to design specifications.
8. Record air delivery amounts at terminal device(s).
9. Adjust terminal device vanes for room air patterns.
10. Measure all heating, cooling, and air cleaning conditions.
11. Measure and record all system data (static, volumes, velocities, temperatures, and pressures).
12. Measure and record final (operating) fan motor voltage and amperage.

Hydronic System TAB
1. Continually vent all air from the system during test. Be sure all air is removed from system while doing TAB work.
2. Dead-head the pump and record the operating curve.
3. Adjust pump to design flow.
4. Adjust boiler, chiller, cooling tower, and so forth to design flow and temperature.

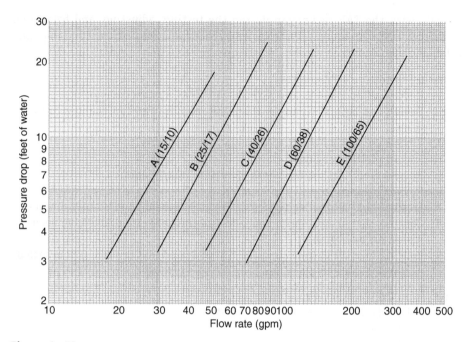

Figure 2–43
Pressure drop chart for a heat exchanger.

5. Measure pressure drop across each heat exchanger (Figure 2–43).
6. Adjust all flows to design specification.
7. Check pressures, amperage, and voltage of pumps.
8. Set bypass valve to 90% of maximum flow; repeat until heat exchangers are within design specifications.
9. Measure and record all final data (pump conditions, boiler conditions, etc.).

DOCUMENTATION AND FORMS

All testing and balancing information is recorded. TAB records are used when the building is commissioned (opened for the first time) to check and verify that systems are operating to design specifications. These records are also important for facilities engineers, who maintain the system and make changes to it as needs change in the future.

Written Records

Written records also consist of printout information from computer systems. In the early days, everything was recorded by hand. TAB forms were made up by the TAB service company to record information in their own format. Standardized forms are now available and produced by AABC, ACCA, TABB, and NEBB. However, for large commercial applications, handwritten records are fast being replaced by computer entry screens.

Computer Databases

Computers are increasingly being used to record TAB information. Computer databases are software programs that can capture and display related information. There are many TAB computer programs; all use a database program to record and store information about testing and balancing. The useful aspect of databases is that the information recorded can be related and displayed in many ways, which makes comparison of system operation easy and understandable. Charts and graphs of system operation can be generated from the data. If desired, for example, two different systems in two different buildings could be compared.

Data is entered into the computer using laptop computers and PDAs (personal digital assistants). Minicomputers capable of accepting keyboard, pen, or voice input also being used. Some system components incorporate wireless communication, which allows the operating data of the device to be entered into the database with minimal human intervention.

Report Forms

Whether they are handwritten or recorded via computer, several forms must be filled out during TAB. Each of the following forms may have several subforms for specific parts of the system:

- System diagrams
- Device test form
- Coil test form
- Heating system test form
- Cooling system test form
- Duct system test form
- Water/fluid system test form
- Terminal device test form
- Instrument calibration test form

SUMMARY

This chapter on TAB provides an overview of testing and balancing. The procedures, tools, equipment, and systems are similar to other aspects of work in the HVAC field. TAB technicians work on the testing and balancing of specific systems. At times, they also troubleshoot the system to determine why some part of it is not meeting specifications. They rarely are involved with repair or maintenance operations.

TAB work is exacting, and it requires that the technician be orderly and precise in operation of equipment and the use of testing tools. Everything that is done is recorded. Everything that is recorded is verified. This means that others will be looking at the data and using it to recreate the operation that was recorded by the TAB technician. This data will also be scrutinized, analyzed, and evaluated for flaws. The TAB technician must be accurate above all.

Here are some of the most important things to take from this chapter.

- TAB tools, meters, and instrumentation must be calibrated before they are used for TAB work.
- Systems with many balance points (valves and dampers) are easier to balance and to operate.
- Flow control for both air and water systems can be accomplished in many different ways.
- Pressure drop measurement across valves, dampers, and heat exchangers relate to flow or volume.
- Increasing the number of fluid flow measurements will lead to more accurate results.
- Pressure and volume have two types of relationships depending on the presence of openings that are fixed versus varied (by valves).
- Water pumps and blowers are similar in design, operation, and function.
- TAB is a systematic approach to system operation.
- All data from a TAB procedure is recorded.

REVIEW QUESTIONS

1. Describe the role of a system's terminal device.
2. List two meters or test instrumentation used for TAB
3. What is a manometer, and how is it used?
4. Describe the different types of manometers.
5. How are velocity and volume related?
6. What purposes does a flow control serve in an hydronic system?
7. What is traversing?
8. What is the first TAB step for an "air" system?
9. What are TAB methods?
10. How is TAB data recorded?

CHAPTER 3

Energy-Efficient Mechanical Systems

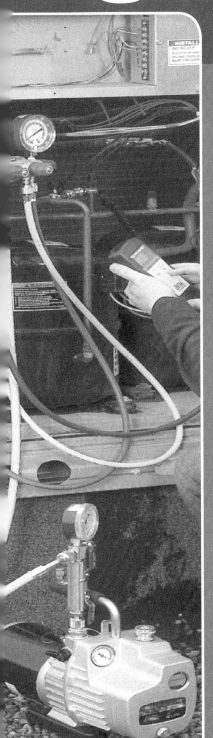

INTRODUCTION

There is much available information about energy efficiency. As energy prices increase, there is renewed interest in energy consumption and energy savings. This chapter will: emphasize the most commonly used energy efficiency ratings; review energy-efficient equipment; look at equipment and delivery system specifications; delve into some of the upgrades that can be made to existing equipment; and look at energy-efficient buildings.

This chapter will give some insights into how energy efficiency relates to systems, system components, and buildings that are built with energy efficiency as a goal. The thread running through the entire chapter is energy efficiency and how it is used from a technician's point of view.

Field Problem

The customer was concerned that one room of her house never seemed to be warm during winter. It always seemed to be colder than other rooms in the house. She had moved the balancing damper handles in every configuration, but nothing seemed to help.

As the technician listened, he made a mental list of the possible problems and those things he wanted to discuss with the customer. When the customer finished, the technician asked the following questions:

1. Has this condition occurred since the system was installed? A: Yes.
2. What have you done besides adjusting the dampers? A: Used an electric heater.
3. Are you satisfied with the temperature in the other rooms? A: Yes.
4. Do you think your heating bills are average? A: No—maybe a little too high.
5. When it gets very cold, is this room as different as when it is mild? A: Colder.

It was time to check a few things. Calling the office, the technician asked:

- Did the office have a record of this installation; and did we install it? A: Yes; and yes, 5 years ago.
- Was a heat load calculation done? A: Yes, 78,500 BTUH.
- Was a test and balance conducted? A: No.
- When was the last time the combustion efficiency was checked? A: Last year, 83%.

A check of the system yielded the following information:

- Forced air furnace, "80 Plus" unit with stainless vent
- Input BTU: 100,000 BTUH

- Output BTU: 80,000 BTUH
- All balance dampers were wide open

So far, everything seemed to be set to make this system function. The plan was to (1) check the temperature rise, which for the system should be about 65°F; (2) measure air volume at the central return; and then (3) start setting the dampers based on some basic cfm calculations. The office had given him the BTU requirements for each room, so after measuring the total cfm, the technician should be able to split it up where it belongs.

Temperature rise was too high, indicating a low volume of air. A check of the fan speed showed that it was on the lowest speed tap. Moving it to the medium speed tap reduced the temperature rise to about 57°F. After measuring the return cfm and then dividing the total according to the percentage needed by each room, he set all of the room balancing dampers. For example, the total cfm measured 1,000 and, according to data from the office, the living room needed 7,850 BTUH. Dividing room load by the total house load, given as 78,500 BTUH, the technician determined that 10% of the cfm should go to the living room (7,850 BTUH ÷ 78,500 BTUH = 0.1 or 10%). So the living room needed 100 cfm, or 10% of 1,000 cfm.

The last room on the run was the problem room. The duct disappeared through a batt of insulation at what appeared to be the rim joist. Pulling out the batt, the technician felt a lot of warm air coming back in his face. The floor of this room was cantilevered and hung out over the basement wall. Using a flashlight and peering into the floor joist, he saw the problem: the duct had become disconnected from its boot. With a little ingenuity, the technician was able to reconnect the duct to the boot and mechanically seal the joints in the process. The customer was satisfied and comfortable with this solution to the problem.

EFFICIENCY CALCULATIONS AND RATINGS

Energy efficiency calculations are used for two purposes: (1) to establish the ratings of equipment and (2) to determine the energy requirements of the application (home or business). Manufacturers will test equipment and provide energy consumption information along with system specifications. Designers and contractors are encouraged to provide calculations of what the energy consumption levels might be for the so-called envelope (walls, floor, and ceiling) and for the energy use within the envelope (lights, equipment, and other activity that uses or generates energy). This calculation of energy efficiency may result in the building receiving an Energy Star or LEED rating.

The energy efficiency ratio, or EER, is a "steady-state" calculation of energy usage. The equipment's design conditions are used to compare the energy that it consumes or moves (heating or cooling) to the amount of energy required to operate the equipment for a short period of time (not a season). The steady-state efficiency is called EER, which may appear on equipment labels along with the SEER rating. The formula is EER = BTUH ÷ Watts used.

An example would be a central split air conditioning system with the following rating:

Equipment	Specification
Central Air Conditioner	14 SEER/11.5 EER

Tech Tip

The EER reports steady-state efficiency. This rating is always lower than the SEER, because the EER requires the cooling system to operate continuously in a 95°F outside ambient environment. The SEER is higher because average outside temperatures are actually lower than this.

Energy efficiency ratings are crucial for determining energy consumption. The single-number SEER is a consumer benefit, facilitating their energy comparisons among equipment manufacturers. To further protect and guide consumers, the government has adopted energy standards for cooling equipment. According to the U.S. EPA's Department of Energy (DOE):

> As of January 2006, all air conditioners sold in the United States must have a SEER of at least 13. ENERGY STAR qualified Central Air Conditioners must have a SEER of at least 14, and an EER of at least 11 for single package equipment and 11.5 for split systems.

SEER

The seasonal energy efficiency ratio, or SEER, is applied to cooling equipment and heat pumps in the cooling mode. It is a rating that results from a calculation by the manufacturer. The SEER indicates, in a single number, the energy use of cooling equipment over an entire cooling season.

The calculation is a ratio: the (estimated) energy produced over a season compared to the total electrical energy consumed. The higher the SEER number, the more efficient the equipment at consuming energy in relationship to the amount of work that is done. The formula is SEER = Seasonal BTU output ÷ Seasonal watt-hours used.

HSPF

The heating seasonal performance factor, or HSPF, is applied only to heating systems. It is used to rate the heat output of heat pumps. HSPF is similar to SEER in that it takes into consideration energy requirements for an entire season of operation. In the case of a heat pump, both the compressor and the electric heater are rated. The formula is HSPF = Seasonal heating output BTU ÷ Seasonal watt-hours used.

Just as with SEER, higher HSPF ratings correspond to units that are more efficient. And just as with the SEER, the HSPF number can be used to make other comparisons. To change the HSPF into a percentage, divide it by 3.41 (the number of BTU in one watt). An HSPF of 7 ÷ 3.41 = 2.05, or 205%. This means that, for every 1 BTU of electrical energy used, 2 BTU of heat energy are moved. An 8 HSPF moves 2.34 BTU for every BTU of electrical energy used. More BTU of heat energy moved per BTU of energy consumed translates into higher efficiencies.

Tech Tip

The HSPF number can be used to compare heat pumps to other heating systems, but always consider the price of the fuel.

Costs of different fuel types may vary significantly.

When selecting an air-to-air heat pump in warmer climates, the SEER is more important. In northern climates, HSPF becomes more important. As of January 2006, EPA Energy Star requirements are that any heat pump sold in the United States must have a HSPF of at least 7.7.

Tech Tip

The coefficient of performance, or COP, is a rating for heat pumps that is less often used. It indicates how an air-source heat pump will operate at a steady-state

condition of 47°F outside ambient. The formula is COP = BTU of heat produced ÷ BTU of electricity used.

AFUE

The annual fuel utilization efficiency, or AFUE, is applied to fuel-burning equipment. It is a rating that results from a calculation by the manufacturer. AFUE compares the amount of energy used (seasonal input expressed in BTU) to seasonal output of BTU and shown as a percentage. The higher the AFUE, the more efficient the unit is at converting energy. If a system is rated at 90% AFUE, it means that, for every BTU, 0.9 or 90% of the energy consumed is used within the building and only 0.1 or 10% is wasted. The following calculations compare a 90% AFUE unit with a 78% AFUE unit:

$$100,000 \text{ BTUH} \times (1.00 - 0.78 \text{ AFUE}) = 22,000 \text{ BTU}$$
of wasted heat energy/hr of usage

$$100,000 \text{ BTUH} \times (1.00 - 0.90 \text{ AFUE}) = 10,000 \text{ BTU}$$
of wasted heat energy/hr of usage

Suppose you paid $1.50 for one therm (100,000 BTU) of natural gas, which is the input of the unit in this example. Using it in each of the two units then yields the following comparison:

$$1 \text{ therm} \times (1.00 - 0.78 \text{ AFUE}) \times \$1.50 = 33 \text{ cents}$$
of wasted heat energy/hr of usage

$$1 \text{ therm} \times (1.00 - 0.90 \text{ AFUE}) \times \$1.50 = 15 \text{ cents}$$
$$\text{of wasted heat energy/hr of usage}$$

Tech Tip

AFUE is not the same as combustion efficiency. An AFUE may be close to but is usually of lower efficiency than a combustion analysis. The combustion analysis involves steady-state efficiency and does not take into account changing outdoor temperatures, system cycling, and other building loads.

ROI

Return on investment, or ROI, is a calculation performed to compare choices. It is also referred to as the "rate of return on an investment" and is usually presented as alternatives. An example would be to compare the cost of one heating unit that burns fuel at a lower AFUE to a more expensive unit that features a higher AFUE. Using our previous example of comparing a 78% AFUE unit with a 90% AFUE unit, a simplistic ROI might look like this:

$1,500 cost of unit ÷ 10 years of expected life = $150 equipment cost/year

$0.11 waste/therm × 2,000 hrs of operation/season = $220 efficiency cost/year

$150 equipment cost + $220 efficiency cost = $370 total cost/year

Compare this result with the following:

$2,000 cost of unit ÷ 10 years of expected life = $200 equipment cost/year

$0.05 waste/therm × 2,000 hrs of operation/season = $100 efficiency cost/year

$200 equipment cost + $100 efficiency cost = $300 total cost/year

These calculations show that buying the more efficient but higher-priced unit will generate savings of $70 per year, or $700 over 10 years. If the efficient unit cost $500 more to purchase, then the operating savings would recover the extra cost in about 7 years (7 × $70 = $490). Therefore, the ROI is approximately 7 years.

Application of Efficiency Calculations

When the efficiency calculations are used, other aspects of equipment cost—including equipment use, fuel cost, availability of fuel, transportation costs of fuel, and more—must be figured into the calculations in order to make comparisons

Tech Tip

When it comes to selecting a system, listen to the customer. If they have not experienced anything other than a forced air gas furnace, they may only be comfortable purchasing the same. Selecting something new may require that they have a sense of adventure.

Table 3–1 Energy Efficiency Calculations: Additional Considerations

Consideration	Additional Calculations or Costs
Fuel costs	Determine cost per BTU for comparison.
Fuel availability	Is the fuel easily available? For example, is there an extra charge to bring LP to this location, or will there be an installation cost to bring natural gas to this location?
Distribution system	Determine the cost of piping or ducting.
Lifestyle	Determine if the customer prefers: • Air or hydronic systems • Radiant heat • Air cleaning • Humidification • Cooling • High or medium efficiency systems • The latest technology or conventional systems • Length of ownership • Resale enhancement of building
Reliability	Check for parts availability and maintenance requirements.
Longevity	Expected operational lifetime.
Warranty	Length of warranty and what parts are warranted.
Contractor	Reliability and customer service.
Existing system	Determine if: • The delivery system needs to be replaced or updated • The new equipment is compatible (delivery system capacity) • Major modifications will be necessary • There are redundant or backup systems
Zoning	Optimize for comfort in each room while attempting to maximize energy savings.

between units or to make the best selection of system and system components. Table 3–1 summarizes important considerations when using energy efficiency calculations.

APPLICATIONS AND HIGH-EFFICIENCY SYSTEMS

Higher-efficiency equipment is being produced every day. The specifications on new equipment always tell us how much more efficient equipment is becoming. Learning about the new equipment is like going to school every day!

Heat Pumps

Heat pumps are considered to be one of the most efficient heating and cooling system combinations. The reason why a heat pump is so efficient is that it uses one unit of energy to move many units of energy from one place to another. The higher the HSPF and the SEER, the more efficient the heat pump is at moving BTU.

Electric companies sometimes provide incentives for the installation of heat pumps. During the winter months, much of utility's capacity sits idle. If consumers use heat pumps in the winter, then the utility can sell its extra capacity. When calculating the energy efficiency, be sure to check with the power company about installation incentive programs.

Tech Tip

When you encounter customers that use a heat pump, always ask if they participate in any utility incentive program. If they do not and you know of such a pro- gram, suggest that they participate and help them get in touch with the electrical provider.

There are several categories of heat pump technology.

- *Air-source* (air-to-air)—these systems are the most common of all heat pumps. Comparing the cost of air-source heat pumps to all-electric heat, the heat pump can reduce the cost of heating a home by 30% to 40%.
- *Geothermal* (ground-to-air)—these systems generally use the heat from the ground to heat the occupied space. The annual savings from installing a geo- thermal heat pump (rather than a conventional system) ranges from 30% to as high as 70% in some areas of the country. Installation costs depend on the type of ground and how the heat pump is connected to the ground. Heat can be extracted from the ground in several ways:
  - *Open-loop* (pump-and-dump)—water is pumped from the ground; heat is extracted from the water; and the water is dumped to a pond or res- ervoir. This system may incorporate the water well system or be separate.
  - *Closed-loop* configurations:
    - *Horizontal loop*—piping is trenched 5 to 7 feet below grade level in looping patterns back and forth throughout the field (see Figure 3–1).
    - *Vertical loop*—a vertical well is drilled; then piping is draped in the well so that the piping is looped one or more times from top to bot- tom. This takes less area to install but does require a well driller.
    - *Pond loop*—a closed loop of piping, suspended or attached to a frame, is lowered to the bottom of an existing pond (see Figure 3–2).
    - *Coiled loop* or *overlaid loop* (*slinky*)—piping is layered horizontally 5 to 7 feet below the grade level, as if a Slinky® were pressed flat (while lying on its side) to create spirals of piping (see Figure 3–3).

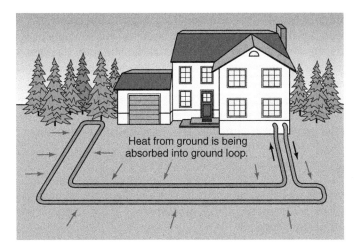

Heat from ground is being absorbed into ground loop.

Figure 3–1
Horizontal loop geothermal system.

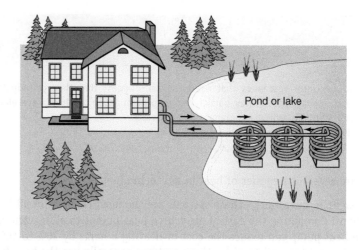

Figure 3–2
Pond loop geothermal system.

Figure 3–3
Coiled loop (slinky) geothermal system.

 o *Standing well*—sometimes considered both an open loop and a closed loop, this system pumps water from the bottom of a deep rock well and returns water to the top of the well. The heat exchange is from the bedrock to the water.
 o *Water-to-water*—this heat pump is similar in operation to a boiler that heats a radiant panel. Just like a boiler in this example, it is unable to use the same delivery system to cool the building. However, there are other delivery mechanisms that can be added to perform cooling in the summer months.
 • *Direct exchange* (DX geothermal)—these systems extract heat from the ground directly with a refrigerant loop. The outdoor coil is in direct contact with the ground: it is buried as prescribed by the manufacturer.

Forced Air

Forced air systems include heat pumps and other fuel-burning equipment. The examples used here will be of gas forced air systems. Similar relationships occur with other types of heating systems.

Three categories of forced air furnaces range in efficiency from standard to high. Each of the three types can often be identified by the venting material. The three efficiency types are:

- *Standard*—these furnaces have standing pilot ignition systems and have the lowest efficiencies (AFUE below 80%). These units vent using class B double-wall venting.
- *Mid-range*—units in this category have an AFUE of 80% to 90%. These units vent using Class B double-wall venting or plastic.
- *High*—these units have efficiencies from 90% to 97% AFUE. Most of these systems use two plastic pipes, one to vent and the other for combustion air. This category includes furnaces that may have two-stage or modulating characteristics that reduce the amount of fuel in order to maximize the unit's efficiency.

With increased efficiencies, the temperature rise across the unit tends to drop. Standard furnaces typically have a temperature rise of between 35°F and 70°F. This is the temperature difference between return air temperature and supply air temperatures. Air moving through the delivery system is hotter for a standard unit. The highest-efficiency furnaces have temperature rises of as low as 30 degrees. This means that high-efficiency furnaces require more airflow because, in order to gain the efficiency, they need to condense the water from the flue gas (see Figure 3–4). Decreased temperature rise means that more air must be delivered to the entire building, and ductwork should be sized accordingly. When a lower-efficiency furnace is replaced with one of higher efficiency, usually the new unit has a lower BTU input rating than the old one.

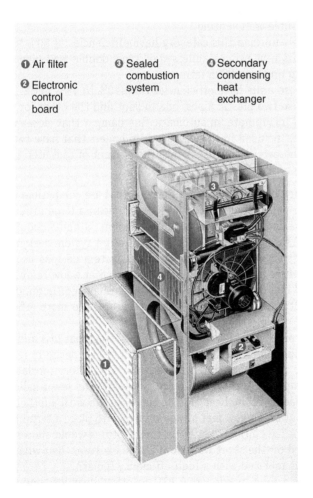

❶ Air filter

❷ Electronic control board

❸ Sealed combustion system

❹ Secondary condensing heat exchanger

Figure 3–4
A high-efficiency furnace with primary and secondary heat exchangers designed to condense water from flue gas.

Hydronics

Hydronics uses water to move heat from the heat source to the building. Heat pumps and other fuel-burning equipment can be included, but the examples given here will be of gas boilers. Similar relationships occur with other types of heating systems.

Three categories of boilers range in efficiency from standard to high. Each of the three types can often be identified by the venting material. The three types are:

- *Standard*—these boilers usually have standing pilot ignition systems and have the lowest efficiencies (AFUE below 80%). These boilers vent using class B double-wall venting.
- *Mid-range*—units in this category have efficiencies of 80% to 90% AFUE. These units vent using stainless steel pipe, double-wall pipe, or plastic pipe depending on their AFUE rating.
- *High*—these units have efficiencies from 90% to 98% AFUE. Most of these systems use two plastic pipes, one to vent and the other for combustion air. They may incorporate an automatic flue damper that closes when the system is off. Included in this category are boilers that have two-stage or modulating characteristics that reduce the amount of fuel used in order to maximize the unit's efficiency.

High-efficiency boilers rely on installations that are compatible with the boiler's efficiency. A few percentage points in AFUE can mean a large difference in savings over a 10-year period. In order to be 98% efficient, a boiler requires 90°F return water and 110°F supply water. Only low-temperature system designs are capable of boiler efficiency such as this. Low-temperature system designs usually are radiant panel systems, where wide areas of floors are heated at a low temperature to offset the building BTU loss (see Figure 3–5). Keeping this concept in mind, it is clear that the lower the return and supply water temperature, the more efficient the boiler will be.

A boiler is typically installed when there is a large heat loss and a long distance between the boiler and the area to be heated. There is a desire to reduce the size of piping needed to supply heated water to a space. This often translates into a higher supply-water temperature and a smaller pipe size to send fewer gallons of water per minute to the space needing heat. Couple this scenario with another: a space with a large heat load but only a few feet of wall on which to place a terminal device (baseboard, radiator, or fan coil). Higher water temperatures would allow smaller terminal units to be placed on the short wall. So if all this is true, then what happens when a standard unit is replaced with a high-efficiency boiler?

Suppose that a room needs 5,000 BTU to offset heat loss and that a standard boiler sends 1 gallon of 160°F water per minute to a fin tube terminal unit, 12 feet long, that is designed to deliver 420 BTU/ft/hr (see Table 3–2). This relationship works fine. But if the boiler were changed to a high-efficiency unit and the highest

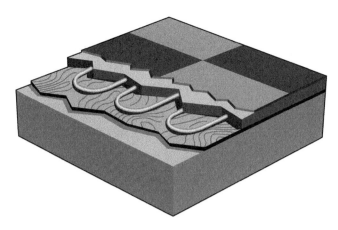

Figure 3–5
Cutaway of a thin-slab radiant floor panel.

Table 3–2 Fin Tube Specification (BTUH per linear foot at flow of 1 gpm)

Water Temp.	100°F	110°F	120°F	130°F	140°F	150°F	160°F	170°F	180°F	190°F	200°F
BTU/hr per foot	90	140	190	240	290	350	420	480	550	620	680

efficiency was desired, what would happen? If only the supply water temperature changed to 100°F, then the room would receive only 90 BTU/ft/hr at the terminal unit or 1080 BTU/hr—not nearly enough energy to meet the 5,000 BTU requirement. Increasing the water flow would not help much because the fin tube is rated to carry only a specific amount of volume at a certain pressure drop, and the water needs time to give up the heat. The fin tube would need to be lengthened to over 55 feet to meet the heating need, which is more than 4 times the original length. The result of this mismatch would be a system that cannot function without major modification and, at the very least, an unhappy customer. The lesson here is that the boiler efficiency must match the delivery system efficiency in order for the boiler to provide what the manufacturer specified. Designers need to ensure that a replacement unit will be compatible with the existing system.

MEETING SYSTEM SPECIFICATION REQUIREMENTS FOR INSTALLATION

When a system is new or has been retrofitted, all system components must be evaluated. Each component of a system has a set of specifications, and each component must be engineered to match and work with all other components. A complete heating and cooling system can be divided into two general groups of specifications: equipment specifications and system delivery specifications.

Equipment Specifications

Equipment specifications (also called manufacturer specifications) accommodate a wide range of operation. Because the manufacturer does not know the conditions under which a piece of equipment will operate in the field, a "range" of specifications is given so that the engineer can match the conditions to the specification. The most general of these is the example of a high-efficiency boiler. The manufacturer may provide a range of boiler output within the same class of boiler. The example to follow involves a high-efficiency gas boiler. This class of boiler may be only one of

Table 3-3 Specification Sheet

	Super 100	Super 200	Super 300
Input (max.)	81 MBH	156 MBH	320 MBH
Output (min.)	17 MBH	32 MBH	56 MBH
Heat capacity (DOE)	72 MBH	140 MBH	291 MBH
Net I=B=R	63 MBH	124 MBH	254 MBH
AFUE (DOE)	92.0%	93.2%	93.4%
Low-temp. operation AFUE	98.2%	97.5%	96.1%
Vent size & material	3″ PVC	3″ PVC	4″ PVC
Combustion air intake	3″ PVC	3″ PVC	4″ PVC
Water volume	0.96 gallons	1.27 gallons	2.6 gallons
Weight (empty)	75 pounds	125 pounds	190 pounds

Key: DOE, Department of Energy; I=B=R, Institute of Boiler and Radiator Manufacturers; MBH, 1,000 BTU/hr; PVC, polyvinyl chloride.

several classes of products within the gas boiler line. The specifications show model numbers and their related input; output ratings by several testing methods; the manufacturer; DOE capacity, and net I=B=R values (see Table 3–3). It also shows the AFUE rating (for low-temperature conditions) as calculated by DOE and by the manufacturer. The designer would use this information to select a boiler that would match the system delivery specifications, using the heating output capacities as a guide.

Example: High-efficiency Boiler—Super Model

Description: This boiler has—as standard equipment—spark ignition; stainless heat exchanger; two-position gas valve with matching two-speed vent motor; electronic control and touch-pad control interface; outdoor reset controller; circulator pump; and plastic connections for combustion air and venting.

The rest of the specifications in this example would pertain to planning and selecting installation materials. The size and material type for the vent and intake determine what material, fittings, and hangers are needed. The location of the vent termination and of the intake are identified on the specifications printout. The system's weight must be known in order to plan for any rigging equipment that may be needed to put the system in place. If the equipment is particularly large, a specialized contractor may be hired to move it into place. The manufacturer can usually provide additional information that describes various parts of the equipment in greater detail.

Tech Tip

Whenever in doubt about a model number or about system output or input, consult the manufacturer's website or call to verify specifications.

A product manual would show additional specifications. In our example, the boiler manual provides information about the circulator pump mentioned in the

Table 3–4 Flow Rate and Temperature Rise Based on Pump Specifications

Flow (gpm)	Temp. Rise (°F)	Pipe Size (in.)	Boiler Loss (ft w.c.)	DHW Loss (ft w.c.)	Pump D	Pump F	Pump C
5.5	41	1	3.8	1.5	6.1	13.4	19.8
9.1	24	1.25	10.1	2.1	N/R	7.1	12.1
12	18	1.50	17.2	2.8	N/R	N/R	2.7

Key: DHW, domestic hot water; ft w.c., feet of water column; gpm, gallons per minute; N/R, not recommended.

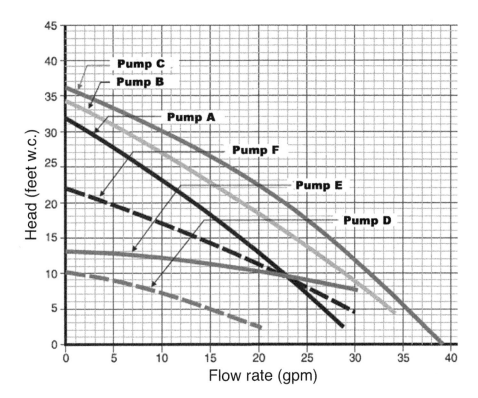

Figure 3–6
Pump curve chart example.

description. The information provided would allow the engineer to pick flow rates and temperature rise based on the pump specifications (see Table 3–4). In this case, the pump that comes with the boiler is highlighted. To provide additional help, the manufacturer may also provide a pump curve chart (see Figure 3–6).

System Delivery Specifications

Delivery specifications involves that part of the system that moves the heat to and from the space. Continuing with our boiler example, delivery would include the terminal devices (fin tube in this example) and all piping runs, pumps, valves, diverters, flow controls, headers, gauges, pipe hangers, and so forth (see Figure 3–7).

Table 3–5 lists some delivery specifications to be considered when selecting equipment for hydronic or forced air. All of these choices will affect the delivery system's efficiency, either directly or indirectly in terms of how efficiently the entire system functions.

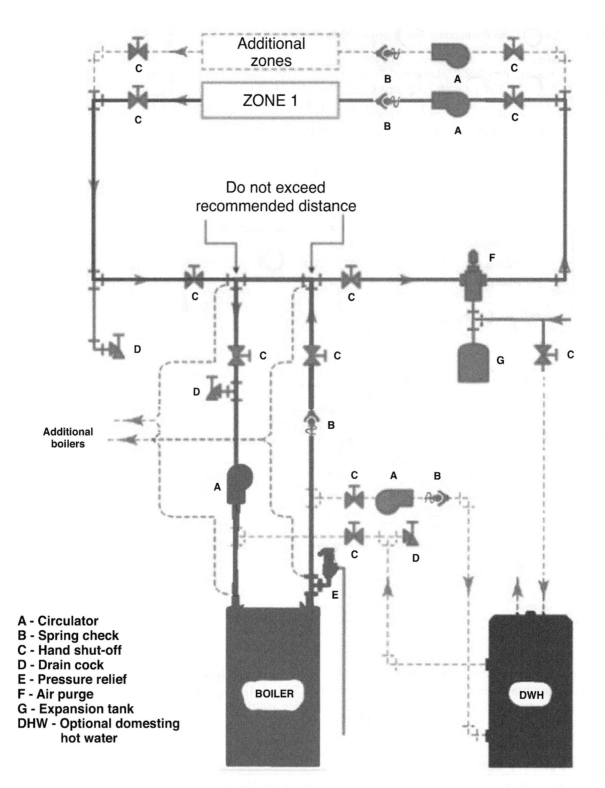

Figure 3–7
System layout of a high-efficiency system with multiple delivery zones.

A - Circulator
B - Spring check
C - Hand shut-off
D - Drain cock
E - Pressure relief
F - Air purge
G - Expansion tank
DHW - Optional domesting
        hot water

Table 3–5 System Delivery Options

Component/Requirement	Options
Heating and cooling load calculation	• Single integrated system (heating and cooling in one package) • Multiple integrated systems (many systems in separate packages connected together) • Multiple independent systems (separate systems in separate packages)
Heating/cooling runs	• Length of piping or ducting runs • Number of elbows and fittings • Fitting connections/takeoffs • Ease of installation (low-cost installation)
Humidity	• Integrated humidifier • Independent humidifiers
Filtration	• Integrated or independent filtration system • Multistep filtration and/or HEPA (high efficiency particulate air) filters
Zoning	• Number of spaces that need to be controlled by separate thermostats • Additional system components to accomplish all zoning
Domestic hot water	• Integrated with the heating system • Independent system
Temperature setback	• Whole building setback • Individual room setback
Terminal devices	• Coverage patterns • Size • Style • Resistance (pressure drop) • Efficiency
Maintenance	• Number of valves/dampers • Location of devices; ease of access • Filters/strainers/flush valves (to maintain system cleanliness) • Signals/alarms/lights (to indicate a maintenance problem)
Thermostat	• Single thermostat • Multiple thermostats • Multistage thermostat • Readability and ease of use • Features to fit application (setback, multi-day programmable, indicator lights, backlit screen, etc.)

UPGRADES OF EXISTING SYSTEM

There are more existing systems than there are new systems being installed. It has been said that if all of the existing system efficiencies were increased by even a small percentage, then the savings in fossil fuels would be huge. There are many HVAC installations older than 10 years that are still functioning. Well-maintained systems can easily operate for 10 years and still be as energy efficient as they were when installed. However, the achievable efficiencies were not as high 10 years ago as they are today. System upgrades can greatly increase efficiencies and comfort while reducing energy consumption.

Cleaning

Resistance to the flow or water or air in piping and ducts leads to increased static pressure and reduced flow. Keeping the system clean maintains the original flow characteristics of the delivery system. Delivery systems can be cleaned, but it is far better to keep the system clean with good filtration (for air systems) or good water chemistry (for water systems). It is far less efficient to allow systems to degrade slowly and then add the cost of cleaning. Regular checks of the filtration system or water chemistry will help to keep a system delivering the designed amount of heat.

Pumps and fans need to be clean and running efficiently. Dirt, dust, or corrosion will affect the capacity of these devices to deliver the fluid. Broken or dirty blades, bad bearings, and overheating motors will all reduce efficiency. Regular checks of the voltage, amperage, and working parts will insure that these fluid pumps work efficiently throughout their mechanical life.

Tech Tip

When inspecting a system, look for any improvements that would increase its efficiency—for example, removing dirt or scale buildup to improve heat exchanger and pump efficiency.

Combustion Analysis

For fuel-burning systems, combustion analysis is a test that ensures the burner is converting the fuel efficiently. Properly used, combustion analysis documents the condition of a fuel-burning system as it is first encountered. As adjustments are made to the system, the process can document incremental changes in efficiency as improvements are made. At the end of any repair, maintenance, or improvement work, combustion analysis is used to document the final, steady-state efficiency of the unit. The steady-state efficiency of the maintained system should be close to or better than the original efficiency.

Combustion analysis takes into consideration these factors:

- *Carbon monoxide*—need low or zero CO emissions; CO results from incomplete combustion
- *Carbon dioxide*—need high CO_2; 10% for gas and 13% for fuel oil
- *Free air* (O_2)—enough free air to create CO_2; 5–10% for gas and 5–20% for oil
- *Nitrogen oxides* (Nox or NO_2)—need low Nox; occurs more with oil than with gaseous fuels; also related to flame temperature
- *Sulfur oxides*—need low SO_2; these compounds occur more with oil than with gaseous fuels
- *Particulate matter* (smoke)—need low soot; soot results from uncombined carbon
- *Hydrocarbons* (unburned fuel)—need low hydrocarbon content; this is fuel that escapes the controlled burn
- *Flue gas temperature*—need low temperature; indicates that heat is being used
- *Adequate draft*—need enough draft to move products of combustion without expending or using a lot of energy

Good burner efficiency depends on the right combination of all of the combustion analysis considerations just listed. Here are some products that will improve overall efficiency and can be retrofitted to an existing system.

- *Vent dampers*—automatically open during combustion and close after combustion.
- *Stack economizers*—remove heat from the flue gas before it is vented. The heat is used to augment other parts of the system or add heat back into the delivery system.
- *Vent fans*—improve venting or make venting more dependable. Vent fans can also eliminate off-cycle heat from moving up a natural draft chimney.
- *Reset controller*—boilers can have their water temperature reduced (to minimum allowable temperatures) during milder weather. Reducing water temperatures too much might cause condensation in the heat exchanger and the flue.
- *Intermittent pilot*—can be added to systems with standing pilots.
- *De-rating*—a process of reducing a system's fuel input to "correct" for an oversized condition. (*Note:* This process may not be allowed in some areas and will void manufacturers' warranties.)

Duct Sealing Methods and Materials

Leaking boiler pipes should be repaired, and it's no different for air ductwork. Forced air systems can have a very high leakage rate. Leaking supply or return duct systems account for huge losses in the delivery system. Leaks affect the comfort of conditioned spaces as well as efficiency and utility costs. Duct leaks in crawl spaces, attics, and cantilevered floors exposed to the outside are examples of wasted air. Leaking air can contribute to other spaces not receiving the proper percentage of air needed to condition them.

Duct systems that leak need to be sealed. Most of these systems are made of sheet metal and leak at every joint and takeoff. Visual inspection is sufficient for many systems. Larger systems use a "plug and pressurize" method to find and correct leaks. While the system is operating use your hand to find large leaks and tissue paper to find the small ones. Both the supply and return need to have their ducting sealed, though in many cases it's the return system that creates more of a problem. Large losses can occur where building materials are gapped or not properly filled in. Many return systems use building structures, such as floor and wall joists, to create return runs back to the furnace. Large gaps and cracks in the return can be filled with foam products or by added building materials. Complete installation and sealing information can be found in the ANSI-recognized ACCA HVAC Quality Installation Specification, which can be downloaded at no cost from www.acca.org.

The most effective way to seal sheet-metal systems is to use duct mastic (see Figure 3–8) where connections are curved and metal duct tape (Figure 3–9) where seams are straight. Table 3–6 summarizes the properties of common duct sealing products.

Maintenance Records

Good maintenance records are important! The difference between maintenance records and good maintenance records is the level of detail. Good record keeping can save time and money. In fact, good maintenance records can make money!

Figure 3–8
Duct mastic is used around all joints.

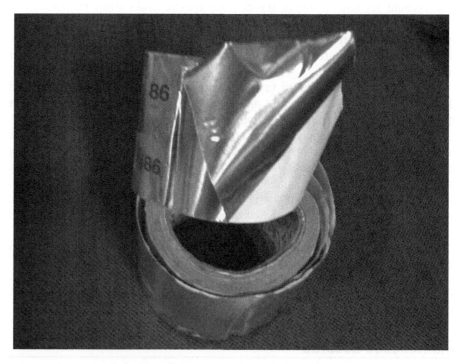

Figure 3–9
Pressure sensitive metal duct tape is used only for clean, straight runs.

Table 3–6 Products and Methods for Sealing Ducts

Product	Method
Metal tape—SMACNA (Sheet Metal and Air Conditioning National Association) or UL (Underwriters Laboratory) rated metal-backed duct tape	Supply ducts—Seams to be sealed must be straight and clean. Tape will not stick to dirty surfaces.
Mastic	Supply and return ducts—Applied with a brush or glove. Cracks wider than $\frac{1}{8}''$ require rated fiberglass mesh.
Caulk (urethane or silicone)	Supply and return ducts—Applied using a caulking gun. Difficult to apply to marginally accessible areas.
Nonexpanding foam	Return ducts—Applied from the can to large openings or gaps in building construction.

When a service call comes in from an existing customer, the dispatcher can easily see from the records the following information:

- Installation dates and age of the equipment
- Prior service records
- Prior maintenance records
- Parts or system components that have been replaced
- Any difficulty that the customer has had in the past
- Who worked on the system last
- Model numbers, voltages, amperages, pressures, etc.
- Combustion efficiency tests results
- Testing and balance information
- Location/address
- Customer contact information
- Anecdotal information supplied by the service technician

What can good maintenance records do?

1. Help to spot recurring problems
2. Identify pending maintenance
3. Estimate system and component life cycles
4. Estimate time and costs for maintenance or repairs
5. Assist in troubleshooting system problems
6. Maintain a record of safe operation
7. Maintain a record of combustion and system efficiency
8. Maintain a record of testing and balance data
9. Help keep systems operating at peak performance and keep a satisfied customer

Record keeping is becoming more digital. All of the information just listed can be loaded into a main computer database program. When needed, data can be downloaded to a portable computer or PDA (personal digital assistant) device and used in the field. Bar-code readers can be used to identify systems and system components without having to type in names, addresses, or phone numbers. The records can be added to and amended while in the field. Once the technician returns to the office, the data can be uploaded to the main computer where it is stored for future use.

ENERGY EFFICIENCY OF BUILDINGS

The efficiency of buildings continues to increase as building techniques and materials change. Buildings built today are more energy efficient and tight. Energy efficient means that buildings have more insulation and use strategies to retain or reflect heat. Tightness means that buildings leak less; hence drafts are reduced and heated air is retained. Tight buildings require the design and installation of a ventilation system.

Green Building

"Green building" is the practice of spending increased time on building details and using materials that conserve energy. Spending more time and money while the building is being constructed is expected to have a positive effect on energy conservation. Each of the building systems is evaluated for ROI (return on investment), and only those that yield the best ROI are chosen.

Green building takes into account all aspects of the building: thermal envelope; heating, cooling, and air delivery; water use; electrical use; lighting; and impact on the environment (fuel, waste, and pollution).

USGBC Certification and LEED

USGBC (U.S. Green Building Council) has created a program called LEED (leadership in energy efficient design). The program began as a guide for the design and construction of new buildings. That program also has an "existing" building guide entitled "LEED for Existing Buildings: Operations & Maintenance." The main thrust of this program is to develop a process of reporting, inspection, and review of the entire building and its systems. The program is designed so that the process can be applied for the life of the building. Owners and companies subscribing to the program are awarded points that lead to recognition for their efforts in retrofitting to increase the efficiency of and reduce the waste in existing buildings.

Specific information and instructions can be found by visiting the USGBC website. Here is an overview of the program:

- The building must be open for a minimum of 12 months to qualify.
- There is an application process and a fee to register.
- Re-registration is required every 5 years.
- Levels of recognition are certified, silver, gold, and platinum.
- A plan must be developed and executed to reduce energy or improve the building.
- When the plan is complete, a report must be submitted.
- Review is conducted by LEED-certified individuals.

SUMMARY

This chapter covered the most popular energy efficiency rating systems being used today. The use of single-number ratings for equipment has reduced consumer confusion and increased trust in the rating system. Customers and contractors both can use the rating numbers to compare and contrast different systems. One thing to remember is that these numbers alone do not make for a complete comparison. The best job of comparing systems requires that other aspects of the system be taken into account. Fuel types, application, and location are a few of the considerations used in addition to the rating numbers.

The chapter discussed the main types of systems and their energy efficiencies. Heat pumps continue to be of the most interest for those who seek the most efficient heating systems. A wide variety of heat pump systems is available for almost every application. Forced air and hydronic systems each have efficiency advantages when used in certain applications. Hydronic radiant floors are able to heat at low temperatures and are very comfortable. Forced air systems and hydronic systems that are retrofitted with new, higher-efficiency units must have their delivery systems evaluated. Mismatches can easily occur, as when a high-efficiency heating unit is coupled with a low-efficiency delivery system. The result is that the system operates at the lower efficiency.

Existing systems account for most of the energy being used or wasted. Enhancing, upgrading, and retrofitting a system will reduce energy consumption and waste. There are a number of ways to enhance an existing system, but the most effective is by cleaning. Combustion analysis is the next most important maintenance solution for saving energy.

Finally, the USGBC program called LEED was discussed. Originally designed for new buildings, LEED now has an existing building program that is designed to guide operations and maintenance personnel in retrofit and upgrade operations. The program is designed to continue for the life span of the building and to recognize reductions in energy and waste.

REVIEW QUESTIONS

1. What does it mean if one system has an SEER of 15 and another has an SEER of 13?
2. Can SEER and HSPF be compared? Why or why not?
3. What is the purpose of energy efficiency ratings?
4. What makes the heat pump an interesting system for energy efficiency?
5. Why are gas forced air systems chosen for energy-efficient delivery systems?
6. Why are hydronic systems often found in energy-efficient homes?
7. Why do equipment manufacturers supply specifications?
8. Why are system delivery specifications developed?
9. Describe one upgrade that qualifies as an energy-efficient retrofit.
10. What is meant by "green building"?

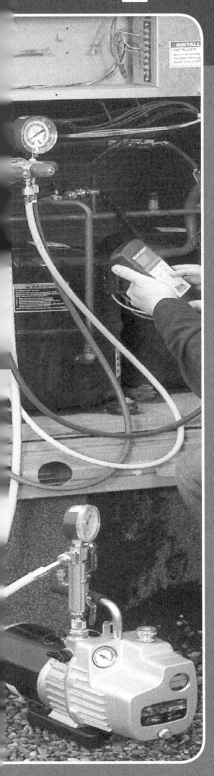

CHAPTER

4

Fluid Handling Systems

LEARNING OBJECTIVES

The student will:

- Recognize heat transfer fluids
- Measure fluid pressures
- Relate fluid pressures to the operation of equipment

INTRODUCTION

All HVAC systems function to handle fluids. Both water and air are considered fluids. Because liquids and vapors are fluids, all vapors like air and all liquids like oils are considered to be fluids. This chapter discusses the handling of fluids to heat and cool structures. Air, water, and other heat transfer fluids are featured. Pumps, fans, and blowers and their functions in handling fluids are identified. The chapter also presents terms that are used to describe how systems operate and are connected. Such terms as *open loops* and *closed loops* are defined. *Static, frictional head,* and *pressure head* are also defined.

This chapter provides a fundamental understanding of fluids and fluid handling systems. How fluids flow through an HVAC component and how they are handled by what type of component are important aspects of HVAC operation. The concepts in this chapter can be applied to many different types of HVAC systems.

Field Problem

The maintenance manager sent his top service technician to meet the pump manufacturer's service representative with the following instructions: he should not let the service representative leave until the source of the system's noise was found and eliminated. (*Note:* Pump noise is often a signal that some problem is occurring with a fluid handling system. The level of noise, location, and type of noise each indicate various types of problems. At this point you should write down all of the possible causes related to noise problems. Use your list and strike through those that do not seem to apply or are have been checked during this service problem.)

The manufacturer's service representative greeted the technician and listened as he explained that the pump had been getting noisier—a high-pitched vibration. He also explained that there had been a history of work on this pump. The pump had been disassembled three times, but the source of the noise was never found. The pump's bearings, impeller, and housing measurements were all within specifications. The motor-to-pump coupling had been laser aligned in an effort to quiet the pump. All of this work seemed only to make the problem worse. The pump seemed to be noisier each time maintenance was

completed. The service representative thanked the technician for the information and began to check the pump and system while the technician watched.

Using a long-shaft screwdriver, which was handier than the mechanic's stethoscope back in the truck, the service representative listened to various parts of the pump by placing the screwdriver on a part and holding the handle of the screwdriver to her ear. She moved from the back of the pump housing to the discharge, the pump bearing, the mechanical seal, the motor drive bearing, the motor stator, and the motor rear bearing. As she listened to each part, she asked the technician how the service to the pump was done. He explained that each time it was done the process was hurried, because the pump was an important and needed device. Each time, only the pump had been removed while the motor remained mounted.

After sounding a few more points on the motor, the service rep was fairly sure about what was needed. Pulling a $\frac{3}{4}$" box-end wrench from her tool kit, she tightened one of the motor mounting bolts and the noise went away.

For an explanation of the solution to this field problem, see page 120.

APPLICATIONS

Water

Fluid handling systems are typically associated with water-based systems. These systems use a circulator to move water throughout the system. Sometimes referred to as circulator pumps (or sometimes just pumps), circulator pumps are not positive-displacement pumps, like piston-style pumps. Circulator pumps are designed to move a certain amount of water against a determined head pressure.

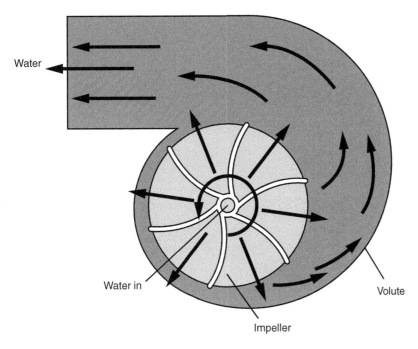

Figure 4–1
Water entering the center of the impeller is thrown to the outer edges, where it is captured and directed toward the outlet.

"Head" (or head pressure) is the frictional resistance to water movement. Circulator pump design is based on the centrifugal movement of water using an impeller. The impeller throws water from the center to the outside, where it is gathered in the scroll (the housing of the impeller); see Figure 4–1. The scroll defines the shape and direction of the water stream as the water is forced out of the circulator. These types of pumps are also called *centrifugal pumps* because they create a centrifugal force on the water to get it to move.

Conversion of gauge pressure (psig) to head pressure allows a technician to read a pressure gauge (in psig) and then convert the reading into head pressure that can be used to measure the total pressure exerted by a pump (see Figure 4–2). The conversion factor is 1 psig = 2.31 feet of head. If a technician reads 4 psig on the suction side of the pump and 8 psig on the discharge side of the pump, the difference in pressure is 4 psig. Therefore, the pump head is the pressure difference of 4 psig × 2.31 = 9.24 feet of head.

To estimate the head of a system, measure the length of pipe (in feet), including the radiation, in the system and multiply by 1.5, which accounts for the fittings. That result multiplied by 0.04 will give an approximation of feet of head. Pump head is equal to system head loss.

Circulator pumps are specified or replaced for a particular application based on the required flow in gpm (gallons per minute) at a particular head pressure (measured in feet of water column). See Figure 4–3. It is important for the service technician to replace the circulator pump—or, if rebuilding, its various components—with the exact same parts. Deviation from the original pump or parts will greatly affect its operation.

If the pump is used in a closed system (recirculating water), then it can be replaced with a cast-iron pump. If the pump is used in an open system (one that circulates new water), then it must be made of more expensive materials, such as bronze, because of corrosion considerations.

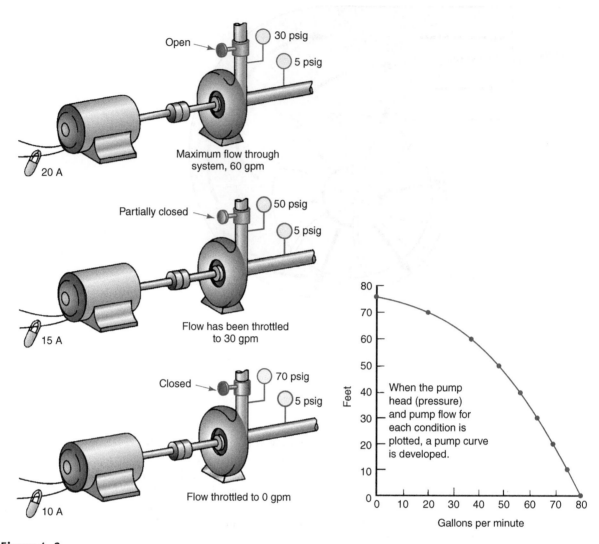

When the pump head (pressure) and pump flow for each condition is plotted, a pump curve is developed.

Figure 4-2
As the flow is changed, the pressure in psig changes. The total head is the addition of both the suction and discharge of a circulator. Also, note that as the pump is throttled, pump head goes up while motor amps go down. This occurs because the pump is doing less work (moving less water).

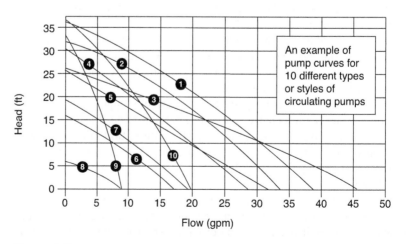

An example of pump curves for 10 different types or styles of circulating pumps

Figure 4-3
Pump curves chart.

Air

Air is a fluid, too! We commonly put air in a different category, but it acts and flows like water and it is actually considered a fluid. Air is handled in much the same way as water. An air circulator or blower is used to move the air by overcoming friction (also called static pressure), as measured in inches of water column, and delivering the flow, as measured in cubic feet per minute (cfm). The blower acts in the same way as the water circulator. The blower wheel (also called the squirrel cage) throws the air from the center to the outside, where it is gathered in the scroll (the housing of the blower wheel); see Figure 4–4. The scroll defines the shape and direction of the air stream as the air is forced out of the blower.

Blowers and drive motors are specified (or replaced) based on their original design conditions. The horsepower of a motor that is coupled to a given blower (also called an air handler) will deliver a certain volume of air under particular static pressures. The total external static pressure is the difference between the blower discharge static pressure and the blower inlet suction pressure. The amount of air conforms to the characteristics shown in a fan curve (see Figure 4–5).

Other Heat Transfer Fluids

Water and air are considered to be heat transfer fluids, but there are other substances that work well under certain conditions. Antifreeze solutions are a mixture of water and glycol; they are used when temperatures are below the freezing point of water. Silicone is used when conditions are very hot. Essentially, any fluid that can move heat from one location to another is considered to be a heat transfer fluid. Circulator pump selection for these fluids is based on the characteristics of the heat transfer fluid and the application. The replacement or repair of a circulator pump should be done such that the pump is restored to nearly original operation. (See Chapter 10, "Water Treatment," for more information on concentrations, burst protection, etc.)

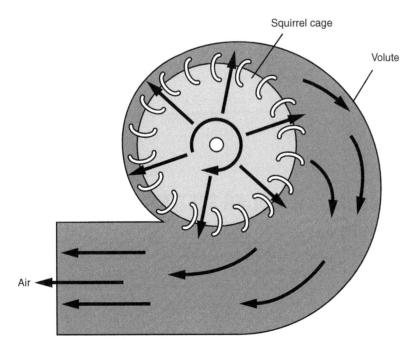

Figure 4–4
Air in the center of the squirrel cage is thrown to the outside of the wheel, where it is captured and directed toward the outlet.

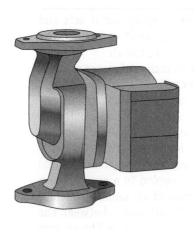

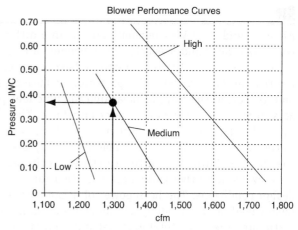

Blower Performance Data

UNIT SIZE	BLOWER SPEED	EXTERNAL STATIC PRESSURE									
		0.10	0.15	0.20	0.25	0.30	0.35	0.40	0.50	0.60	0.70
WX009	LOW	320	310	300	290	275	—	—	—	—	—
WX009	HIGH	380	370	360	350	335	—	—	—	—	—
WX012	LOW	350	335	320	305	290	—	—	—	—	—
WX012	HIGH	400	385	365	350	335	—	—	—	—	—
WX019	LOW	560	550	540	530	520	—	—	—	—	—
WX019	HIGH	665	650	635	620	605	—	—	—	—	—
WX025	LOW	845	825	805	780	750	—	—	—	—	—
WX025	HIGH	895	875	850	825	795	765	—	—	—	—
WX031	LOW	1000	985	970	950	920	885	—	—	—	—
WX031	HIGH	1040	1030	1010	995	970	940	910	—	—	—
WX033	LOW	1010	980	960	930	890	850	—	—	—	—
WX033	HIGH	1080	1050	1010	980	940	900	870	—	—	—
WX036	LOW	1125	1110	1090	1070	1050	1030	—	—	—	—
WX036	HIGH	1200	1180	1160	1145	1115	1090	1060	—	—	—
WX041	LOW	1215	1210	1205	1200	1190	1180	1170	—	—	—
WX041	HIGH	1395	1385	1375	1365	1350	1335	1310	1260	—	—
WX049	LOW	1320	1315	1310	1300	1290	1275	1260	1220	—	—
WX049	HIGH	1650	1625	1600	1575	1545	1510	1475	1425	1340	—
WX059	LOW	1765	1750	1735	1720	1700	1680	1655	1660	1550	—
WX059	HIGH	1960	1930	1905	1880	1855	1825	1790	1730	1670	1600

Includes allowances for wet coil and for filter. All units factory wired for high speed.

Figure 4–5
As the amount of static pressure (measured in water column) decreases, the pump will deliver more cubic volume of air. (Charts courtesy of ACCA).

TYPES OF FLUID HANDLING SYSTEMS

Circulator Pumps

In HVACR work, circulator pumps are generally viewed as devices that move water or waterlike heat transfer fluids. Specifically, the most common circulator pump used in HVAC is the centrifugal pump. Two general styles of centrifugal pumps are "wet rotor" (see Figure 4–6) and "mechanical seal" (Figure 4–7). Mechanical seal style pumps are also referred to as "standard" and "in-line." Wet rotor pumps are unique in that the rotor of their electric drive motor spins in the heat transfer fluid being pumped. Because the rotor is inside with the fluid, it does not need a mechanical seal to contain the fluid. Each type of pump has its advantages and disadvantages.

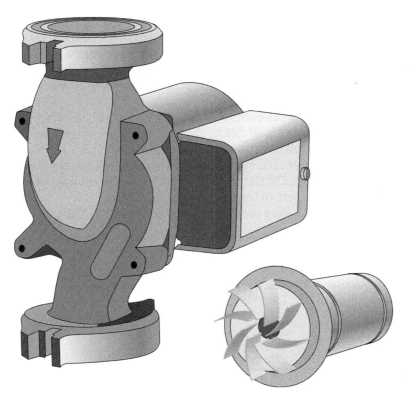

Figure 4–6
Wet rotor pump with replacement rotor cartridge.

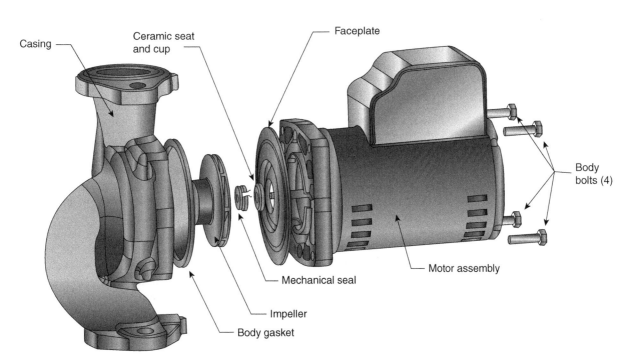

Figure 4–7
Exploded view of a typical in-line pump, showing the mechanical seal.

Circulator pump use is also based on the type of fluid. In the case of water, cast-iron housings can be used if the water is being recirculated. Recirculated water has lower amounts of oxygen, and in many cases the water is treated. Where raw (or new) water is being circulated, pump housings must be made of bronze or other corrosion-resistant materials to eliminate the effects of corrosion and corrosion products.

It is important to note the position of the circulator pump: vertical or horizontal. The pump's positioning is related to the bearings that are used and how they are lubricated. Some small circulator pumps can be used in either position. As the pump size and horsepower increase, the need to cool and lubricate the bearings becomes more important. Circulator pumps must be selected in light of their proposed position with respect to the pump and motor shaft.

Circulator pumps can have some or all of these additional features:

- More than one impeller, or multistaged operation
- Magnetic drive that eliminates a mechanical seal and allows the motor to be on the outside of the system
- Housings and other components made of various plastic-type materials for acidic fluids

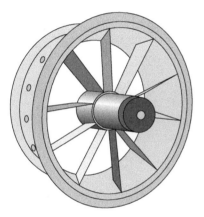

Figure 4–8
Duct booster fan.

Fans

Fans are propeller devices that move air. Fans can be thought of as devices that look like ceiling fans in a house or commercial building. These are also referred to as axial fans. An axial fan moves air *in parallel* with the fan's axle (the motor or fan shaft). Axial fans are designed for high volumes and low static pressure. They are commonly used as duct boosters (see Figure 4–8), but not as the primary air mover in a duct system. In this application, the fan is fitted inside of round ductwork and is used to boost the air volume.

Axial fans are used for exhaust applications when gases or vapors must be moved from the interior. Such applications include laboratories, welding facilities, and mechanical rooms. Exhausting operations require large volumes of air to be moved with very little static pressure resistance.

Blowers

Blowers are also referred to as radial fans and air handlers (see Figure 4–9). A radial fan moves air *perpendicular* to its axle (shaft). This type of fan has a squirrel cage driven directly or indirectly by a motor. Radial fans are used where higher static pressures are encountered and higher volumes of air are needed. The primary air-moving fan in a ducting system is a radial fan or blower. A blower will usually run more quietly than an axial fan and will move more quantities (cfm) of air than a axial fan of the same horsepower.

Air handlers such as furnace blowers use radial fans to move large quantities of air through heat exchangers, filters, and ducted delivery systems. All of these system components present resistance to air flow; this resistance is referred to as static pressure. Radial fans are designed to overcome higher static pressure than axial fans, making them ideal for ducted air delivery systems.

A blower wheel (also known as a squirrel cage) can have blades that are forward-curved or backward-curved with respect to the blower rotation (the blades can also be straight). Each of these characteristics of blower wheels gives the blower certain advantages that designers can exploit. Each type has its own fan curve characteristics as shown in Figure 4–10 and Figure 4–11 for a blower of the same horsepower.

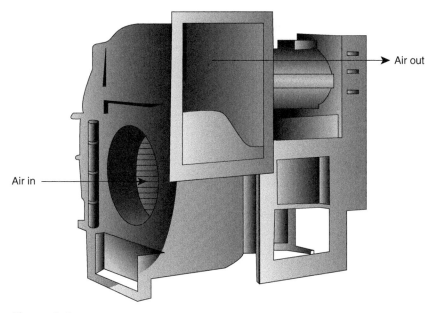

Figure 4–9
Radial blower or air handler.

Backward-curved radial fan blades pull air into the center of the blower wheel. Air is pulled into the back of the blade and slung into the scroll (the curved outer housing). As the pressure in the scroll increases, the air is pushed out of the scroll in a pattern defined by the scroll.

Forward-curved radial fan blades work in a similar fashion to the backward-curved blades: they pull the air into the center of the blower wheel and pressurize the air moving into the scroll; and the scroll defines the pattern of the air leaving the scroll. The difference is that the forward-curved blade scoops the air from the center of the wheel instead of pulling it into the back of the blade, as in the case of the fan with backward-curved blades.

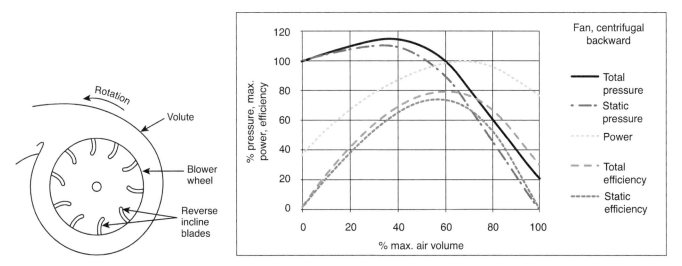

Figure 4–10
Effects of backward-curved blades on the blower wheel.

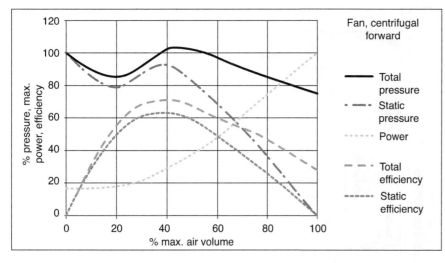
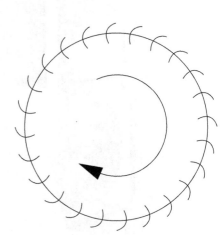

Figure 4–11
Effects of forward-curved blades on the blower wheel.

SERVICE AND INSTALLATION

General Service and Installation of All Fluid Handling Systems

When installing any mechanical equipment, it is very important to read and understand the installation and operation manual that comes with each component or device. An HVAC system will not operate as designed, provide comfort, or operate efficiently if not installed properly. Recommendations found in these textbooks will guide the installation and will also describe service requirements. Maintenance and service recommendations should always be heeded, because every piece of equipment will need maintenance and service. Some of the most disappointing installations are where the removal of shafts, for instance, require cutting a hole in an adjacent wall to obtain the required clearance. Look in the installation and operation manual for:

- Safety warnings
- Service clearances
- Component removal space requirements
- Drain connections
- Mounting/support requirements
- Lifting points
- Inspection
- Start-up
- Removal of packing material and transportation supports
- Piping and ductwork connections
- Operating restrictions
- Electrical connections
- Operational limitations (temperature, moisture, dust, etc.)
- Lubrication, if applicable

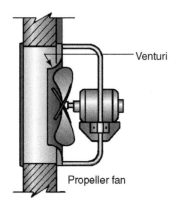

Figure 4–12
An axial fan moves air *parallel* to the propeller shaft. This fan is being used as an exhaust fan.

Air-based Systems

Radial blowers are different than axial fans (see Figure 4–12 and Figure 4–13). Blowers are typically part of air handling systems that may include some or all of the following elements:

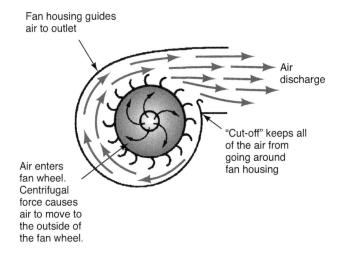

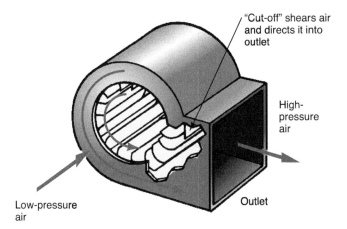

Figure 4–13
A radial fan (centrifugal blower) moves the air perpendicular to the blower wheel (squirrel cage).

- Refrigerant coils
- Heating coils
- Ductwork
- Dampers
- Drain connections
- Straightening louvers
- Filters
- Electrical and control connections
- Methods of controlling airflow (dampers, variable pitch blades, frequency drives, intake dampers, etc.)

Air handling systems are designed and specified long before a technician arrives to install or service the system. However, it is important for the technician to proceed carefully with the installation. Many problems can be averted by paying close attention to details. Read the prints (blueprints) as well as the installation and operation manual from the equipment supplier. Here is a brief list of some important things to check before proceeding with any installation:

- *Connection sizes and locations*—be sure that the sizes of the fluid connections (air and water) are as specified on the print; locate the connection (openings) with the existing equipment and compare these with the required clearance for the equipment.

- *Connectors and fasteners*—be sure that the connectors and fasteners used for connections match those that are called for by the manufacturer.
- *Vibration eliminators*—be sure that mounting pads and vibration eliminators are the right size and quantity for the installation; check that mounting bolts are of the right size and quantity.
- *Wiring*—be sure that wiring access points are in the right location for connection to existing wiring as specified on the prints; that field wiring is the right size for the amperage of the unit; that wiring connectors (wire nuts and crimps) are the right size for the number of conductors; and that electrical access panels are not obstructed.
- *Service clearance*—be sure that the minimum clearances are available as per the manufacturer specifications found in the installation materials; look over the unit for other service situations and ask yourself if the largest component could be removed and replaced easily.

Water-based Systems

Water-based fluid systems consist mostly of centrifugal circulator pumps. Pumps supply heating and cooling coils found in air handlers with hot or chilled water. Pumps can be connected to:

- Hot-water coils
- Chilled water coils
- Drains
- Cooling towers
- Converters (steam to hot-water coils)
- Boosters in existing systems
- Boilers and chillers

Pumps can move many types of waterlike heat transfer fluids, including freshwater, conditioned water, glycol fluids, silicone fluids, and many other fluids designed to move heat. Whenever a pump is specified for an application, the type of fluid that is to be moved is of primary importance. Freshwater systems (sometimes called open-loop systems) require that pump impellers and housings be made with materials (like bronze) that will not rust. Conditioned water systems (sometimes called closed-loop systems) are able to use less expensive materials (like cast iron). The pump must be compatible with the fluid it is pumping. There are several things to look for when installing water-based fluid handling systems:

- Fluid and pump compatibility
- Dielectric connections (separates dissimilar metal piping; Figure 4–14)
- Filters and strainers
- Piping connection material (copper, cast, galvanized, plastic, etc.)
- Mounting location and pump mounting requirements
- Service access
- Leak containment

As with air handling systems, water-based systems have been designed and specified for an installation. And as emphasized previously, the technician must pay close attention to the installation and refer to all available documentation. Important items to check before an installation include the following.

- *Pipe connections*—be sure to use dielectric fittings when connecting dissimilar metals to the pump, such as iron to copper; use pipe sizes as called for in the prints; and use reducers only as specified by and acceptable to the manufacturer.
- *Mounting*—be sure that the mounting and positioning are within the manufacturer's specifications; that mounting bolts are of the quantity

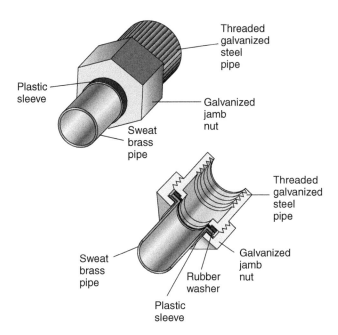

Figure 4–14
The dielectric union separates two different metals through isolation sleeves and rubber gaskets.

and size required; that supports for the pump or the mounting base are of sufficient size and type as specified in prints or the manufacturer's literature; that piping supports are adequate to reduce stress on pump connections and housing; and that vibration eliminators are used to reduce sound transmission in the pipe.

- *Electrical connections*—be sure that wiring access points are in the right location for connection to existing wiring as specified on the prints; that field wiring is the right size for the amperage of the unit; that wiring connectors (wire nuts and crimps) are the right size for the number of conductors; and that electrical access panels are not obstructed.
- *Service clearance*—be sure that the minimum clearances are available as found in the manufacturer's installation specifications; look over the unit for other service situations and ask yourself if the largest component could be removed and replaced easily.

Steam-based Systems

Fluid movement in steam-based systems can be induced by gravity and the pitch of the piping. As steam rises to heat exchangers, the steam condenses and travels back to the boiler to be heated and begin the process again. In the industry, these systems are called *one-pipe* and *two-pipe systems*. See Figure 4–15 and Figure 4–16. More complicated steam systems use pumps to return the condensate to the boiler. Condensate pumps (also called condensate recovery pumps) come in two styles, electric and mechanical.

Mechanical condensate pumps are specialized centrifugal circulating pumps that are designed for the level of heat encountered at the pump. These specialized pumps work on pressure difference to move trapped condensate back to the boiler.

As with other systems, the installation and service technician should check the manufacturer's installation and operation manuals. Make sure that the pump, still in the box, matches what's called for in the prints. Some other things to check before installation are listed below.

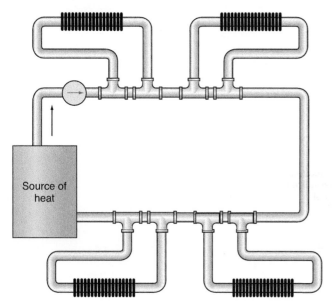

Figure 4–15A
Simplified one-pipe hydronic system.

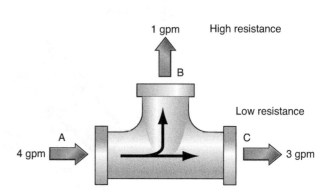

Figure 4–15B
The high resistance at point B lowers the flow through that branch, while the lower resistance at point C results in greater flow through that branch.

Figure 4–15C
Diverter tee for use on one-pipe systems.

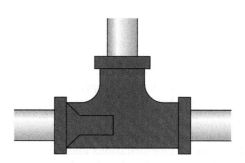

Figure 4–15D
Cross-sectional view of a diverter tee.

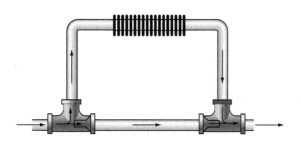

Figure 4–15E
Two diverter tees are recommended if the radiator is located above the main loop and there is significant resistance in that branch.

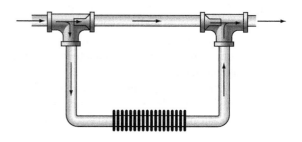

Figure 4–15F
Two diverter tees are recommended if the radiator is located below the main loop.

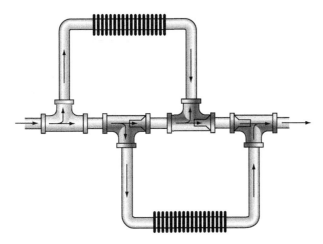

Figure 4–15G
Alternate the diverter tees if there are terminal units both above and below the main loop.

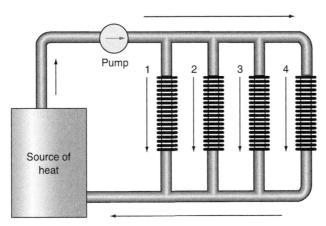

Figure 4–16
Simplified two-pipe hydronic system.

- *Volume and lift*—be sure that the condensate pump is no more than the maximum distance and height from the boiler or the amount of lift in psig.
- *Connection size and type*—be sure to check for threaded or flanged connections and for proper sizes of connections.
- *Pressures*—be sure to check the maximum allowable pump pressure compared with system pressures.
- *Operating pressure* (system pressure)—be sure to check that the pump has the necessary pressure to operate at capacity.
- *Capacity* (measured in lb/hr of steam)—be sure to check the system specifications or requirements against the capacity of the condensate pump.

Measurement

Measurements are critical to a proper installation. Installation is followed by a process of testing, adjusting, and balancing (TAB), which is meant to ensure that the installation is complete and that the device or system functions as designed and specified. Measurements of system operation are required to determine if the device or system is operating within specifications. Table 4–1 describes

Table 4–1 System Measurements (TAB)

System Type	Measurement	Process
Air-based fluid systems	Static (duct) pressure in inches of water column	Static pressure can be measured with a pitot tube and inclined manometer (Figures 4–17 and 4–18).
	Velocity in feet per minute (fpm)	Velocity is measured with a manometer (Figure 4–19) or with a pitot tube in a process called "traversing," where many readings are taken and then averaged (see Chapter 3).
	Volume in cubic feet per minute (cfm)	Volume is determined by multiplying velocity (in fpm) by the size of the duct opening (in square feet).
Water-based fluid systems	Pump head pressure in feet of head	Measured with a differential pressure gauge or two pressure gauges and converted; 1 psi = 2.31 feet of head.
	Pressure drop in psig	Measured with two gauges (Figure 4–20) or a differential pressure gauge from the inlet to the outlet of the pump.
	Flow per minute in gpm	Can be estimated by plotting the pressure drop on the manufacturer's pump curve (Figure 4–3).

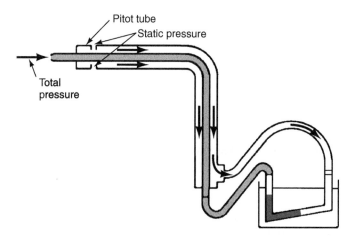

Figure 4–17
A pitot tube used with an inclined manometer to measure velocity pressures.

measurements that may need to be taken during the TAB and compared with the operating design specifications.

Loop Type

There are two types of fluid system loops: open and closed. An open loop is one that moves fluid to an open sump or basin (e.g., a cooling tower or a building air discharge). The closed loop does not have a connection or opening to the outside atmosphere. All fluid systems, whether air based water based, can be fit into one of these classifications.

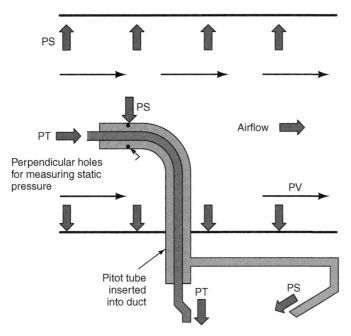

Figure 4–18
The pressures exerted in a pitot tube: PT is total pressure to the manometer; PS is velocity pressure to the manometer.

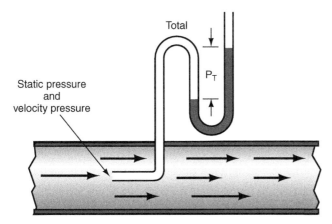

Figure 4–19
Velocity pressure in a duct system being measured with a manometer.

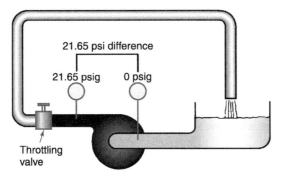

21.65 psi difference × 2.31 feet of head = 50 feet of water column

Figure 4–20
Pressure difference being measured on both sides of the circulator. The pressure drop or difference can be measured by subtracting inlet from outlet.

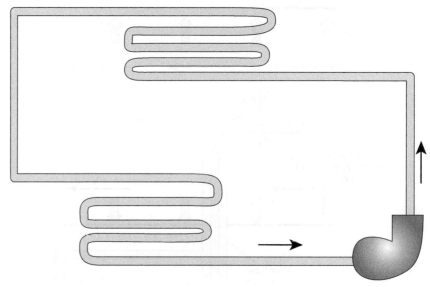

Figure 4–21
Closed-loop system: all piping is sealed from the pump discharge back to the intake.

Closed Loop   Closed-loop systems (see Figure 4–21) move fluids within ductwork or piping. The piping is connected from the pump or blower discharge; through heat exchangers, ductwork, and rooms; and back to the return or intake side of the pump or blower. For instance, water is sometimes returned vertically back to the pump. In this case, the pump does not have to pump against the vertical weight of the water because the water has the same weight coming down. Because the weight of the water is the same going up as going down, the pump does not have to pump water weight. Instead, pumps in closed-loop systems need only pump against the frictional resistance of the piping and fittings.

Open Loop   An open-loop system is a pumping arrangement that pumps the fluid outside of the piping or ducting system. For instance, water may be pumped to the top of a cooling tower, sprayed to the atmosphere, collected in the sump, and returned to the pump inlet. In this example, the pump must move the weight of the water vertically to the top of the cooling tower. When the water gets to the spray rail, it needs to have enough extra pressure to force water through the nozzles. Pump head in an open-loop system is the head due to friction loss to the piping (same as in a closed-loop system) plus the head in feet, and is equal to the difference in the height of the levels of water between where water is being pumped from and being pumped to.

SOLUTION TO THE FIELD PROBLEM

You may already know that vibration is one of the biggest sources of aggravation. What's more, vibration and noise can be signals that mechanical systems are going bad. Bad bearings are the most common source of noise, which indicates that they need to be replaced. Bad bearings can cause mechanical seals to wear and leak. Bearings can go bad if the motor shaft and coupling are putting stress on the bearing because they are misaligned. In the service scenario described in the Introduction, the technician had done all of the right things to check the pump bearings and align the shafts. Checking the impeller is important too, because the impeller may have lost a blade or developed a crack or could be rubbing against the pump housing. The manufacturer's representative assumed that these me-

Table 4–2 Troubleshooting Chart for Circulating Pumps

Problem	Possible Causes
Vibration	• Alignment • Flexible coupling/drive is bad • Impeller missing a blade • Cavitation caused by insufficient net positive suction head, or temperature of fluid too high coming into the pump • Resilient motor mounts deformed
High amp draw on motor	• Too much flow • Motor too small for load • Adjustable-type packing too tight • For single-phase pumps in parallel, backflow could be causing the pump to spin in the wrong direction prior to start-up because of check valves that are missing or stuck open
Low flow	• Motor too small • Blades missing on impeller • Restricted flow or buildup on inlet, outlet, or impeller • Worn impeller or impeller wearing rings • Pump driver not up to full speed • Pump rotation incorrect (it will still pump, but at a lower flow rate) • Plugged screens • Excessive piping restrictions
Leak	• Mechanical seal is worn • Packing is old or burned • Housing is cracked • Gasket has failed • Bearing is bad • Packing gland may need adjustment • Shaft may be worn or corroded • Field water supply lines may be clogged, causing premature failure of mechanical seal • Alignment is off

chanics were competent and had disassembled and reassembled the pump according to manufacturer's specifications. Since the pump was the only component removed, the service representative concentrated her search on the connected electric motor. When she asked the technician how the service work was done, she discovered that the motor had not been moved. This means that the pump was aligned to the motor, which is not the correct way of aligning shafts but rather an expedient way of completing the job. When she touched the screwdriver to the motor, it seemed to be noisier than the pump. Touching the screwdriver to the base and each of four mounting bolts; it was apparent where the noise was coming from. Tightening one mounting bolt made the noise go away.

However, upon further inspection, the service technician found that the mounting pad was worn from the vibration and needed to be replaced. The bolt was also worn and was replaced with thread locker. The technician was instructed to check all of the other bolts and realign the motor to the pump after tightening the bolts. It was found to be within manufacturer specifications.

Table 4–2 provides troubleshooting guidelines for circulating pumps.

It should be noted that the same thing that happens to circulator pumps can happen to blowers. If blowers are not aligned properly, bearings suffer from extra stress and can fail prematurely.

SUMMARY

This chapter covered the basic attributes of fluid handling systems. Air and water are both considered fluids and behave in generally the same way under the influence of blowers and circulators, respectively. Fan and pump curves graph the relationship between pressure and flow. Measurement of pressure is called *static* in air systems and *head* in water-based systems. The similarity between these two fluids is striking. If you understand one, it will be easy to understand the other.

REVIEW QUESTIONS

1. Describe the function of a heat transfer fluid.
2. Identify the heat transfer fluids that are used to heat a room with a boiler.
3. Describe how air pressure is measured.
4. What are the units of measurement for air pressure?
5. What is velocity pressure, and how is it measured?
6. Describe how to convert velocity pressure to feet per minute.
7. Describe how to convert gauge pressure to Feet of head.
8. Describe the tool that is connected to an inclined manometer and used to measure pressure in a duct.
9. Describe what happens to a fluid when the resistance to flow is increased (air or water can be used, and a curve may be used to diagram).
10. Describe what will happen in a hydronic system if additional pipe and fittings are installed in a loop and the circulator is unchanged.

CHAPTER

5

Applied Electrical Problem Solving

INTRODUCTION

This chapter focuses on electrical problem solving in practical ways. It begins with safety and describes the types of safety clothing that must be worn. A section on "arc flash" is included that demonstrates the need for proper clothing. Proper tools and their correct use also protect the health and safety of a technician. Reading electrical prints provides valuable information on system wiring, device placement, and electrical and mechanical sequences. With this information you will be able to conduct the tests necessary to find electrical problems.

Electrical troubleshooting techniques will be discussed along with how electrical tests are conducted. Understanding control wiring and controls is essential to your understanding of any system operation. The motor serves as the heart of the HVACR system. Single- and three-phase motor starting, monitoring, and protection will be discussed, with an emphasis on testing. The different ways that single- and three-phase motors are started will also be discussed.

Field Problem

The four-story office building was intimidating to the inexperienced technician. She was just a week into this new job, but the supervisor had asked her to go to this building without the mentor technician to whom she had been assigned. The air-cooled condenser was on the roof, the chiller was in the basement, and there were air handlers, thermostats, and water coils on every floor. The thought of assessing all this equipment was starting to unnerve the technician until she remembered her instructions: don't be overwhelmed by size of the system; start with the reason for the service call and troubleshoot from there.

The customer reported that only part of the second floor was cool. The first, third, and fourth seemed to be cooled normally. The building control prints, located in the basement next to the chiller, indicated that each floor of the building was supplied by its own air handler with a cooling coil and a circulating pump.

Next the technician went to the second-floor air handler. She checked and found the pump not running. The pump and control circuit were suspect. Locating the pump, control relay, and thermostat, she had all of the information needed to determine the cause of the inoperative pump.

After checking the wiring schematic (see Figure 5–1), the technician set her multimeter for 24 volts to test the input to the relay coil 2C. It read 24 V, which is a normal reading. Next the output was tested. Line voltage (208 V) was found across the relay contacts, which meant that the contacts were open when they should have been closed. Deenergizing the line power and disconnecting one of the power leads, she tested with an ohmmeter and confirmed that the contacts were open with the relay coil (2C) reenergized; the coil should have closed the contacts but did not.

The relay was a push-on type and easily replaced from the stock on the service truck. The technician completed the service call and headed back to the shop.

ELECTRICAL SAFETY

Enough cannot be said about electrical safety. Every day someone has an incident with electricity, either out of ignorance or because of carelessness. If you work with electricity long enough, your risk of having an incident will increase. An incident can be anything from nearly coming in contact with electricity to being electrocuted. Severe electrical burns and fatal electrocution may occur when less than 5 milliamps (5/1000ths of an amp) of electricity flows through the body, and the ability to let go may be lost. Being aware of the electrical risk and taking all proper precautions is necessary at all times.

Personal Protective Equipment (PPE)

Personal protective equipment (PPE) is a strict requirement of the U.S. Occupational Safety and Health Administration (OSHA). A technician may be required to

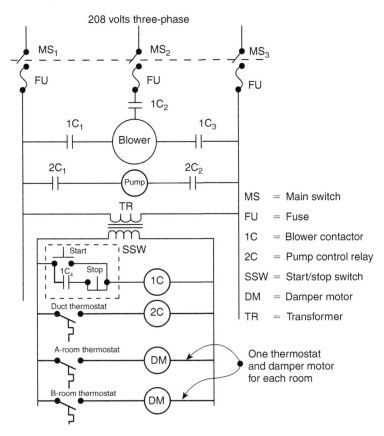

208 volts three-phase

MS = Main switch
FU = Fuse
1C = Blower contactor
2C = Pump control relay
SSW = Start/stop switch
DM = Damper motor
TR = Transformer

One thermostat
and damper motor
for each room

Figure 5–1
This is a schematic for the air temperature and delivery of a "constant volume" commercial air conditioning system. The SSW (start/stop switch) controls the blower. When the start button is pressed, the blower runs continuously until the stop button is pressed. The SSW is considered to be a hold-in or latching circuit. 1C4 closes on start and remains closed unless the power to 1C is lost. Cooling is accomplished through the circulator and room control circuits.

attend or watch a training program on safety before working on company premises; there may also be a checklist of PPE required. Typically, the minimum equipment includes safety glasses, hearing protection, steel-toed shoes, and a hard hat. When it comes to electrical work, there may be more stringent requirements imposed. The type of work and the level of activity affects the requirements for PPE. According to OSHA, it is the employer's responsibility to require that employees wear appropriate, employer-provided PPE in all operations involving exposure to hazardous conditions.

Table 5–1 presents a suggested list of standard PPE for HVAC technicians and when it should be worn. Note that job-site, employer, and OSHA requirements may supersede this chart.

Additional PPE may be required for other types of operations. Always check with your employer or local OSHA office for help in determining potential risks and for advice on PPE.

Table 5–1 Standard Personal Protective Equipment

PPE Item	To Be Worn . . .
Safety glasses	At all times while on the job
Hearing protection (ear plugs, muffs, etc.)	Whenever in noisy environments and around operating equipment
Hard hat	Whenever there is potential for objects to fall from above or for electrical shock from above
Work shoes or safety boots	Whenever working with objects that could fall from waist high or where feet could come in contact with rotating machinery
Goggles	Whenever there is a possibility of dusts or liquids (such as refrigerants) entering the eyes
Gloves (friction, liquids, electrical)	Whenever there is a risk of physical injury (sharp edges), hazardous liquids, or line voltage electrical contact
Clothing	Clothing must be made from cotton (see following section entitled "Arc Flash")
Face shield	Whenever there is a risk of physical or chemical exposure
Respiratory equipment—masks, respirators, or SCBA (self-contained breathing apparatus)	Whenever there is a risk of dusts, vapors, toxic agents, or asphyxiation (as when entering a confined space)
Belts and lifelines	For proper work positioning
Harnesses and lanyards	For fall protection or fall arresting

Arc Flash

Arc flash results when there is a rapid release of energy due to an electrical fault (as in a short circuit). The resulting energy discharge can cause eye damage and electrical burns from the heat and molten metal as well as hearing damage from the concussion. Although arcing can occur at any voltage, levels of 120 volts or more can sustain an arc. It is important to know the industry standards to prevent an arc flash and to be in compliance with tools and clothing (PPE) required for the job.

Potential hazards are grouped into five classification categories. Each category requires specific PPE, as shown in Table 5–2.

Table 5–2 PPE Classification Categories

Category #	Clothing Required
0	Untreated cotton, wool, rayon, silk, or blend (1 layer)
1	Flame-retardant shirt and pants or coverall (1 layer)
2	Cotton underwear *and* flame-retardant shirt and pants or coverall (2 layers)
3	Cotton underwear *and* flame-retardant shirt and pants *plus* flame-retardant coverall (3 layers)
4	Cotton underwear *and* flame-retardant shirt and pants *plus* multilayer flash suit (3 or more layers)

Arc flash hazards are to be identified with signage that warns of the danger, states the category number, and lists the PPE required. In addition to this knowledge, the technician must be able to identify a potential arc flash hazard and reduce or eliminate this potential. For instance, deenergizing a circuit is enough to eliminate the possibility of an arc flash. If the circuit must be in operation for testing, use tools that reduce a potential arc flash. Standards that apply to arc flash include:

- OSHA 29 CFR (Code of Federal Regulations), Part 1910, Subpart S
- NFPA 70 National Electrical Code
- NFPA 70E Standard for Electrical Safety Requirements for Employee Workplaces
- IEEE Standard 1584 Guide for Performing Arc Flash Hazard Calculations
- ANSI Z535.4 Product Safety Signs and Labels

An example of a warning label is shown in Figure 5-2.

Safe Testing Techniques

The best safety practice is to test electrical devices with the power off. However, in HVAC work this is not always the most productive or practical thing to do. Often the system needs to be tested under operating conditions. This means that the technician must adopt safe testing practices to prevent any electrical incidents.

Safe testing techniques that isolate or insulate will keep the technician from becoming part of a circuit. Becoming part of the circuit means that your body (or parts of your body) become a load, just like a light bulb or any other load. To keep from becoming part of the circuit, always employ the safety techniques described in Table 5-3.

Safer Tools and Equipment

Standards organizations have developed standards to guide the testing and rating of electrical test meters. These standards are known as CAT (categories). A higher CAT number indicates that the meter is protected from higher transient voltages (voltage spikes). Table 5-4 gives a brief description of how the CAT numbers apply.

Safety Tip

Anytime the technician is working on equipment that does not need to be energized for troubleshooting, etc., local lockout/tagout procedures should be followed. Usually this will consist of the technician using a padlock (for which he has the only key) and tag warning that the equipment is not to be energized and applying them to the disconnect so that the equipment cannot be reenergized (on the type with the handle, the handle should be in the off position; for the type with the plug, the plug should be removed and the disconnect cover locked shut).

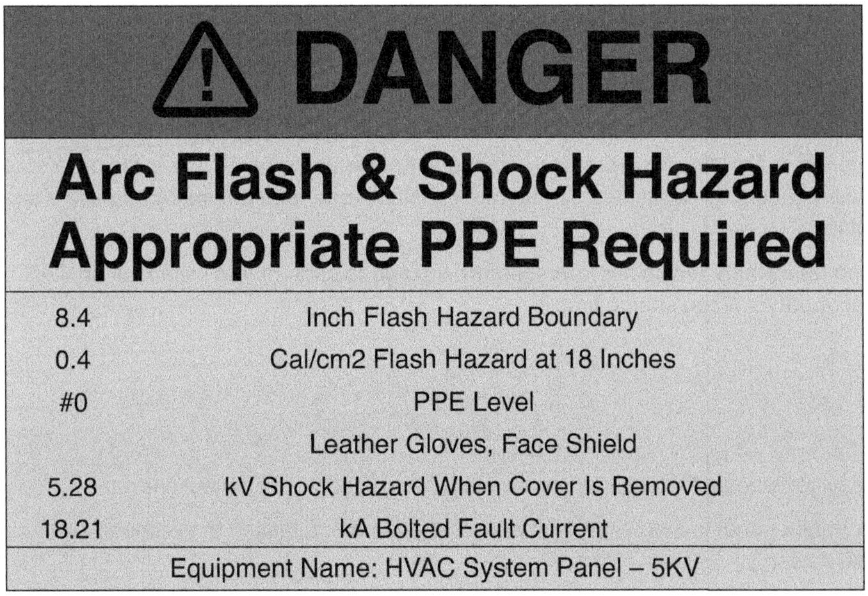

⚠ DANGER

Arc Flash & Shock Hazard
Appropriate PPE Required

8.4	Inch Flash Hazard Boundary
0.4	Cal/cm2 Flash Hazard at 18 Inches
#0	PPE Level
	Leather Gloves, Face Shield
5.28	kV Shock Hazard When Cover Is Removed
18.21	kA Bolted Fault Current
Equipment Name: HVAC System Panel – 5KV	

Figure 5-2
A warning label.

Table 5–3 Safe Techniques for Working with Electricity

Technique	Application
Deenergizing	Turn off the power; lockout and tagout if necessary; perform a voltage test to confirm the power is off; and use an ohmmeter to test for continuity. Electrical testing to determine if switches are closed (or circuits are open) can often be performed with the ohmmeter.
Testing for ground fault	Connect one lead of the voltmeter to a ground source and test with the other probe connected to the meter, using just one hand to the voltage point. Line voltage and three-phase voltage will be read as half of the applied voltage, because tests are being conducted from ground. *Note:* Tests for applied voltage must be taken across the line voltage connections.
Testing for ground potential	Connect one lead of the voltmeter to a ground source (*other* than the cabinet of the HVAC machine) and then test the cabinet, compressor, or other devices to verify ground potential. *Note:* If voltage is read, then the machine or device has a ground fault; it is not safe to continue energized testing or working on a grounded system.
One-hand testing	Only one hand enters an electrical access panel. When testing electrical devices, one hand is used to hold the test instrument and to make electrical test lead connections. The other hand is essentially "behind your back" and is not touching the cabinet, panel, piping, or any other grounded object. This practice reduces the chance of a fatal electrical incident by keeping your heart out of the circuit.
Replacing covers	Replace the electrical cover after completing a test and before continuing with other tests. Electrical covers will protect against accidental contact during other electrical tests.
No jewelry	Remove all jewelry, watches, or metal objects.
Voltage test (three-step measurement)	1. Test your voltmeter on a known line voltage source to be sure that your meter is working before testing an unknown source. 2. Use the voltmeter to verify that equipment is deenergized. 3. Once again, verify your meter on a known line voltage source.
No water	Do not test for voltage or work with live power while in a wet environment (e.g., rainstorm) or where there is heavy condensation. *Note:* An exception can be made if the right PPE for the work and environmental condition are used.
Alligator clips	Insulated alligator test leads are available to fit all test meters. The alligator clip maintains the electrical connection without slipping. If two probes are being used, one should be an alligator: you can watch only one probe at a time *and* two probes may require two hands inside the electrical panel—an unsafe practice.
Neutral or ground first	Always connect a test meter to the neutral or ground connection first. In this way, the other meter test lead will be at ground potential. Connect to the power last and disconnect from the power first.
Positioning the meter	Set the meter on a stable surface or use the manufacturer's mounting device (possibly a magnet). The meter should be positioned so that the technician can read the meter and use a single test probe.
Test the lowest voltage	When testing for voltage, test to the lowest known voltage to confirm that voltage is applied. Test to a secondary voltage source first.

Table 5–4 Standards Categories

Rating	Description
CAT IV	Used by utility companies for testing three-phase power on outdoor connections
CAT III	From the entrance panel to machinery and electrical equipment; testing three-phase and single-phase power
CAT II	Household outlets (receptacles) and connected loads
CAT I	Within the equipment; as in the case of electronics, where tests are done of the power and transformer power sources

In addition to standards that can help guide our choice of meters, there are a number of safety features and tools available to the technician for work on and around electricity. These include:

- Meter leads that screw or lock into the meter to prevent pulling out
- Shock coverings for meters to help protect the meter if dropped
- Insulated screw drivers and other tools to prevent arc flash (high-amperage arc that occurs when terminals are shorted)

Look for safety features and ratings on all meters and test equipment.

Reading Nameplate Data

Before conducting electrical tests, read the equipment nameplate (see Figure 5–3). There is no standard for an equipment or device data plate, so read it thoroughly. The plate may be located in one of several places and contain various

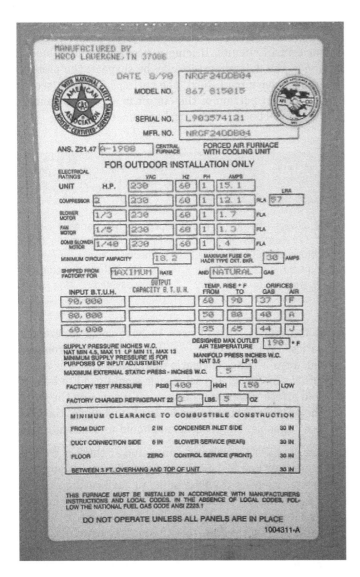

Figure 5–3
This nameplate has a large quantity of information about the system. The technician can use this information as a reference while conducting electrical tests. (Photo by Bill Johnson)

types of information. Most often you will find general information of the following sort:

- Electrical information (voltage, amperage, phase, minimum and maximum fuse or circuit breaker sizing, etc.)
- Capacities (BTU, maximum pressures, refrigerant and charge, etc.)
- Minimal clearances (to combustible construction or to work space)
- Testing agencies (UL, AHRI, etc.)
- Model, serial, and manufacturer numbers

READING ADVANCED ELECTRICAL PRINTS

Reading any electrical schematic takes time and understanding. The important thing to understand is that a technician reads only part of a wiring schematic when troubleshooting. When a system is malfunctioning, the first job of the technician is to determine if the problem is mechanical or electrical. If it is determined that the problem is in the electrical system, the technician determines in which part of the electrical system the problem exists. For instance, if an air-source heat pump is not working and the problem is in the electrical system, the technician will try to determine if it is in the reversing valve circuit, the defrost, or the compressor control circuits. If icing is occurring and is blocking the outdoor heat exchanger, the defrost system may be suspect; the problem may lie in the defrost reversing valve circuit. Next, the wiring schematic (see Figure 5–4) is consulted to see how that portion of the circuit is wired and to identify related controls and read the electrical/mechanical sequence.

Tech Tip

Schematics are always drawn in the powered-off state. Contacts may be shown in either the closed or open state, depending on how the control functions—in other words, NO (normally open) or NC (normally closed). A good way to think about the components of a schematic in the powered-off state is to think of them as being disconnected from the system.

At this point the technician will need to have some experience in reading electrical schematics. Unlike Figure 5–4, which shows the problem of contacts being stuck closed, a real wiring schematic would show only the electrical system in the powered-off state. The wiring schematic would need to be read for all the switches that could affect "reversing the system" (operating the reversing valve), which controls the heating or cooling mode. In this example, the system will not operate in the heating mode. The technician should understand that:

- The secondary power (24 volts) is connected at the R and C terminals, where C is one side and R the other side of the 24-volt power supply.
- The thermostat operates the heat pump in the heating mode by closing from R to Y; the technician must imagine a switch within the thermostat that closes between these two terminals.
- R to O also closes (another internal thermostat switch) at the same time R to Y closes for the heating mode and changes the heat pump from heating to cooling. Keep in mind that this unit opens the circuit to O to go from heating to cooling.

Knowing this and remembering that the system is having problems going into heating mode, we can read Figure 5–4 as follows.

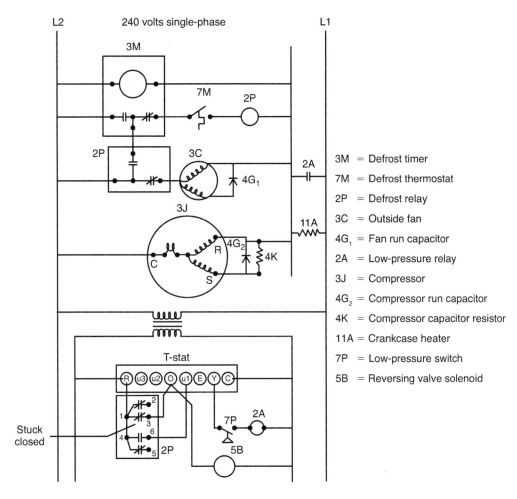

L2   240 volts single-phase   L1

3M = Defrost timer
7M = Defrost thermostat
2P = Defrost relay
3C = Outside fan
$4G_1$ = Fan run capacitor
2A = Low-pressure relay
3J = Compressor
$4G_2$ = Compressor run capacitor
4K = Compressor capacitor resistor
11A = Crankcase heater
7P = Low-pressure switch
5B = Reversing valve solenoid

Figure 5–4
The contacts in relay 2P are closed when they should be open.

1. Because the system did not go into heating mode, start with an electrical component that is common to both: the reversing valve.
2. The reversing valve is operated by closing a switch from R to O (inside the thermostat). Following the 24-volt wire at O to the load, we find the symbol 5B, which identifies this as the reversing valve solenoid. Look for another switch that might be operated by 5B—it would also have a 5B label.
3. There is no other switch with a 5B label, so we can be fairly sure that this is the reversing valve solenoid and not a relay, motor, or some other inductive load.
4. Read the schematic back to O and then to the defrost relay 2P (a portion of this relay is in the 24-volt circuit, and another piece is in the line voltage circuit). We find that there is a switch that is supposed to be open (shown stuck closed). If closed, this switch conducts from R to O. If the defrost relay (2P) was not operating, this switch would be open. We need to find the operator for the defrost relay to determine if it is receiving power (remember, we cannot see the stuck switch and still have no idea that it is stuck closed).
5. The defrost relay is labeled 2P. We need to find the operator for this defrost relay, so we look for another load symbol (circle symbol) labeled 2P.
6. We find the 2P load (operator coil) in the upper portion of the line voltage diagram.
7. Setting the voltmeter for line voltage (240 volts), we measure the voltage applied to 2P and read zero voltage. Hence the relay operator is not

powered, and the contacts should not be in the operated state. Instead, the contacts should be in the non-powered state.

8. Setting the meter for 24 volts, we test from terminal 3 to terminal 1 on the defrost relay. The switch should be open; we should read 24 volts, but the reading is zero. We have determined that the switch is closed when it should be open.

9. To confirm that these contacts are closed, we turn off the power, set the meter for ohms, and test from 3 to 1. We will find that the meter reads continuity, so the relay contacts are confirmed to be stuck closed. The relay needs to be replaced.

Another experienced technician might say that we could have started by testing from 1 to 3 on the gang relay, and he would be (partly) right. He would have confirmed that the relay contacts were closed, but he would still have to check that the operator (load 2P) was energized in the line voltage circuit. Either way, both of these things must be tested and the wiring schematic must be interpreted to find where to test.

In this example, the technician would not need the rest of the wiring schematic. It's not necessary to know about the compressor circuit, the outside fan circuit, the indoor fan circuit, the backup heat circuit (not shown), or the defrost circuit. It is the same way in the field. Technicians should remember that troubleshooting requires reading only one or two small circuits to troubleshoot problems for what may be large and complicated electrical circuits.

Field Problem

A technician was called to service an inoperable heat pump. The customer complained that the outdoor unit was not running.

Symptoms:
- Emergency heat on
- Outdoor unit not running

Possible Causes:
- Refrigerant leak
- Inoperable compressor
- Reversing valve stuck
- Pressure switch
- Defrost control problems

Upon arrival, the technician checked the thermostat to see that it was actually calling for heat and not in the emergency heat mode. The blower was found to be running, and the discharge air was warm because the second-stage electric strip heat was on but the outdoor unit was not running. Next, the technician went to the outdoor unit and checked for 24-volt input, which was present. Referring to the unit schematic, he checked the control circuit and found a reading of 24 volts across the high-pressure safety cut-out, indicating that the switch was open. The technician knew that it could be open only if something occurred during the run cycle to make it open. So even though it may have been the reason the unit was not running now, this fault was probably the result of some other problem. Before obtaining a new switch, the technician decided to attach a gauge manifold to the system and test the operation. In this way, he was ready to shut off the system if pressure got too high. Perhaps the outdoor coil fan was inoperable. He checked the outdoor fan motor for worn bearings and free spin. Next, he bypassed the high pressure switch and reenergized the outdoor unit. The compressor and outdoor fan both started normally; pressures looked normal and were not rising anywhere near where the high pressure switch would have tripped. Next, he checked the fan motor amperage, which tested below nameplate FLA rating. All looked normal to this point, so he proceeded with testing the defrost cycle. This unit had a set of terminals on the defrost circuit card. When jumped across these terminals, the unit went into defrost. All seemed normal at first. The four-way valve shifted to the cooling mode to defrost the outdoor coil, and the outdoor fan shut off to allow heat to warm the outdoor coil for defrost. After a few minutes, however, the compressor discharge pressure began to rise quickly, and the technician shut off the unit before it exceeded the 425-psig setpoint of the high pressure switch. He discovered that the defrost termination thermostat, which should have closed when the outdoor coil was warm enough (indicating that it had defrosted), failed in the open position. This made perfect sense. When the defrost termination thermostat could no longer close, the unit would run until it cycled off on the high pressure switch. With the unit off, the pressure would fall, causing high pressure switch to reset and allow the system to run again, repeating the cycle until the auto reset–type high pressure switch wore out and failed open. The open high pressure switch discovered initially was just a symptom of the actual problem: the defrost termination thermostat. The technician replaced both components and again tested the system in all modes of operation to verify that the problem had been corrected.

Logic Wiring

Logic wiring is another name for ladder schematics (also called across-the-line schematics). Many times these schematics are part of or additional to the point-to-point wiring diagram, as in Figure 5–5A and 5–5B (where the schematic is printed to the right of the legend). The schematic helps the technician understand what switch operates what load (remember, every circuit diagram contains only loads and switches) and understand and read the electrical–mechanical sequence.

Here is how both schematics would be used if the gas valve (GV) were not operating (i.e., for a complaint of no heat).

1. In the schematic, find the GV symbol (round symbols indicate that they are loads).
2. The GV is connected at wire #11 and wire #12. (*Note:* Wire numbers are shown along a wire; terminals are shown as a point; and connectors are shown as double arrows.) Follow #11 to C and stop. This is where GV is connected to one side of the power supply, so there is no need to go any further this way. Now we need to follow the circuit to the other side of the 24-volt power source.
3. Follow wire #12 to K1 and then another K1. These are internal switches to the IFC (integrated furnace control), an integrated circuit that is designated by the dark rectangle. These switches close on a call for heat, which is an input on W on the IFC. The connection comes back out to wire #13 through the NPC (read the legend for what this is), which connects to #14.
4. Follow through the IFC and back out (without going through any switches) to wire #15, where it connects to MRLC (a thermostatic switch) and then to another MRLC and wire #16.
5. Follow through the IFC again and back out to wire #17; then through the LC to wire #18 and the ALC (both thermostatic switches) to wire #19.
6. Follow through the IFC to terminal X and out to optional SD (smoke detector; which we will say is wired to this system) and terminal 6; through the closed switch to terminal 4, where the other side of the 24-volt power supply connects.

Once we know what switches are controlling the gas valve, we can test them one by one with the power on in order to see which is open. Starting with the gas valve (and reading the upper left of the left, point-to-point, diagram), we connect the multimeter (set to read 24 volts AC) with an alligator clip to terminal 4 on the SD. This is the other side of the 24-volt system from the gas valve connection. Next, follow the steps below.

1. With the other meter test lead, test to terminal PL3 for voltage (have voltage; gas valve has continuity).
2. Next, probe PL6 (have voltage; both K1 switches are closed).
3. Next, probe PL5 (have voltage; NPC is closed).
4. Next, probe PL7 (no voltage; MRLC is open). We have found the problem and need to determine why the lockout has tripped. The MRLC uses wire #15 and wire #16; locating them in the point-to-point diagram, we see that they have the color code of Y (yellow). These two yellow wires will lead us to the MRLC (manual reset limit control).
5. Verify that one or more of these switches is open by turning off power, isolating them from the rest of the circuit (unplug the yellow wires), and then using an ohmmeter to check continuity (or visually check for manual indicator).

What we have done here is read the schematic for system function or the logical operation of the system. Another way to read this schematic is to read from

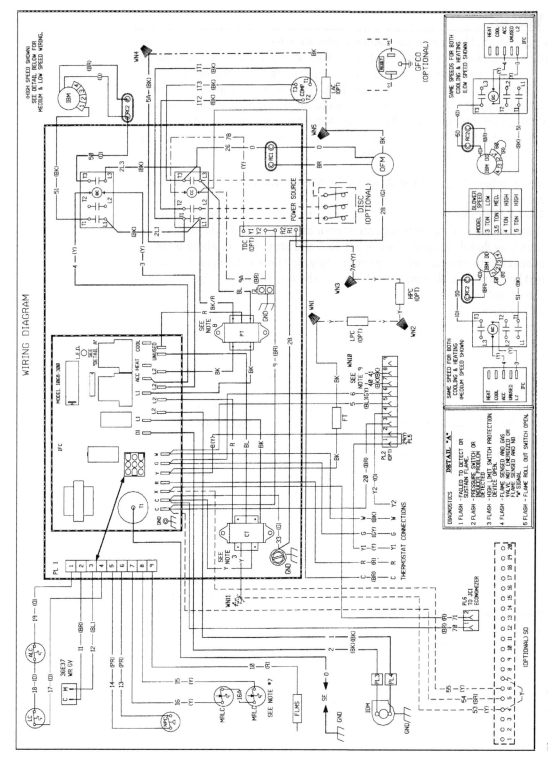

Figure 5–5A
Electrical diagram.

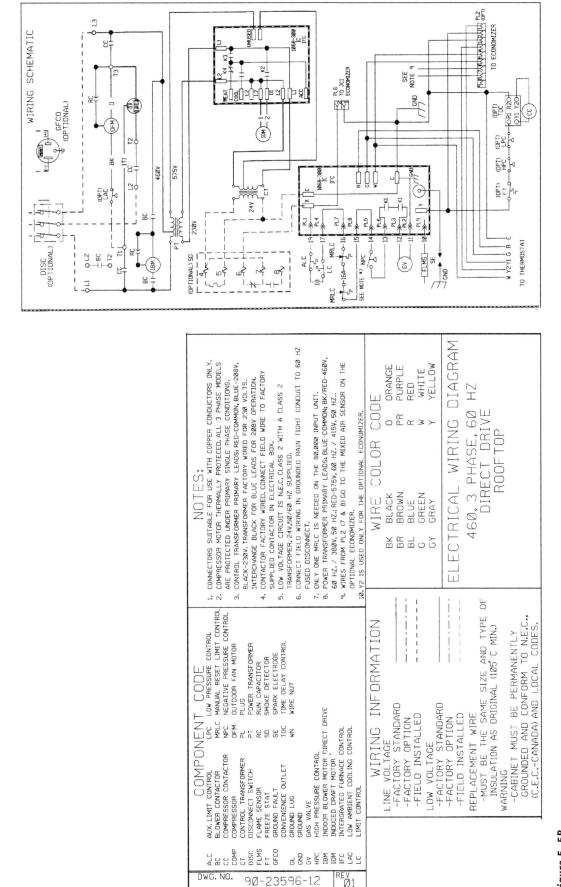

Figure 5–5B
Schematic that might be found on the inside of the electrical access panel for a small commercial rooftop packaged unit (unitary system).

Table 5–5 Logic Circuits

Logic	Meaning
AND	Switch 1 AND switch 2 must close to operate the load (switches wired in series)
OR	Switch 1 OR switch 2 must close to operate the circuit (switches wired in parallel)
NAND (not AND)	Switch 1 AND switch 2 must close to turn off a load (two *series* switches connected to a relay operator that will open a switch in the relay when *both* switches are closed and the operator is powered)
NOR (not OR)	Switch 1 OR switch 2 must close to turn off a load (two *parallel* switches connected to a relay operator that will open a switch in the relay when *either* switch is closed and the operator is powered)

terminal 4 on the SD and say: If SD and LC and MRLC(s) and NPC and K1(s) are all closed, then the GV will operate. This is an "AND" logic control circuit. A brief listing of basic logic circuits is given in Table 5–5.

TROUBLESHOOTING TECHNIQUES

Troubleshooting techniques can be learned, but they must also be practiced. In the preceding sections, examples were provided to describe how to use wiring schematics. Both types of wiring schematics were used to find electrical problems. Here we discuss how to follow the electrical/mechanical sequence and then pinpoint the problem. Some of the most popular techniques are:

- Follow the electrical sequence.
- Use a systematic approach: analyze the customer's complaint and symptoms; then list possible causes and start by checking the most likely.
- Rule out systems and devices that are working.
- Keep in mind cause and effect as part of the systematic approach.

Following the Electrical/Mechanical Sequence

With all HVAC systems, the technician needs to understand the mechanical working of the system. If what needs to happen mechanically is understood, then the job of the electrical system is simply to make it happen. On the other hand, the converse is also true. If the technician can read the electrical schematic, then the mechanical system should function accordingly. How do these two ways of thinking work? They constitute what is known as the "electromechanical sequence," which works something like this:

- A technician with lots of experience with one type or model of HVAC equipment will know how the mechanical system needs to operate in order to provide comfort conditioning. The electrical system follows a particular wiring logic that makes things happen in an electromechanical sequence.
- If the technician encounters a new or enhanced HVAC system with different controls and a different way of operating (perhaps to meet a special purpose), she can read the wiring schematic to deduce how the electromechanical system should operate to provide comfort conditioning.

If we are familiar with a packaged rooftop air conditioning system, then we know that the cooling system should operate when the thermostat calls for cooling. Sending control power from R to Y (refer to Figure 5–5) through the thermostat should start the compressor and condenser fan. But if an economizer is attached, then electrically we see that it may or may not allow the mechanical cooling to operate. It depends on the economizer—the control wire from Y1 goes to the economizer, but this schematic does not show how that will occur. We will need to read the electrical schematic inside the economizer or obtain the electrical schematic and operational instructions from the economizer manufacturer to determine how the economizer should function with the packaged system.

Following the electrical/mechanical sequence means physically determining what is or should be running based on the electrical information. If the technician sees that the compressor is not running, then the electrical schematic is consulted to see what controls the compressor. From electrical to mechanical, each step proceeds to the next in the electromechanical sequence.

Pinpointing the Problem from the Listed Possible Causes

The technician must be systematic in pursuit of the possible causes in order to pinpoint the problem. In the previous sections, a systematic approach was demonstrated. If this approach had not been used, then it may have taken longer to pinpoint the problem, or the technician may have become confused and not found the problem. Do not bump or disturb the system to force system components to operate. Care should be taken when using jumpers to force systems to run, because they may mask the actual problem. In general, every system should be tested "as is"—without forcing or manipulating controls or wiring. Forcing a system to operate may (temporarily) eliminate the original problem, but it usually leads to a return visit to fix the original problem or its cause. Callbacks do not generate income for the company, and they put the technician in an unfavorable light.

For example, suppose that a cooling system was operating but then suddenly stopped. The technician climbed on the roof and carefully opened the access panel. After looking over the wiring schematic, he measured the voltages at the low-voltage terminal block: C to R had 24 volts; C to Y had 24 volts; and R to Y had 0 volts. In other words, the secondary power was good and the thermostat was calling for cooling. Next he checked the operator terminals at the contactor (the coil); 24 volts. Checking across the contacts of the contactor he found 240 volts; the contactor was open. The technician gently took the protective cover off the

Case Study

A split air conditioning system would not cool. The customer explained that it was working fine and then just quit. The indoor fan was not operating.

Symptoms:
- No cooling
- Was cooling before

Possible Causes:
- Control system problem
- Compressor failure
- Blown breaker or fuse
- Thermostat set incorrectly

The technician checked the thermostat and found it to be set correctly. Setting the indoor blower switch to manual did not bring on the system blower. Because the blower did not work, the technician concentrated on the low-voltage control system. Accessing the control panel, his first check was for secondary voltage—there was none. Checking the primary of the transformer for voltage produced 120 volts. To verify the cause, the technician turned off the power, disconnected the secondary terminals of the transformer, and checked for continuity. There was none. The control system was checked for a short circuit, and the transformer was replaced.

contactor to expose the contacts and to visually confirm that the contactor was stuck open; it was. Further verifying the condition of the contacts, the technician found that they were lightly pitted but not enough to prevent the contacts from conducting electricity. The technician replaced the contactor.

TESTING

There are many tests that can be conducted to find a problem. Once that problem is found, another test is required to verify or confirm that the problem exists at that point. Several of those tests will be described in this section.

Evaluating Input and Output

Imagine a black box connected to five wires. Suppose that there is no wiring schematic but that you need to know what is inside. Everything is sealed, preventing access, but you are told that it is a 24-volt relay. What can you do?

Using an ohmmeter, you start testing, but it gets confusing quickly. You need a systematic way to determine what the input and output terminals are for the relay. The best way to do this is to draw a "truth table" to help keep track of the readings you get with the multimeter. Table 5–6 logs a series of possible readings, and Figure 5–6 depicts the result of these values.

Between terminal 1 and 2 you find 1 ohm, and between terminals 4 and 5 you read continuity or 0 ohms. A 1-ohm value (resistance) signifies a load, while a 0-ohm value (no resistance) signifies a switch. The rest of the tests show an open circuit or infinity. We know now that (a) the coil (operator) is between terminals 1 and 2 and (b) there is a switch that is NC (normally closed) between terminals 4 and 5. But there is still a question: What is connected to terminal 3, or is it just a dummy terminal? To answer this question, we need to apply input voltage to the operator and measure the output at terminal 3. (*Note:* The operator is the coil. Only the coil is powered, so all other power to the contacts must be eliminated!) That measurement would find no voltage (testing from terminal 1 to 3 and from 2 to 3 with a voltmeter), so terminal 3 is not connected to the input power. Testing with the ohmmeter would reveal continuity (0 volts) from terminal 3 to 4 and an open circuit from terminal 5 to 4. The truth table is now as shown in Table 5–7.

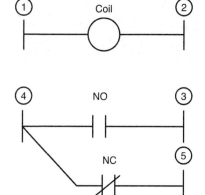

Figure 5–6
A drawing of the "black box" relay that results from the truth table values.

Table 5–6 Test Results in Truth Table Format, Coil Power Off

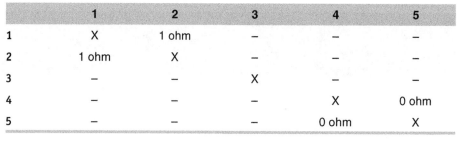

	1	2	3	4	5
1	X	1 ohm	–	–	–
2	1 ohm	X	–	–	–
3	–	–	X	–	–
4	–	–	–	X	0 ohm
5	–	–	–	0 ohm	X

Table 5–7 Test Results in Truth Table Format, Coil Power On

	1	2	3	4	5
1	X	–	–	–	–
2	X	–	–	–	–
3	–	–	X	0 ohm	–
4	–	–	0 ohm	X	–
5	–	–	–	–	X

Because the second test shows that terminal 4 to 3 is closed and terminal 4 to 5 is open, we know that the switch is energized (moved) by the operator—a coil between terminals 1 and 2—and that terminal 4 is common to both 3 and 5. The input and output measurements indicate that the switch is a single-pole, double-throw relay switch.

This procedure can be used to deduce the input and output for any "black box" device. In most cases, however, the manufacturer will stamp a wiring schematic on the relay that shows the contact positions when the operator is not powered. Another way to check a circuit for power is to measure amperage. Both input (primary) and output (secondary or 24 volts) could be measured to determine if 24-volt loads (solenoids, relays, etc.) were operating. This would prove that 24 volts is being provided, contacts are closed, and loads are being operated in the control voltage circuit.

Open Testing

The most common "open" test that can be done is to ensure that there is no voltage from a machine to ground. This open test is performed using a voltmeter. If no voltage is measured on the lowest scale, then there is an open circuit (no volts or potential) from the source (power) to ground.

The next most common test for "open" is performed using a multimeter. The ohms setting of the multimeter is used to power the circuit (using the meter's internal battery) and then detect either an electrical path or a closed circuit. If the meter reads zero (no ohms of resistance), then there is a complete path for electricity and the circuit is not open. The same is true if the ohmmeter shows high resistance. The higher the resistance, the harder it is for electricity to flow, but it still means a closed circuit. Only when the ohmmeter reading is infinity (an infinite amount of resistance) is the circuit considered to be open.

A voltmeter is used to check for an open circuit between two points, since the probes of the meter are touching the opposite sides of the potential (voltage in the circuit). An ohmmeter used to check for an open circuit will show either no resistance or more resistance than can be measured (infinity). The check with the voltmeter should be done with the power on, whereas the check with the ohmmeter should be done with the power off.

Ground Tests

The term "ground source" is used when looking for a good ground, and the term "grounded or ground fault" is used to describe a path to ground.

When testing for a ground *source,* the objective is to find a good earth ground from the device or system that generates a value of 2 ohms or less. The test can be conducted using an ohmmeter. One probe is placed on a known ground, and the other probe is placed on the point to be used for ground.

When testing for a ground *fault,* a multimeter is used on its highest scale. One test probe is placed on a known ground, and the other probe is placed on the equipment, motor, or cabinet to be tested. Safe operation requires that there be more than 100 megohms of resistance.

Megohm Testing

A megohmmeter is a testing instrument that applies a voltage of 50 to 500 volts DC to a circuit in order to test for leakage to ground (see Figure 5–7). The leaking voltage is measured in megohms by the meter, where 1 megohm equals 1,000,000 ohms.

The megohmmeter is used to test compressor winding resistance to ground. The test is to discover if the insulation of the windings has deteriorated to a point where it is unsafe to operate the compressor. The megohmmeter is also used to monitor the condition of the winding of a compressor during the entire life of the compressor. By taking periodic readings during scheduled maintenance,

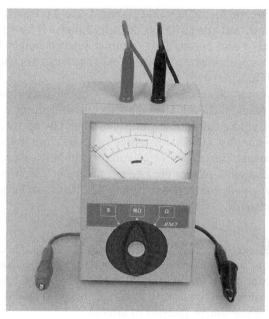

Figure 5–7
Megohmmeter for checking ground potential of motors, circuits, and systems. (Photo by Bill Johnson)

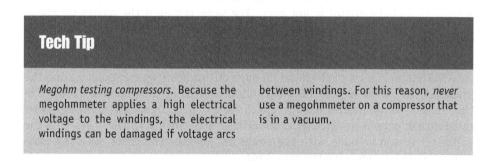

Tech Tip

Megohm testing compressors. Because the megohmmeter applies a high electrical voltage to the windings, the electrical windings can be damaged if voltage arcs between windings. For this reason, *never* use a megohmmeter on a compressor that is in a vacuum.

deteriorating conditions of the windings can be detected before the compressor burns out.

It should be noted that the megohm readings can be affected by temperature, moisture, and the condition of compressor oil. Guidelines for reading megohms can be obtained from the manufacturer of the compressor.

Documentation of Tests

When tests are conducted of a system or compressor, the data are recorded (see Figure 5–8). Records of data are kept to compare the operation of the system or compressor over the lifespan of the unit. The data are reviewed and compared with previous data to determine if there is any change in the equipment's operating characteristics. If a compressor is used as an example, the following electrical data would be obtained and documented:

- *Voltage*—to see if applied voltage is consistent.
- *Amperage*—to see of the amperage is fluctuating and if it is within tolerance for the compressor at various conditions.
- *Megohms*—to monitor leakage to ground and leakage from winding to winding.
- *Phase*—voltage differences and amperage per leg are compared to determine if phasing angles and power are within tolerance.

ROTARY SCREW LIQUID CHILLER LOG SHEET			CHILLER LOCATION										
			SYSTEM NO.										

Date												
Time												
Hour Meter Reading												
O.A. Temperature D.B./W.B.		/	/	/	/	/	/	/	/	/	/	
Motor	Volts											
	Amps											
Cooler	Refrig.	Suction Pressure											
	Liquid	Inlet Temperature											
		Inlet Pressure											
		Outlet Temperature											
		Outlet Pressure											
		Flow Rate — GPM											
Condenser	Refrig.	Discharge Pressure											
		Corresponding Temperature											
		High Pressure Liquid Temperature											
		System Air — Degrees											
	Water	Inlet Temperature											
		Inlet Pressure											
		Outlet Temperature											
		Outlet Pressure											
		Flow Rate — GPM											
Compressor		Oil Level											
		Oil Pressure											
		Oil Temperature											
		Suction Temperature											
		Discharge Temperature											
		Filter PSID											
		Slide Valve Position %											
		Oil Added (gallons)											

Figure 5–8
An example of a basic system log sheet on which information about the system, including voltage and amperage, is recorded on a regular basis.

Case Study

A forced air, natural gas furnace was short-cycling, and the blower was not coming on. The customer said that this problem seemed to start occurring slowly. The furnace would turn on and off two or three times and then start working. Now it just turned on and off and didn't produce heat. See Figure 5–9.

Symptoms:
- No heat
- Burner is short-cycling
- Blower does not operate

Possible Causes:
- Blower motor is inoperable
- Blower temperature control is faulty
- Problem in the control circuit

The technician started with the thermostat and could not make the blower operate on the manual switch setting. Going to the blower motor circuit, he checked for power at the blower line voltage connections (with the thermostat switch still on manual blower). There was no voltage to the blower (voltmeter test 1). Checking from one side of the blower to the opposite side of the blower temperature control (voltmeter test 2), he still found no voltage. Testing a third time from one side of the blower to the opposite side of the BR1 contacts (voltmeter test 3), he read line voltage. This indicated that the blower relay was open. Resetting his meter for low voltage, he checked for power to the blower relay (voltmeter test 4) and found 24 volts. The relay was powered but not pulling in. Turning off the power, disconnecting the relay coil leads, and setting his meter for ohms, he checked for continuity through the relay coil (ohmmeter test 1) and did not find continuity. The relay was replaced.

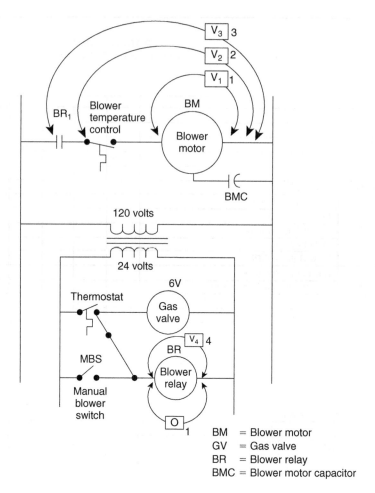

Figure 5-9
Schematic for a furnace. Notice the test instruments (V = voltage and O = ohms) and test locations used to diagnose the system.

BM = Blower motor
GV = Gas valve
BR = Blower relay
BMC = Blower motor capacitor

CONTROLS AND WIRING

Control wiring can be very simple or very complex. As systems become larger and more expensive, additional controls are added to protect and customize their operation. Every compression refrigeration system has four system components: compressor, condenser, metering device, and evaporator. Basic compression refrigeration systems also have no fewer than four electrical devices: compressor motor, temperature or pressure control, motor starting control, and motor safety/ limit. It is the same for both small and large systems, although larger systems have many more motors, limits, safeties, pressure controls, and user interfaces (e.g., thermostats). In this section we will look at some of the generic controls, components, and control systems encountered in the field.

Relays

Relays (see Figure 5-10) are used in many applications throughout the HVAC industry. The most common use of relays is for creating control sequences that meet the needs of the application and/or the customer's requirements. It should be mentioned that there is no *operational* difference between a relay and a contactor (see Figure 5-11); the only difference is the amount of amperage that can be switched. The dividing line between relays and contactors seems to be about 20 amps. If higher amperage can be handled by the contacts then the unit is

Figure 5–10
A double-pole, double-throw relay commonly used as a fan relay. (Photo by Bill Johnson)

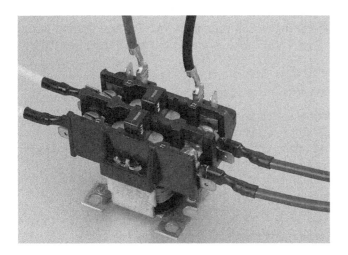

Figure 5–11
A contactor used to control a compressor. (Photo by Bill Johnson)

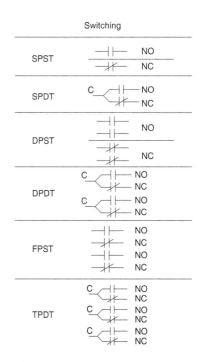

Figure 5–12
These symbols and open or closed designations may be printed on the relay.

called a contactor; if only lower amperage can be handled, it is considered a relay (the distinction is not a rigid one). The point is that, electrically speaking, relays and contactors operate in the same way.

There are two major classes of relays, line voltage and low voltage (also known as control voltage). Within these two classifications, the technician must understand that line voltage can be anywhere between 50 and 600 volts and that low voltage ranges from 50 volts down to millivolts. (Thus, 50 volts is the arbitrary divide between low and line voltage in HVAC work.) Line voltage means that the power being supplied is being used to power the operator (coil) of the relay. Low voltage means that a step-down transformer has been used to lower the voltage from line voltage to a voltage that can be used for control.

Within the two major classifications of line and low voltage, relays are further categorized by their switching action (see Figure 5–12):

- Single-pole, single-throw (SPST)
- Single-pole, double-throw (SPDT)
- Double-pole, single-throw (DPST)
- Double-pole, double-throw (DPDT)
- Three-pole, single-throw (TPST)
- Three-pole, double-throw (TPDT)
- Four-pole, double-throw (FPDT)

Next, switches are designated as either NO (normally open) or NC (normally closed). When the manufacturer of the switch stamps this on the relay or states this in the packaged instruction sheet, it means that the switch is in this position when the operator (coil) is *not* energized. Each of the switch configurations can come with NO or NC designations. For instance, you can purchase a relay that is low voltage, single-pole, single-throw, and NO or one that is low voltage, single-pole, single-throw, NC. It depends on whether the relay is being used for an AND logic (NO; turning something on when the operator is powered) or NAND logic (NC; turning something off when the operator is powered); refer to Table 5–5.

Finally, the relay is rated for the amount of load that each switch can handle. The load rating for each switch is listed in terms of:

1. Amperage rating of the contacts
2. Voltage applied
3. Power applied (AC or DC)

An example of a relay specification might be:

Enclosed Power Relay, Coil Voltage 24 VAC, Contact Current Rating Resistive @ 28 VDC 13 Amps, Contact Current Rating Resistive @ 300 VAC 25 Amps

Here's what this description tells us:

1. The relay is sealed (enclosed).
2. The operator (coil) is rated for low voltage (24 volts AC).
3. The first rating of the contacts is for DC power at 28 volts, which provides up to 13 amps of capacity.
4. The second rating of the contacts is for AC power up to 300 volts, which provides up to 25 amps of capacity.

Both of these ratings are "resistive"; this means that they are rated for lights and heaters, not for inductive loads like motors and solenoids.

Tech Tip

Inductive loads draw more power when first energized. If a motor is to be operated by a relay or contactor, the amperage of the motor must be matched to the contactor rating. Remember, the contacts must handle the amperage not only for a short time when closing but also, if the motor is stalled (locked), until some safety device signals to open the contacts—and then the contacts must be in working order to open. Look for "full load" ratings for contactors.

An example of a contactor specification might be:

Definite Purpose Contactor, Inductive Full Load Current 40 Amps, Resistive Full Load Current 50 Amps, Number of Poles 3, Power Rating @ 120 VAC 3 HP 1 PH 3 HP, Power Rating @ 240 VAC 1 PH 7.5 HP, Power Rating @ 240 VAC 3 PH 10 HP

Here's what this description tells us:

1. The operator (coil) can be changed to any voltage value and is purchased separately (this is why it's not mentioned).
2. The contactor is designed for a particular use or specification ("definite purpose").
3. The first rating of contacts is for an inductive full load of 40 amps.
4. The second rating of contacts is for a resistive full load of 40 amps.
5. There are three poles or three contacts for three separate circuits.
6. The horsepower rating of the contacts is for a 3-HP, 120VAC, single-phase motor, a 7.5-HP, 240VAC, single-phase motor, or a 10-HP, 240VAC, three-phase motor.

Transformers: Step-up and Step-down

Transformers change the value of voltage and work only with AC circuits. An easy way to remember this is that power in equals power out. If the voltage goes down in the secondary, ampacity increases. The changing polarity of alternating current allows magnetic energy to commute power from the primary (input) to the secondary (output). Most often, transformers are used to step down voltage (see Figure 5–13) from line voltage to control voltage. In some cases, transformers are designed to step up the voltage (Figure 5–14) in order to overcome the resistance of an air gap or to meet the requirements of a spark ignition control system (see also Figure 5–15 and Figure 5–16).

Transformers are rated VA, which stands for (theoretical) voltage times amperage. If the VA of a transformer's secondary is 24 VA and if the voltage output is

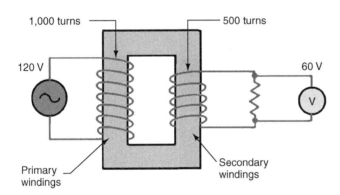

Figure 5–13
A step-down transformer.

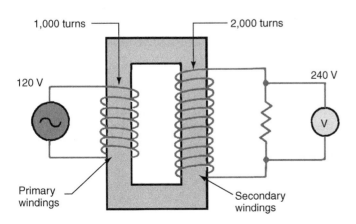

Figure 5–14
A step-up transformer.

Figure 5–15
A multi-tap transformer for use on multiple primary voltages (120 V, 208 V, or 230 V). Unused primary wires should be insulated and capped. This transformer has a 24-volt secondary. (Photo by Bill Johnson)

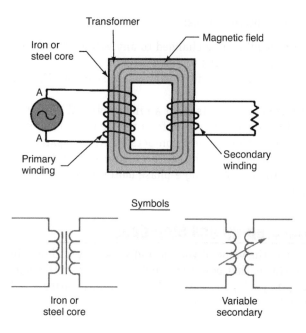

Figure 5–16
The symbol for a transformer always shows an iron core between two coils. If additional wires are drawn from the coil, these are called "taps" that represent different voltages.

24 volts, then the amperage output is 1 amp (24 V × 1 A = 24 VA). This means that transformer power is available at 24 volts and up to 1 amp for operating 24-volt relays, solenoids, and heaters. The VA rating is the rating of a transformer's secondary, whether the transformer is a step-up or a step-down type.

Safety Tip

When testing transformers with an ohmmeter, keep in mind that the ohmmeter applies a small voltage (battery voltage) to the circuit. If you are testing a step-up transformer on the primary side, the secondary (step-up) side could produce enough voltage to cause a spark or an electrical shock. The same is true if you are checking the secondary on a step-down transformer.

Tech Tip

Technically speaking, a step-down transformer is also a step-up transformer (winding to winding relationship). However, it is not designed to be turned around in a circuit—the result would be a burned out (and possibly smoking) transformer. Be sure of the purpose, rating, and wiring of every transformer.

Case Study

The store owner reported that the heating system was working until two hours ago when it simply quit. The thermostat was adjusted to see if it would come on, but nothing happened. The electric heat rooftop unit would not respond. See Figure 5–17.

Symptoms:
- Thermostat is set correctly for heat
- Heating system is not operating

Possible Causes:
- Blown fuse or breaker
- Control circuit problem
- Thermostat problem

The technician checked to see if the blower would operate with the manual switch, and the blower came on. Because the system was on the roof, the technician continued with troubleshooting at the thermostat. Removing the thermostat from the subbase, he checked for voltage from R to W and read 24 volts. Jumping from R to W made the electric heat system function, and warm air blew from the ceiling diffuser. To verify that the thermostat was the problem, he set his meter to ohms and checked between R and W. There was no continuity with the thermostat in the correct position (level). The thermostat was replaced.

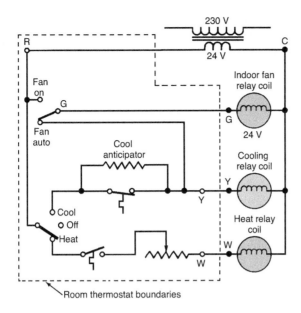

Figure 5–17
Control voltage circuit showing the internal thermostat schematic.

Transformers are tested for primary and secondary using an ammeter and voltmeter. The measured values are compared to the transformer's specifications. If a transformer is suspected of being burned out, standard procedure is to conduct an open winding test with an ohmmeter.

Tech Tip

Use proper antistatic procedures when handling solid-state components. This includes grounding yourself before touching the component.

SOLID-STATE ELECTRONICS

Solid-state electronics is a growing part of HVAC system control (see Figure 5–18 and Figure 5–19). Solid-state devices have no moving parts and thus are less likely to wear out than electromechanical devices. However, solid-state devices are more sensitive to stray current, static electricity, and voltage spikes, and they will burn out if wired incorrectly or accidentally shorted. It is often said that, when testing around solid-state devices, if you see a spark then it's all over. One wrong touch of a stray electrical wire may mean that an entire board must be replaced.

A great deal of study could be devoted to solid-state electronics, but an HVAC technician should mainly be familiar with the following basic principles.

- An ohmmeter sends power (internal battery) to a circuit. Do not use an ohmmeter to test an electronic board except in accordance with the manufacturer's instructions.
- Always follow the manufacturer's testing procedure for an electronic board.
- Every board has a set of inputs and outputs; apply input and measure output strictly for each function of the board.

Figure 5–18
The electronic control board. (Courtesy of Ferris State University. Photo by John Tomczyk.)

- Many boards have self-diagnostic lights or readouts. Use the diagnostic information and the manufacturer's troubleshooting guides to isolate the problem.
- In most cases it is impractical to test or repair individual components on an electronic board. If the board is faulty (as determined by measured input and output), then replace the entire board.
- Many problems with boards involve sensors. Be sure that you test and verify sensor input to the board.

See Figure 5-20 and Figure 5-21.

Sensors/Monitors

Electronic sensors are available in many shapes and sizes (see Figure 5-22). They are used to monitor conditions that can be controlled by HVAC equipment. By monitoring a specific condition, the sensor sends continual information to the controller. When the monitored condition falls below the control setting, an action occurs to correct the condition. Sensors fall into these classes:

- *Temperature*—thermostats, heat sensors, and cooling sensors (Figure 5-23)
- *Pressure*—high pressure transducers and low pressure transducers
- *Electrical*—volts, amps, phase, and hertz
- *Moisture*—liquid level sensors and humidity sensors
- *Motion*—motion, rotation, speed, and vibration sensors

Sensors (see Figure 5-24) send information (input) to a controller (electronic board, or microprocessor) in three ways:

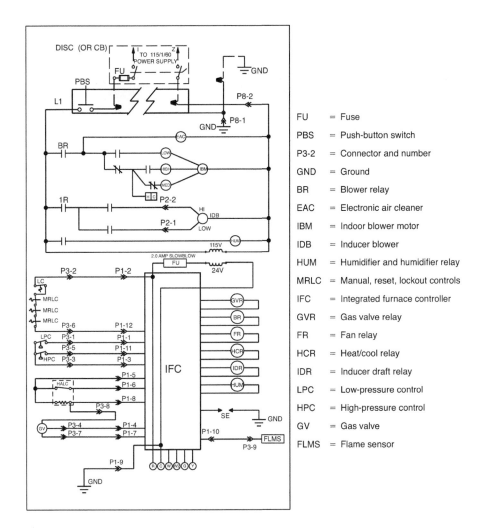

FU	= Fuse
PBS	= Push-button switch
P3-2	= Connector and number
GND	= Ground
BR	= Blower relay
EAC	= Electronic air cleaner
IBM	= Indoor blower motor
IDB	= Inducer blower
HUM	= Humidifier and humidifier relay
MRLC	= Manual, reset, lockout controls
IFC	= Integrated furnace controller
GVR	= Gas valve relay
FR	= Fan relay
HCR	= Heat/cool relay
IDR	= Inducer draft relay
LPC	= Low-pressure control
HPC	= High-pressure control
GV	= Gas valve
FLMS	= Flame sensor

Figure 5–19
Schematic (ladder or across-the-line) of the system's operation.

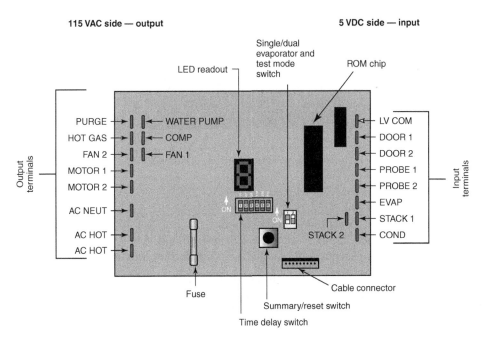

Figure 5–20
An electronic control board (microprocessor) showing a readout, input, output, and test switches.

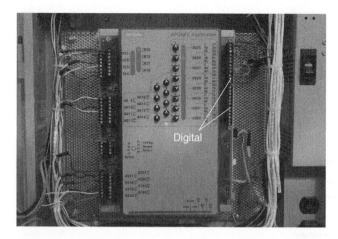

Figure 5–21
A controller (microprocessor), also referred to as a DDC (direct digital control), that can accept both analog and digital input connections. (Courtesy of Ferris State University. Photo by John Tomczyk.)

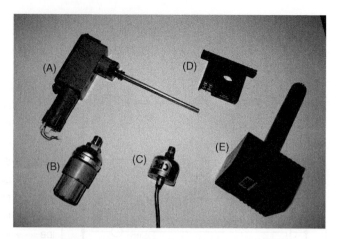

Figure 5–22
There are many different types of sensors: (A) thermistor; (B) & (C) pressure transducer; (D) current sensor; (E) humidity sensor. (Courtesy of Ferris State University. Photo by John Tomczyk.)

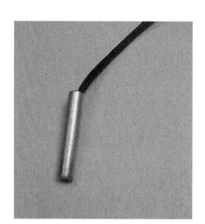

Figure 5–23
A thermistor used as a temperature sensor. (Courtesy of Ferris State University. Photo by John Tomczyk.)

Figure 5–24
Modular, nonadjustable pressure sensors. (Courtesy of Ferris State University. Photo by John Tomczyk.)

1. *Resistance*—measured in units of ohms
2. *Voltage*—measured in units of volts down to millivolts
3. *Amperage*—measured in units of amps down to milliamps

The meter that is used to test these input signals must be sensitive enough to measure small values of voltage and amperage. Loose or dirty connections with the controller or the test meter will make it difficult to obtain good signals for or readings from a sensor. Always be sure that the connections are good for both the controller and the test meter.

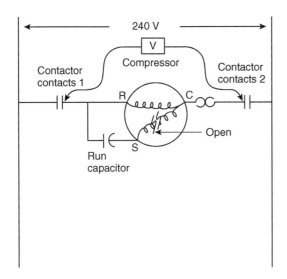

Figure 5–25
Compressor circuit for an air conditioning system showing an open circuit in the start winding.

Case Study

No cooling was the customer's complaint. The system had stopped working the day before when it was very hot outside. The blower was operating, but there was no compressor hum. The outside fan was operating, too. The breaker was not tripped, and the thermostat was set correctly. See Figure 5–25.

Symptoms:
- No cooling, thermostat set correctly, line voltage was supplying to the system
- Indoor and outdoor fans were operating

Possible Causes:
- Safety control problems
- Compressor is inoperable
- Low refrigerant pressure and off on the low pressure control

The technician decided to check the compressor first to see if it would run. Removing the electrical panel cover of the outdoor unit, he checked to see if the contactor was pulled in. It was. Checking for voltage to the compressor at the contactor, he found 240 volts. During the check, the compressor overload clicked. The compressor was hot. Turning off the power, he disconnected the compressor leads and set his meter for ohms. Checking for continuity across the three compressor terminals, he found C to R had continuity, but C to S did not. The continuity check confirmed that the start winding was open. The customer was informed that the compressor needed to be changed out.

ELECTRIC MOTORS

Electric motors are part of every HVAC system. Understanding and being able to test motors is a large part of a technician's job. Motors fall into one of two large categories: single phase and three phase. Single-phase motors generally require some assistance to start. A three-phase motor can start on its own, but if the motor is large then it is started in stages in order to reduce the required amount of starting amperage. Motors may use anywhere from 3 to 20 times their normal operating amperage to start. Large starting amperages match the "locked rotor" amperage, or the amperage that would be used if the motor were stalled and/or the rotor locked. To reduce the size of wire and switching gear, large motors are started and "proven" at lower amperages. If the motor does not start and/or rotation is not established, then limit controls shut off power before the equipment is damaged.

This section covers motors, motor starting, and control devices. There are many different styles of motors used in HVAC. The most common ones may be described as follows.

Single-phase Motors
- Permanent-split capacitor (PSC) motor
- Capacitor-start (CS) motor
- Capacitor-start–capacitor-run (CSCR) motor

Three-phase Motors
- Part-start motor
- Wye-Delta start (or Delta-star start) motor

Single-phase Starting Controls

Small-horsepower motors (3 HP or less) may be started using start relays and capacitors, but larger motors are usually started with a motor starter. The 3-HP dividing line is arbitrary; a motor rated slightly above or below this value could be configured to start either way. This is only given as a general guideline for identifying motors that will need higher amperage and, as a result, a form of starting control in which contacts can be replaced as they wear out.

Figure 5–26
A current magnetic relay, showing the operator made of heavy wire. (Photo by Bill Johnson)

Tech Tip

When replacing motor-starter heaters, always use the manufacturer's selection table and the motor data plate information. Replace heaters as a set, not one at a time. When heaters have been tripped a number of times, they may need to be replaced because they become weak. Check the amp draw and compare against heater load rating.

Current magnetic relays (see Figure 5–26) are used for smaller (fractional-HP), low-amperage motors. Current magnetic relays sense the total amp draw of the run windings. When high amp (locked rotor) conditions are sensed, the relay closes the switch to the start windings. As the amperage drops and the motor reaches 85% of its rated running speed, the relay drops open the switch and deenergizes the start windings (see Figure 5–27). This type of relay is NO (normally open) and position sensitive—it must be mounted in the upright position (as identified on the relay). Use an ohmmeter to test the contacts in the deenergized condition.

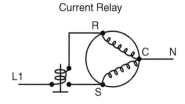

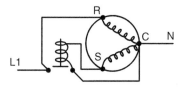

Figure 5–27
Current and potential magnetic start relays, shown wired without start or run capacitors. Notice that the current relay is a NO switch and the potential relay is a NC switch. The potential relay operator (coil) is wired in parallel with the start windings.

Tech Tip

Current magnetic operators (coil) seldom burn out. These are used on fractional-horsepower motors. There is usually no need to test the operator. Contacts usually burn up or get stuck. The best way to check contacts is to remove the relay and turn it over (upside down). Contacts will fall together and can be checked with an ohmmeter.

Potential magnetic relays monitor the voltage (potential) of the start windings. As the motor starts, the start winding acts as an alternator, creating a voltage that is opposite to the voltage being used. This is referred to as *reverse voltage* (or *counter emf,* electromotive force). For example a 240-volt motor starts and produces a counter emf in the start winding, which in turn creates a magnetic field in the potential relay. The magnetic field is not strong enough to open the start switch unless the voltage reaches a counter emf of 340 volts, after which it closes the switch when the counter emf drops to 290 volts. Both the operator (coil) and the switch are susceptible to damage. The operator should be checked for continuity, and the contacts (which are NC) can be checked using an ohmmeter in any position but with no power applied. Potential relays are generally applied to single-phase motors (from fractional to 3 HP) with high starting torque and/or amp draw. Compared with current relays, the advantage of potential relays is that the normally closed (NC) contacts avoid contact arcing on start-up.

Solid-state motor starters have no moving parts, and one relay may fit a number of compressors over a range of horsepower values (see Figure 5–28 and Figure 5–29). When the solid-state motor control is cold, there is very little resistance. As amperage flows through the device to the start winding, the control begins to heat up. Heating of the thermistor in the solid-state motor starter increases resistance and thus blocks the flow of electricity to the start winding, acting in effect like an open circuit.

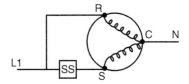

Figure 5–28
The SS (solid-state) relay is wired in series with the start winding.

Capacitors

Capacitors used with motors are classified as either start or run capacitors (see Figure 5–30 and Figure 5–31). Start capacitors are "short duration" capacitors and have no means of dissipating heat. They are packaged in plastic and are typically round in shape. Run capacitors are designed to remain in the circuit for as long as the motor runs. These capacitors are packaged in metal containers, which are oval or rectangular, with an internal heat transfer liquid that allows the capacitor to give up heat to the metal package.

Tech Tip

Capacitance testers are available and inexpensive. Many multimeters include them as one of their features. Look for test instruments that meet your needs on the job.

Hard-start Kits

A hard-start kit typically consists of a relay and a start capacitor connected together in one package. The package is designed to be wired to a compressor and act as a capacitor start (see Figure 5–33). Motors that do not have a start relay are PSC (permanent-split capacitor) motors. The hard-start kit is connected from the line directly to the start terminal on the motor (see Figure 5–34). When the motor is powered, the solid-state relay allows power to the start winding with an extra boost from the start capacitor. As the motor reaches 85% of its rated running speed, the solid state relay stops the flow of electricity and thereby turns off the hard-start kit. Motors of PSC type are generally found in low-horsepower (3 HP or less) a/c applications with low starting torque requirements and equalized pressures at start-up.

Safety Tip

Always handle capacitors as if they were fully charged. Capacitors can pick up stray electricity and may carry a charge at any time. Always bleed off energy using a 20,000-ohm resistor. A resistor may be attached to the capacitor for this purpose (see Figure 5–32). Never use a screwdriver to discharge a capacitor; it may be damaged by shorting.

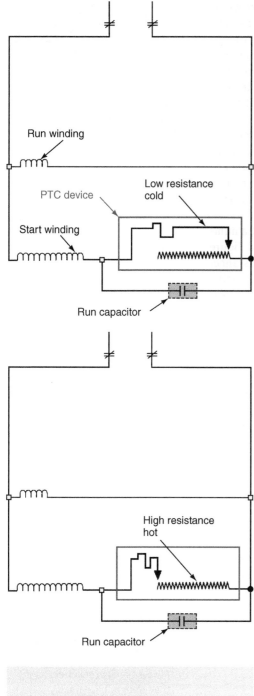

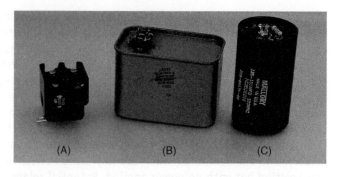

Figure 5-30
Three motor-starting components: (A) potential relay; (B) run capacitor; (C) start capacitor. (Photo by Bill Johnson)

Figure 5-29
One type of solid-state motor starter: a PTC (positive temperature coefficient) device or thermistor. (Photo by Bill Johnson)

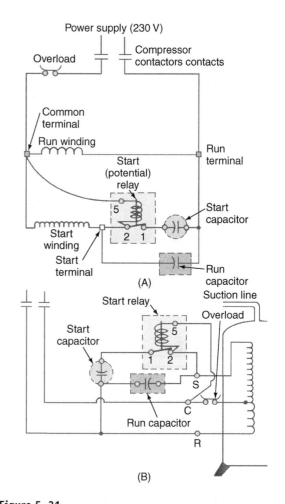

Figure 5-31
(A) A ladder diagram showing the function of the start and run capacitors. (B) An actual (or point-to-point) diagram showing how capacitors are correctly wired.

Figure 5-32
A start capacitor with bleed resister permanently attached.
(Photo by Bill Johnson)

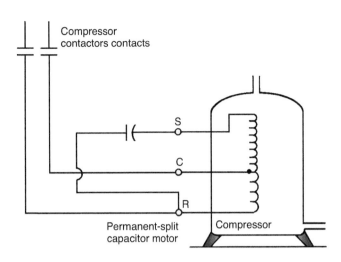

Figure 5-33
A pictorial wiring diagram that shows the silhouette of a PSC
compressor without a hard-start kit.

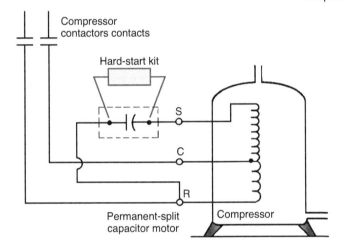

Figure 5-34
The hard-start kit is installed in parallel with the run capacitor.

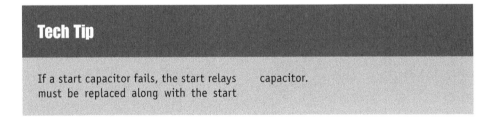

Tech Tip

If a start capacitor fails, the start relays   capacitor.
must be replaced along with the start

Overloads

Motor overloads (see Figure 5-35) measure amperage. Depending on how they
mount to the motor, some are affected by motor and ambient heat. Most overloads
are of the *external* type. This class of overloads includes those that attach to the
motor housing to measure internal temperature as well as amperage.

Motor starter overloads are external. They are housed within the motor starter
that is separate from the motor itself. Motor starter overloads, called "heaters,"
can only monitor motor amperage (see Figure 5-36).

Another external overload, one that is not affected by heat, is the magnetic
overload (see Figure 5-37). Like the current magnet relay, this overload operates
only on amp draw. When amp draw is too high, it will open the circuit and require
a manual reset to close the switch.

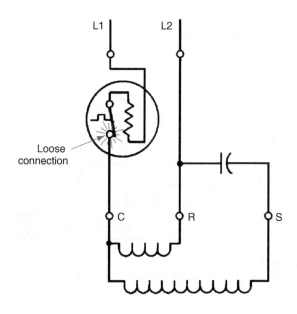

Figure 5–35
Overload shown in the circuit with a PSC motor. The overload itself is shown tripping because there is a bad connection that is adding heat internally to the overload.

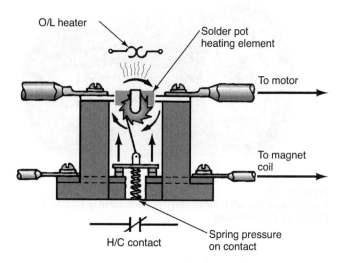

Figure 5–36
A resistive overload that heats as amperage increases. The ratchet wheel and trip lever open the circuit when it heats.

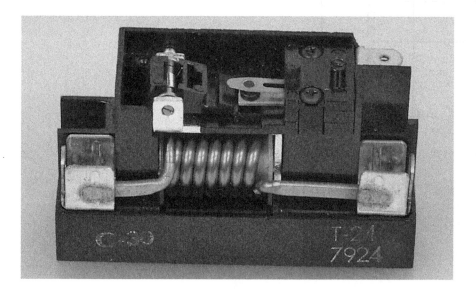

Figure 5–37
A magnetic overload device. (Photo by Bill Johnson)

Internal overloads are located within the motor and are generally embedded in the windings (see Figure 5–38). Figure 5–39 shows the operation of this component.

All overload contacts can be checked with an ohmmeter. A reading of 2 ohms or less indicates that the contacts are good. Overloads that operate in combination with heat cannot be easily tested in the field. Observation and measurement of temperature and amperage while they are in the circuit are the only options.

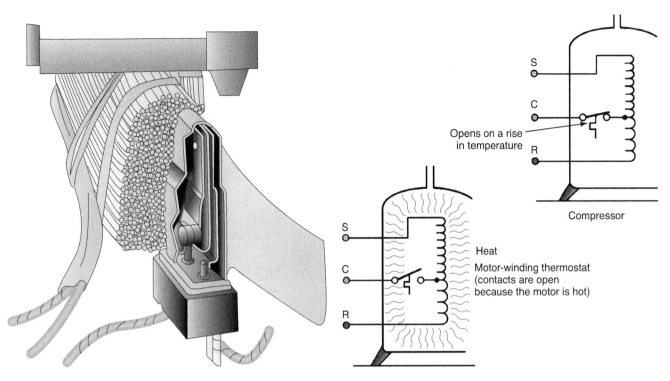

Figure 5-38
An internal overload protector, showing location in the windings.

Figure 5-39
Operation of an internal motor overload.

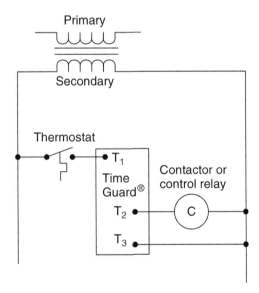

Figure 5-40
A delay timer is wired into the system to prevent the compressor from operating too soon after a shutdown.

Delay Controllers and Applications

Delay relays and monitoring controls are used to delay the restart of motors after shutdown (see Figure 5-40). This is to protect motors from high amp draw while other system components are starting. The delay allows other components to start first and then, after these system parts are running and the amperage draw is lower, the larger motors are allowed to start.

Figure 5–41
A solid-state timer to prevent short-cycling. (Courtesy of Ferris State University. Photo by John Tomczyk.)

A delay timer can be used to prevent short-cycling (see Figure 5–41) and allow systems to equalize system pressures, reduce starting load and voltage drop.

Tech Tip

Time Guard® is a proprietary control that is designed to prevent system components or whole systems from cycling too often. Other manufacturers make similar products. All of these function in some fashion to "time" the amount of time between on and off states. Most new digital electronic thermostats include time delay compressor protection. When this is enabled, make sure the equipment does not have a time delay as well.

Tech Tip

Delay relays come in many different configurations. It is important to know the application before replacing an existing relay or installing a new relay. Of these, two are important to know: delay on make and delay on break.

Delay on make delays the closing of contacts when initially energized. For instance, a thermostat could open and initiate the timer. The timer will not let the thermostat close to operate the system until the timer has timed out. This may prevent a compressor from coming on too soon after shutdown.

Delay on break maintains contacts open for a minimum amount of time at the end of a cycle. For instance, a thermostat circuit will not turn off until the timer has timed out. This may prevent the burner circuit from cycling off before the ignition system completes the sequence.

The delay timer is essentially a clock that will not allow the motor to start for a predetermined or preset length of time. Some delay timers can be adjusted, while others are fixed. A time-delay device can be either a "delay on make" (close contacts when the timer cycles) or a "delay on break" (open contacts at the end of the cycle). Timers that have multiple sets of contacts and perform several functions at the same time are sometimes known as "gang" timers. Gang timers are able to operate several circuits at the same time and depend on the number of NC and NO contacts housed within the timer.

Three-phase Part-start

All three-phase motors will start without external starting devices. The physical nature of a three-phase motor allows it to take advantage of all three phasing angles supplied by the power company (see Figure 5–42).

Large three-phase motors may draw too many amps. If the locked rotor amperage were accommodated, then the installation and operational costs would be extremely high. To lower those costs, some three-phase motors use a strategy known as "part winding start" (see Figure 5–43). What this means is that part of the winding is energized to get the motor turning up to a threshold rpm; when the motor reaches that threshold, the other half of the winding is energized to bring the motor to full rpm.

At the threshold rpm, start amperage is beginning to drop with one set of windings and the amperage draw of the next winding part will not exceed the initial starting amperage. Because the sequence is time delayed, the procedure is generally referred to as "timing" the start sequence. Start systems of this type are usually found with 208–230-volt three-phase systems. Recall that three-phase dual-voltage motors are actually two motors in one: they have windings wired in parallel for 208–230 volts in addition to windings wired in series for 480 volts. To limit inrush current at start-up and the resulting voltage drop on the building electrical system, even low-voltage compressors sometimes employ the part winding start configuration.

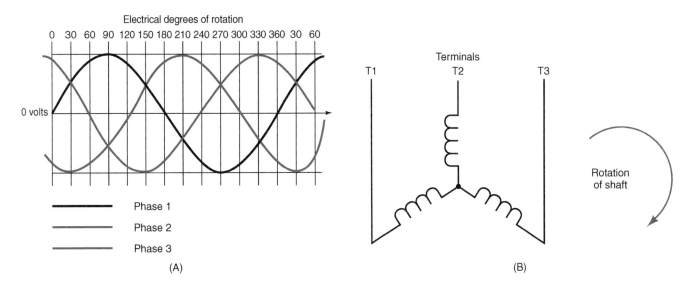

Figure 5–42
(A) A diagram of all three phasing angles; each angle is 120 degrees out of phase with the other. (B) Winding of a three-phase, Wye-wound motor. If T1, T2, and T3 were respectively connected to L1 (phase 1), L2, and L3, then the motor would turn clockwise. Switch any two wires and it will turn counterclockwise.

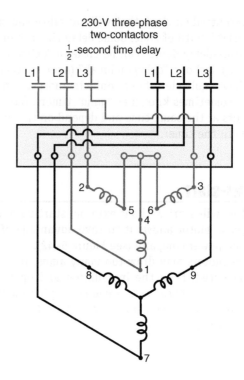

Figure 5-43
Three-phase motor wired for part winding start with a half-second delay timer. The set of contactors on the left starts the motor; the set of contactors on the right takes the motor to full speed.

Three-phase Wye-Delta Start

Much like the part-start, the Wye-Delta (also called star-Delta) system starts the motor from a locked rotor position; it uses a Wye configuration and then switches to a Delta configuration to finish bringing the motor to full speed (see Figure 5-44 and Figure 5-45). The wiring involves three contactors. Two are used to configure windings into a Wye to start, then one is dropped out and the other engaged to form the Delta. These two contacts are interlocked mechanically so that one cannot close until the other is open to prevent a direct short across all three legs.

Three-phase Protection (Single Phasing)

While a three-phase motor is running, all of the windings are energized with separate phasing angles of power. If one leg or phase is lost then the motor is reduced to a single phase, and it will operate on only one phasing angle between two power conductors. In this event, small motors will growl and large motors will burn out in minutes. When one leg of power is lost, the motor circuit needs to be opened to all incoming power sources.

The motor starter offers a degree of protection. When higher amperage is felt on one leg, that heater trips the entire motor starter, disconnecting the motor from the power source (see Figures 5-46, 5-47, and 5-48). For more control, electronic control packages can be used (see Figure 5-49).

Motor starters are simply contactors that have motor protection built in. Motor starters are specified by horsepower. The "heaters" are replaceable devices that trip the contactor open if they sense high amp draw (see Figure 5-50). Motor starters are checked in much the same way a relay or contactor is checked:

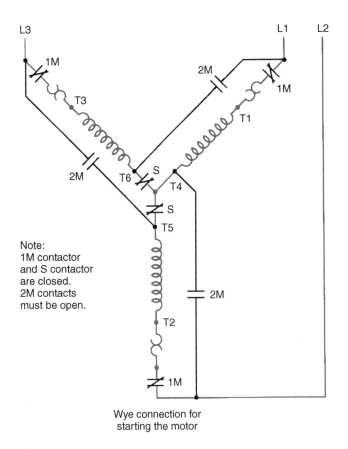

Note:
1M contactor
and S contactor
are closed.
2M contacts
must be open.

Wye connection for
starting the motor

Figure 5–44
Wye-start configuration, showing contactors that are closed.

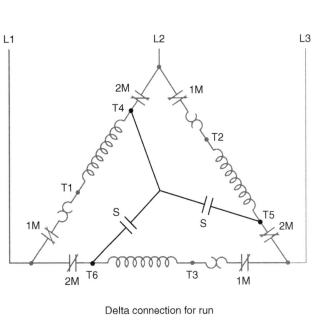

Delta connection for run

Note: 1M and 2M contacts are closed.
S contacts must be open.

Figure 5–45
Delta-run configuration, showing contactors that are closed.

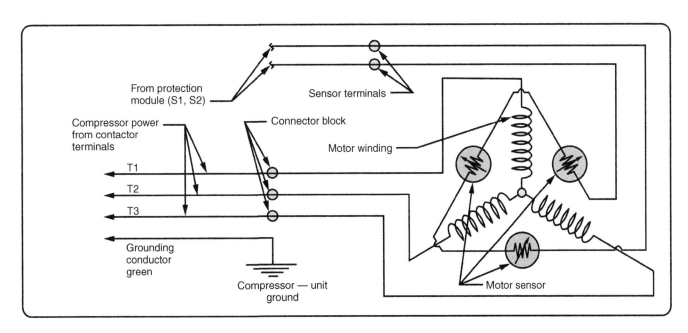

Figure 5–46
Diagram of a three-phase Wye-wound motor with internal thermistor protection linked to an electronic motor protection module.
(Courtesy of Tecumseh Products Company)

Figure 5–47
Disassembled motor starter, showing heaters and contacts at the bottom of the picture.
(Photo by Bill Johnson)

Figure 5–48
Assembled motor starter, showing the location of the overload protection. (Photo by Terry
Miller)

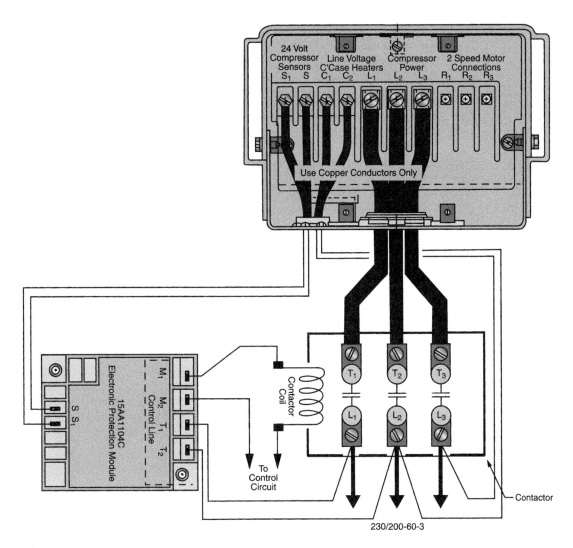

Figure 5–49
Pictorial wiring schematic of the electronic motor protection package. (Courtesy of Tecumseh Products Company)

Figure 5–50
A melting alloy heater. (Photo by Terry Miller)

1. Check the operator for applied voltage and magnetism; with the power off, use a continuity tester or a small screwdriver.
2. Check the contacts by measuring the voltage drop (it should be 2 volts or less) while powered or taking an ohms reading (it should be 2 ohms or less) with the power off.

Check the heaters by comparing the motor starter manufacturer rating chart to the actual FLA rating of the motor being protected.

Tech Tip

Prevent callbacks for single phasing. On larger motor systems that have experienced an overload condition causing one or more fuses to blow, it is good practice—once the short or ground is corrected—to replace all three fuses as a set. One reason for this is that some larger fuses have multiple elements within them. If all elements are not intact, the fuse may ohm-out okay but will no longer be capable of carrying its rated amperage. Hence the fuse may blow, necessitating a callback for another single-phase condition on the new motor.

Variable and Frequency Drives

Variable speed controls are predominantly frequency drives and are a means of controlling the speed of the motor from start-up to full operation. Frequency drives use electronics to smoothly ramp up motor speed, which is sometimes called a "soft start" (see Figure 5–51).

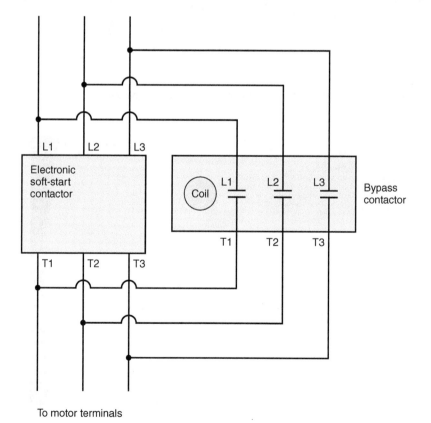

Figure 5–51
The electronic motor control provides speed control from start-up to full-speed operation.

A frequency drive allows only a portion of the power's sine wave of power to the windings when the motor is starting. As the motor starts, the control allows more of the sine wave to reach the windings until the full sine wave is received. Check with the manufacturer for information on troubleshooting these controls.

Case Study

The customer complained that there was no heat. The blower was not operating, but the furnace turned on and off. (Refer to Figure 5–9 on page 142.)

Symptoms:
- Furnace burner short-cycling
- Blower not running

Possible Causes:
- Blower is inoperable
- Blower control circuit is inoperable
- Blower motor starting circuit or capacitor is inoperable

The technician turned the thermostat switch to manual blower, but the blower did not operate. Going to the service panel on the furnace, the technician checked the blower for power and found 120 volts going to the blower (voltmeter reading 1). The motor was humming. The blower motor starting circuit requires the capacitor to start the PSC (permanent-split capacitor) motor. Turning off power, he removed the blower motor capacitor (BMC). Setting the meter to read capacitance, he checked the capacitor and verified that it was open. The motor started and operated within specifications when the capacitor was replaced.

DISPOSAL OF ELECTRICAL DEVICES

Electrical devices should be properly disposed of. Some of the old switches and capacitors have toxic or hazardous materials in them. Thermostats that use mercury should not be thrown into the regular trash. Old capacitors will have PCB (polychlorinated biphenol), which is considered toxic and is regulated by the Environmental Protection Agency (EPA). Some components contain lead.

The EPA itemizes all of these substances on its "P" and "U" lists of discarded commercial products. The list contains many more substances that can be found in spent electrical devices. All replaced electrical parts should be brought to the shop and disposed of properly. Most companies have a hazardous waste policy and/or a contract with a waste disposal company.

SUMMARY

This chapter covered many electrical control devices along with their operation, wiring, and testing. Much of what was presented needs to be put into practice. Safety came first in this chapter for a reason. If safety is not practiced, harm will not be far behind. Work safely and use the safety techniques, PPE, and tools listed to stay safe.

Read nameplates as one of the first steps when looking over a machine and before working on it. Nameplates can give a quick insight into the power supplied and the amperage used. Read electrical prints for what you need: to understand how devices are wired and where to test with a test meter.

When troubleshooting, separate the mechanical system from the electrical system. Pinpoint problems by ruling out those subsystems and devices that are working. Conduct your testing systematically to isolate the problem, and try to test without disturbing system components or forcing systems to function. Use the input and output information that you acquire to determine the possible cause.

Open, ground, and megohm tests can provide valuable information about the system's safety and viability. Conduct ground-fault tests before touching cabinets, motors, and other electrical devices. If any circuit is leaking and you touch the ground path, you could become part of the circuit.

This chapter covered many of the standard devices used in HVAC to operate and control equipment. A thorough understanding of these basic devices will help with any troubleshooting or service work that you do. Relays and contactors both work in the same way; it's just that the former can handle more power than the latter. Transformers change or convert power for controls or spark ignition systems. Solid-state electronics is providing us with new and simpler ways and sensors to monitor and control HVAC equipment.

Electric motors are at the heart of most HVAC systems. Understanding how they are controlled, monitored, started, and protected is an essential part of learning in this field. Single-phase motors have more and varied types of start controls. Three-phase motors start by themselves, but larger motors use high amperage and need to be started in steps or ramped up to speed.

Being a good steward of the resources we have and disposing of spent electrical devices in a responsible way will ensure that we all have a better world to live in tomorrow.

REVIEW QUESTIONS

1. Explain why PPE is used.
2. How is the correct PPE determined?
3. Thoroughly describe one safe testing technique.
4. What does CAT III mean for an HVAC technician?
5. What types of electrical information are available on the equipment's nameplate?
6. How much of an electrical print needs to be read before testing?
7. What is logic wiring, and how is it read?
8. What is AND logic?
9. What is "ruling out," and how does it work as a technique?
10. What is a truth table?
11. How can testing for an open circuit be done with a voltmeter?

12. Describe how to find a good ground to which an electrical ground can be hooked.
13. How does a megohmmeter test show problems with the compressor?
14. Why should electrical tests of equipment be documented?
15. How many classifications of relay switching are there? Name one.
16. What is the output of a transformer (either step-up or step-down) called?
17. To what are solid-state controls more susceptible than electromechanical ones?
18. Why do some three-phase motors require starting controls?
19. What is the difference between *current* and *potential* magnetic relays for starting motors?
20. Why is phase protection necessary?

CHAPTER

6

Applied Refrigeration System Problem Solving

The student will:

- Describe the checks done for system start-up
- Discuss how to rule out working system parts
- Describe how to use cause and effect
- Define scheduled maintenance
- Discuss the reason for cleanliness as related to system repair
- Describe the process of purging
- List the steps in evacuation and recharging
- Describe the process of pumping down
- Describe compressor failure
- Discuss the results of manufacturer compressor failure data

INTRODUCTION

Refrigeration system problem solving requires a good understanding of system operation and components, a systematic approach, and experience. Every system is different, and even when the system is made exactly the same, the application or conditions in which it operates will vary.

This chapter takes a practical look at refrigeration system problem solving.

Field Problem

A commercial air conditioning service call for a new customer involved a complaint that the system was not working on the first day of the new cooling season. The technician listened to the customer and asked a few general questions to determine how well this system worked last year. He then gathered a list of symptoms and possible causes.

The thermostat was correctly set for cooling and the system was running. Feeling the condenser discharge air, the technician noticed that it was not warm. After opening the access panel on the split-system condensing unit, the technician observed that the suction line was warm and the compressor was too hot to touch. He decided to turn off the breaker to allow for additional testing.

Starting with system checks, the technician began with the power off and observed the refrigeration system. One of the first checks is looking for oil, and it wasn't long before he found signs of oil at the bottom of the condenser. A leak check confirmed that something had pierced the fin-tube on the last row of the condenser. The hole was small and the tubing could be brazed after moving the fins back. The technician recovered the refrigerant, made the repair, checked for leaks, evacuated the system, and recharged to the manufacturer's specifications.

Moving to the power-off checks, the technician also found and replaced a dirty filter; cleaned the blower and housing; oiled the bearings; and replaced a loose electrical connector. He finished with the power-on checks and found the system operational. At completion of the service call, he talked to the customer about the repairs and suggested a scheduled maintenance agreement so that the system would be checked annually.

SYSTEM START-UP

After everything is hooked up, a refrigeration system needs to be operated. The first impulse is to throw the switch to see what happens, but the best method is to be sure in advance that it will work. Always follow the manufacturer's start-up instructions, if available. Table 6–1 describes general start-up procedures for refrigeration and air conditioning systems. These system checks apply to both air and water delivery systems and to both refrigeration and air conditioning systems (although some checks will not apply to all systems).

WHAT IS THE PROBLEM?

This question is asked both mentally and verbally by service technicians on a troubleshooting service call. The question is asked at each level of the investigation: as the technician reads the service call slip, when the technician is listening to the customer to determine the customer complaint, and at each step in the troubleshooting process of determining the symptoms (what the unit is or is not doing) and listing the possible causes of the problem. The process of good troubleshooting starts by *ruling out* (as possible causes) those parts of the system that are working properly.

Techniques for Ruling Out

Ruling out is a technique used by all professionals (including medical doctors) to establish the general area or system where the problem exists. One example is a

Table 6-1 General and Initial System Commissioning Checks

Focus	With power off: With the system locked out at the breaker, check the following	With power on: *After* the power-off check, operate the system and check the following
Electrical system	• Tightness of power wires from the disconnect to the system • Look for abraded (missing insulation) wires • Tightness of the control wires • Correct color coding of control wires • Verify that transformer is wired for correct primary voltage	• Operate the crankcase heater before startup based on the manufacturer's recommendations • Check for proper voltage • Check system amperage • Check compressor amperage • Check all motor amperages • Check control circuit amperage

Controls	• Thermostat setting: set for a temperature below ambient operation (not highest or lowest)
	  ○ System switch is set to COOL
	  ○ Fan switch set to AUTO
	  ○ Stage controls are set
	• All resets are locked in
	• Low pressure controls are set
	• Low and high limit controls are set
	• Temperature controls are set
	• Defrost clock is set for time of day
	  ○ Defrost switch is calling for refrigeration
	  ○ Defrost limit is connected in the right location and can sense properly
		• Operate temperature control (wait a minimum of 2 minutes before changing thermostat to induce system operation after a shutdown)
		• Operate the system to cause limit actions that will test the limit controls (once for each type of limit)
		• Cycle through one defrost period (where applicable)
		• Two-stage thermostats must be cycled through the second stage to ensure that second stage systems are working
		• Check for solenoid operation
		• Check for reversing valve operation

Delivery system	• Check dampers for normal operating positions • Check for open diffusers and grilles (covers are removed) • Check blower to ensure that the squirrel cage is free to operate • Check condenser fan for free movement • Check circulator pumps for free movement	• Check blower and fans for proper direction of rotation • Adjust and balance with system dampers • Measure air flow at diffusers and grilles • Check system static pressure • Listen for and investigate any unusual system noises • Check for duct leakage • Check pump differential pressure • Check for fluid leakage
A/c and refrigeration systems	• Check that all valves are in their operating positions (compressor service valves are *not* closed) • Look for oil leakage at all connections • Leak-test all connections • Check all line (tubing) insulation • Check compressor oil level • Check to see that compressor mounts can move freely (remove any packing material) • Connect manifold gauge set and ensure there is pressure in the system (*Note:* If there is no pressure, then a complete evacuation, leak test, recharge, possible oil change, and filter-drier replacement are necessary.)	• Monitor for signs of oil • Check for refrigerant leakage at high-pressure connections • Check refrigerant sight glass • Check compressor oil sight glass (*Note:* Compressor oil level may fluctuate; oil level may drop slightly until normal operating conditions are reached.) • Check system refrigerant pressures and line temperatures • Log start-up data • Adjust refrigerant charge if necessary • Check flow through reversing valve (temperature check) • Calculate superheat and subcooling • Check Delta-T

tripped manual reset on a high-pressure control for a beer cooler. A first step might be to determine whether the system has power (see Table 6-2). Test the circuit (see Figure 6-1) with one probe on L_1 and the other on L_2. Line voltage confirms power. With one probe on L_2, go to the left side of the thermostat and then the right side. Voltage on each side confirms that the cooler thermostat is calling for

Table 6-2 Examples of Ruling Out

Technician Can Rule Out . . .	By . . .
Main power and secondary voltage (except with three-phase systems, where actual three-phase voltage must be verified)	Testing the secondary voltage output of the transformer; there should be 24 volts present.
Solenoid valve power and operation	Measuring the temperature of tubing on both sides; it should be the same for open valves and different for partially closed or restricted valves.
Electrical safety switches	Measuring voltage across the switch or switches in series; a reading of 0 volts indicates that none of the switches are open or there is no voltage at all.
Operating controls	Measuring control voltage at the compressor contactor coil; voltage here means that controls are calling for cooling.
Pumpdown system	Forcing the pumpdown system to operate; pressure drop on the suction side indicates that system is operating independently of the initiator (a temperature control or other device that normally triggers system operation). See the pumpdown system electrical diagram in Figure 6-1 and notice that, when the thermostat opens, the solenoid closes; when the low-pressure safety senses low pressure, the compressor shuts down because the contactor coil is deenergized and so the contacts open.
Defrost system	Forcing the defrost system to cycle; operation means that system and controls are capable of functioning independently of the defrost initiator (thermostat or clock).
Second stage of a thermostat	Setting the thermostat to operate second stage; verify operation of second stage.

cooling. Voltage across the pumpdown solenoid tells us that if the solenoid is okay, it should be open, allowing the low-pressure control (LPC) to call for cooling. With one probe of the voltmeter on L_2 and the other on the left then the right side of the LPC, we get line voltage confirming that the LPC is closed and calling for cooling. Doing the same test on the high-pressure control (HPC), we find that on the right side of the HPC, there is no voltage, confirming that the HPC is open and requires a manual reset.

Electrically, we have confirmed that a mechanical problem exists and that we need to diagnose why the HPC is open. By listing the possible causes, we can then troubleshoot the problem and repair.

Possible Causes:
- Dirty condenser coil
- Overcharge
- Defective condenser fan, wrong rotation, or airflow is being diverted back to inlet of condenser
- Noncondensables in system

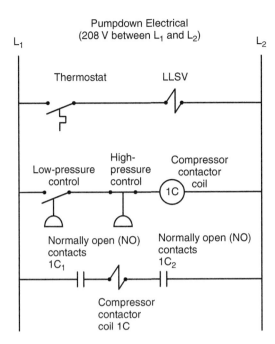

Figure 6–1
Schematic of a thermostat in series with a pumpdown solenoid.

Troubleshooting: Because the system has been off for some time and the condenser appears to be clean, the first check should be checking the high-side pressure with the compressor off. A pressure above the saturation per ambient temperature would indicate the presence of noncondensables. If this was an R-22 system and the ambient was 78 degrees, a pressure above 139 psig could indicate that air (noncondensables) are trapped in the top of the condenser, reducing the effective size of the condenser and causing a high head pressure condition that tripped the HPC. Bleeding the air off (if possible), refrigerant recovery, recycling, deep evacuation, filter change, and proper recharging would solve the problem.

Cause or Effect: Diagnosis and Correction

"Is this the cause or just an effect?" Technicians should ask themselves this question every time they think the problem has been found. The failure of one component can (and often does) lead to the failure of other components. For example, a leaking relief valve may be the result of its being lifted by high pressure. Further investigation reveals that the high pressure was caused by a failed condenser fan. Replacing both will restore the system to service, but the technician who continues to ask, "Cause or effect?" will look a little further and find that the system is equipped with a high-pressure cutout control mistakenly attached to a flare fitting with a Schraeder valve and that the high-pressure control does not have a Schraeder valve depressor. This prevents the high-pressure control from sensing actual system pressure.

Manufacturers sometimes supply cause-and-effect diagrams in their equipment literature. These are generally known as "troubleshooting flowcharts." The troubleshooting chart (see Figure 6–2) describes an effect and asks a question (diamond). The question is answered either yes or no, which leads to another question until the problem is identified. Figure 6–2 is a flowchart for a refrigeration compressor.

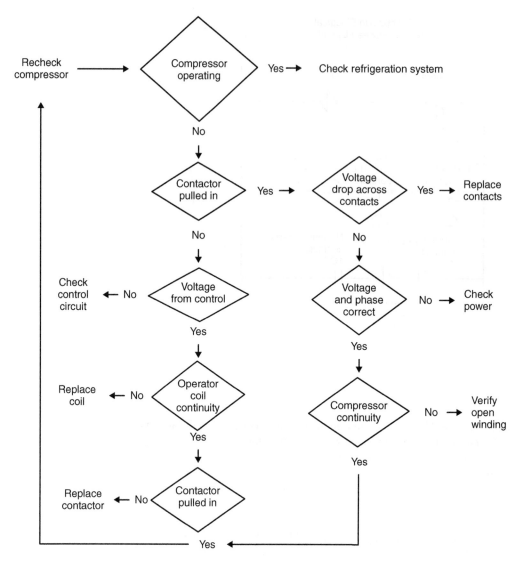

Figure 6-2
A flowchart for troubleshooting a compressor.

SCHEDULED MAINTENANCE

Scheduled maintenance is a process used to prevent unexpected failures, save energy, maintain comfort levels, and increase the longevity of an HVACR system. It is also sometimes known as "corrective maintenance" because corrective action may be applied before a system condition becomes a problem that necessitates a service call. Service contracts are a form of scheduled maintenance: a customer's system is periodically checked and adjusted for optimal operation from season to season.

There are several levels of maintenance:

- *Reactive maintenance*—wait until there is a problem before making repairs.
- *Scheduled maintenance*—maintain system, preempt repair problems by taking corrective action. This may mean replacing parts before they fail completely.
- *Predictive maintenance*—using operating data and manufacturer longevity information to make informed maintenance choices. This means that parts are replaced before they fail, based on known longevity of the parts.

Table 6–3 Scheduled Maintenance Checks

Focus	Scheduled Maintenance
Cleanliness	• Clean heat exchangers • Change the air filter • Clean the blower and housing • Check blower wheel for cleanliness and signs of missing balance weights • Clean equipment cabinets • Clean drain lines, traps, and sumps
Amperage	• Check and record amperage and voltage readings of system motors, including the compressor motor
Condition	• Check all electrical connection points for discoloration and looseness • Check all wiring for discoloration and signs of deteriorated insulation
Air (or water) flow	• Measure the total air flow of the delivery system • Measure temperature rise and/or drop across heat exchangers • Check diffuser air flow for each room • Check pump head pressure (differential pressure) • Check blower static pressure (differential pressure) • Check pressure drop across heat exchangers • Check water chemistry
Safety controls	• Check the operation of overloads, high limit controls, and low limit controls (e.g., low pressure cutout, oil safety cutouts, and freeze stats for economizers and chillers)
Lubrication	• Lubricate all bearings that are not sealed • Check compressor oil level and net oil pressure when possible
Refrigeration system	• Measure system temperatures and pressures; calculate superheat and subcooling • Check for leaks • Look for oil leakage around fittings • During system operation, look for components that may wear against each other owing to system vibration, possibly causing future leaks; take necessary steps to correct

Scheduled maintenance generally involves the items listed in Table 6–3.

REPLACING COMPONENTS

In this section we discuss specific types of HVACR problems that involve procedures, techniques, and system components. Each problem is different, but they are all related to one another. As you read, look for differences and similarities.

Replacement Procedures

Cleanliness is the most important aspect of system component replacement. The system component must be removed and replaced *without* contaminating the rest of the system. For example, the thermostatic expansion valve (TXV) must be replaced without allowing atmospheric conditions to enter and remain in the system. The valve must also be removed and replaced without causing damage to the cabinet, panel, plenum, or other components. Brazing requirements for changing the valve may include heat shields and heat sinks to prevent this damage.

During component replacement, special consideration should be given to covering all open connections on which no work is being done. This will help ensure internal system cleanliness.

Finally, after any repair, the system compartment should be left clean. If, for instance, oil is left in bottom of unit, this will complicate future troubleshooting when it comes time to look for leaks. This oil could also find its way out of the unit and cause damage to roofing materials. Furthermore, it looks unprofessional, which reduces the customer's level of satisfaction.

Line Connection Techniques

Refrigerant line or tubing connections are of great concern in HVAC work. Potential problems include:

- Leaks
- Ruptures
- Contamination
- Flaking
- Oxidation

Proper brazing and connection techniques that a technician must use include:

- Flare connections (Figure 6–3), where system components—such as liquid-line filters—may need to be removed and replaced
- Brazing while purging with inert gas to remove oxygen and eliminate oxidizing (flaking)
- Mechanical compression tubing connectors to eliminate heat and flame from sensitive areas; fittings can be used to connect dissimilar tubing (e.g., copper to aluminum)
- Push-on mechanical fittings for tight areas where tools cannot be used; these fittings can also be used to connect dissimilar tubing types

Purging with an Inert Gas

Inert gas is used to purge contaminants and air from a system. Purging replaces oxygen molecules in the air that enters a system that is open to the atmosphere for repairs. Oxygen in the air causes oxidation and flaking to occur while brazing a refrigerant tubing connection. Brazing heats copper tubing to temperatures that cause rapid oxidation in the presence of oxygen, creating flakes of copper oxide that could be carried by the refrigerant and oil to other parts of the system. Copper oxide (flaking) is a contaminant to the refrigeration system. When inert gas is used during brazing, the inside of the refrigerant tubing remains bright because oxygen is not present. Inert gas is also helpful in purging all other gases,

Figure 6–3
Flare joint; used where system parts need to be removed and replaced. (Photo by Bill Johnson)

residual refrigerant, and air from the system. Because the inert gas has a low moisture content, it can aid in absorbing a small amount of moisture while in the system.

Tech Tip

An inert gas is a gas that does not react with other elements. The periodic chart of elements lists a number of gases that are inert. True, "elemental" inert gases are He (helium), Ne (neon), Ar (argon), Kr (krypton), Xe (xenon), and Rn (radon). Other gases considered to be chemically inert are N (nitrogen) and CO_2 (carbon dioxide). Either of these latter two gases can be used as an inert gas in refrigeration systems. Dry nitrogen is generally used for purging (see Figure 6–4) because it prevents oxidation and copper oxides and also removes moisture from the system.

Safety Tip

An inert gas, such as nitrogen, requires a high-pressure gauge and gas pressure regulator. The regulator is used to monitor and control the pressure entering the system. If the regulator output is connected to a refrigeration system and the regulator is faulty, then pressure in the refrigeration system could build to dangerous levels. To ensure that the system is not overpressurized (typically a maximum of 150 psi test pressure), always turn off the inert gas cylinder valve after adding inert gas to the system.

Filter-driers

Filter-driers are one of the most important safeguards in the system. They help to ensure that refrigerant flowing through the system is both clean and dry. Filter-driers can be placed on both the high and low sides of the system. If filter-driers are to be changed regularly, then provisions should be made to enable changing them without disturbing the rest of the system. Refrigerant valves are placed before and after the filter-drier with access fittings that allow refrigerant to be recovered and vacuums to be placed on the filter-drier before it is placed back into service with a refrigerant charge. Low-side (suction-side) filter-driers are generally applied to clean up a system after a compressor burnout (see Figure 6–5). When the system is thoroughly cleaned, they are commonly removed to prevent pressure drop on the low side and decrease in capacity and efficiency (see Figure 6–6 and Figure 6–7).

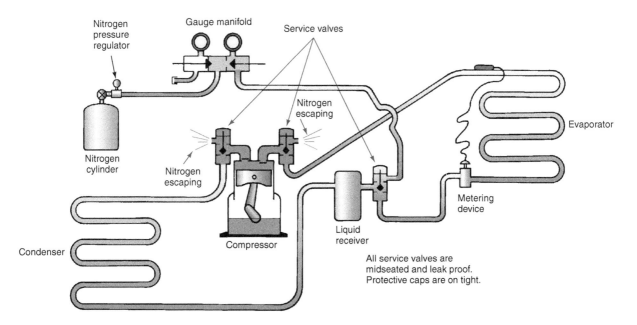

Figure 6–4
Purging the system with nitrogen (at a very low flow rate so that no pressure can build up within the piping) removes all oxygen from the system. Brazing can be done while nitrogen escapes, keeping the copper tubing clean on the inside.

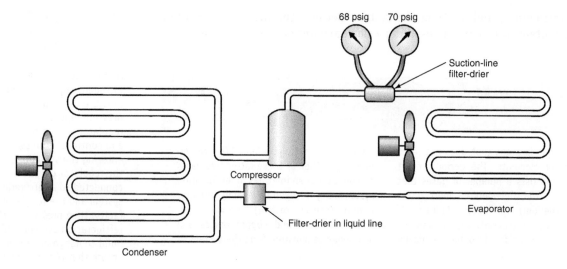

Figure 6-5
Checking the suction filter pressure drop.

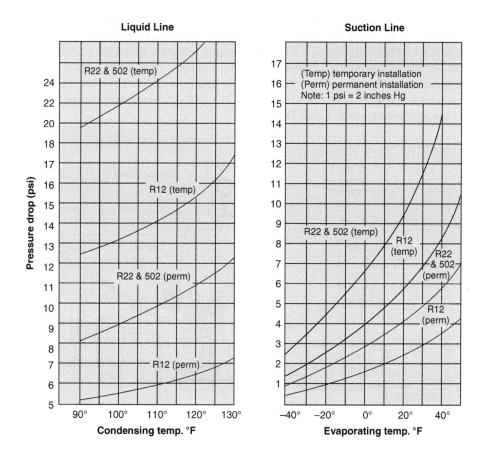

Figure 6-6
Maximum recommended pressure drop for filter-driers. (Courtesy of Copeland Corporation)

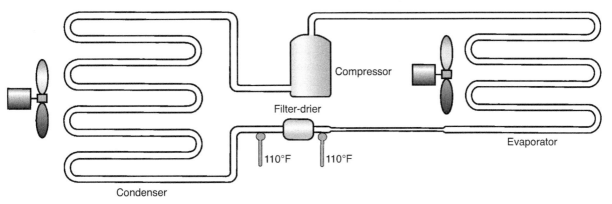

Figure 6–7
Checking the liquid filter temperature drop.

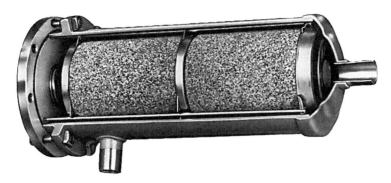

Figure 6–8
Cutaway view of a drier with replaceable core. (Courtesy of RSES)

On small systems, liquid-line filter-driers (see Figure 6–8) are replaced after recovering the refrigerant to the receiver or condenser; this is also known as pumping down the system. Vacuum is broken with an inert gas and then the drier is replaced. Putting the system back into services requires that the portion of the system that was open to the atmosphere be pressurized with inert gas in order to leak-check the disturbed joints and evacuated to a deep vacuum to remove any moisture and noncondensables; the system pressure is then brought up to proper charge with system refrigerant.

Tech Tip

Contamination of isolated refrigerant in a system that has been pumped down may occur if system valves are not leak-proof and inert gas exceeds the pressure of the isolated refrigerant. Care should be taken not to pressurize the system with inert gas to pressures above that of the isolated refrigerant. Higher inert gas pressures would force the inert gas into the isolated refrigerant through leaking valves and thereby contaminate the refrigerant.

Evacuation and Recharging

Evacuation and recharge requirements are critically important for both new and field-repaired systems. Equipment is expected to meet evacuation measurements of 500 microns or less. By attaining 500 microns or less and holding this vacuum level, we are assured that no moisture or system leaks are present.

Evacuation of a system removes predominantly air and moisture (see Figure 6–9). Air and moisture enter the system when the system is opened. Moisture-ladened air enters the refrigeration system, and moisture is condensed on the inside of the tubing in areas that are colder than the ambient air temperature. Air can be removed more easily than moisture (see Figure 6–10).

Charging the system after establishing a good vacuum requires the following steps for an R-410A system using a TXV:

1. Connect a container (cylinder) of refrigerant specified for the unit (check the manufacturer's tag).
2. Purge the hose from the cylinder to the gauge manifold to eliminate air and moisture.
3. Place the cylinder on a charging scale and record the weight with the hoses attached.
4. Determine the weight less the amount of system charge (check the manu-facturer's tag). *Note:* If the manufacturer's tag does not specify the cor-rect charge, the charge must be estimated based on tubing sizes, refrig-eration components (like receiver size), and heat exchanger capacities. Consult the manufacturer.
5. Record the starting and ending weight.

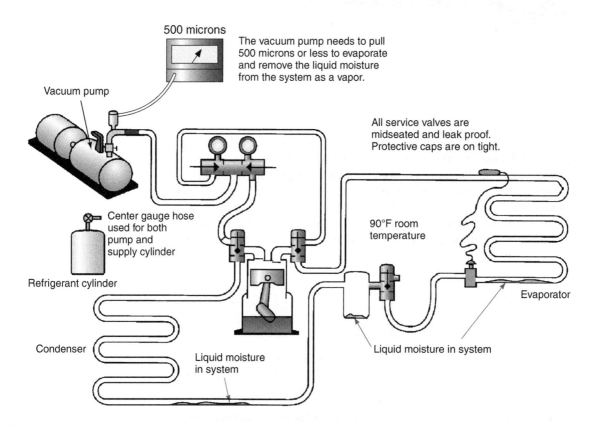

Figure 6–9
Moisture in a system can only be removed by vaporizing, but if the water is in a cold location or under oil then it will not evaporate easily (or at all). Wherever moisture may deposit, heat will need to be applied in the form of a heat light or electric resistance heater; do not use a torch.

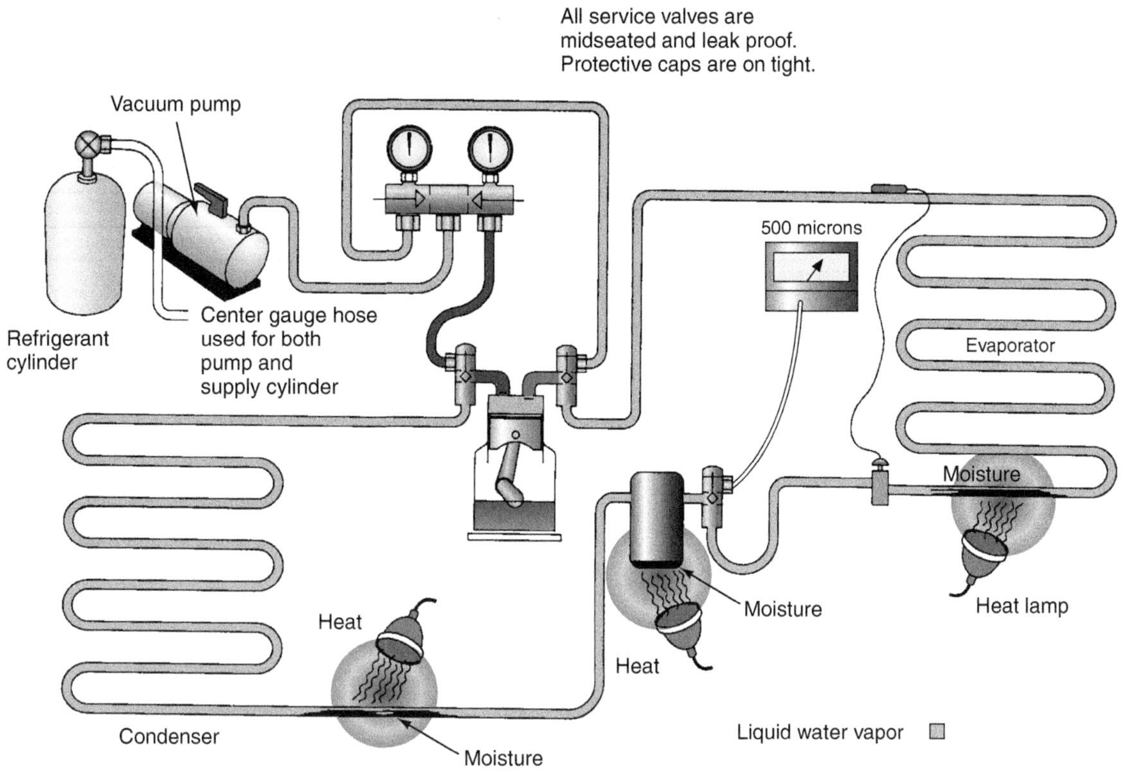

All service valves are
midseated and leak proof.
Protective caps are on tight.

Figure 6–10
Heat is applied to locations of possible moisture. If moisture is suspected in the compressor crankcase, then either the crankcase heater is operated or another heat source is applied.

6. Add all of the charge, as a liquid, at one time through both the suction and discharge connections to the unit.

7. If all of the charge cannot be added at one time, close the manifold gauge valves. Wait for 5 minutes and start the compressor. With the compressor operating, slowly add the remaining amount of refrigerant by opening the low-side gauge manifold slightly. Using the restriction caused by the manifold gauge valve, meter the liquid refrigerant to the system as a saturated gas until the ending weight has been reached.

Tech Tip

In some units, the low pressure switch may need to be jumped to allow the compressor to operate. The pressure may not be high enough to start or keep the compressor running while adding refrig- erant charge. Jump the pressure switch until there is enough pressure to keep the switch closed, and then remove the jumper.

8. Add the complete manufacturer's charge.

9. If the superheat of the TXV is known (for instance, 20°F, as in Figure 6–11, or 15°F, as in Figure 6–12 and Figure 6–13), the superheat of the system can be checked to see whether the charge is close and the system is operating properly.

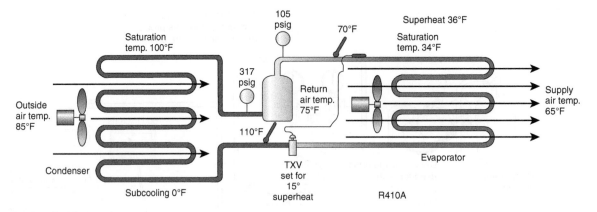

Figure 6–11
Undercharged system.

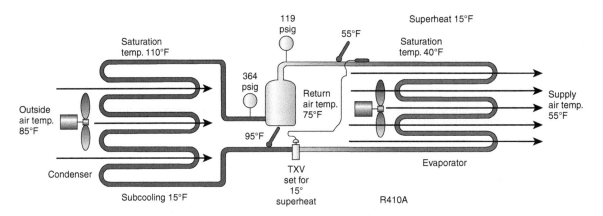

Figure 6–12
Normal system charge.

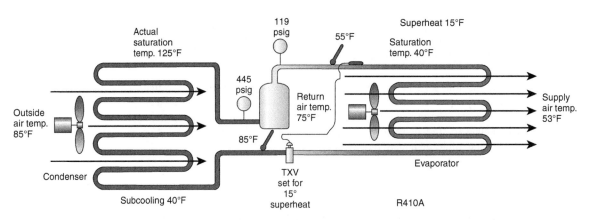

Figure 6–13
Overcharged system.

Tech Tip

Follow these steps to determine superheat:

1. Measure the actual line temperature at the outlet of the evaporator (TXV sensing bulb location) with an accurate electronic thermometer.
2. Take the pressure at the compressor—this is the suction saturation pressure.
3. Use a pressure-temperature chart to cross-reference the refrigerant in the system with the pressure taken at the compressor—this is the saturation temperature of the refrigerant.
4. Subtract the saturation temperature from the actual temperature measured at the evaporator outlet.

5. The difference in temperature is the evaporator superheat.

If the superheat of the TXV (in this example, 15°F) is operating within 2°F immediately after charging, the system may be at or close to the right charge. At this point, it is important to allow the system to reach a steady state or "stabilize." This means that the system is heating up at the condenser and cooling down at the evaporator. The TXV is still trying to find the right metering rate. The technician may also find that the superheat is 12°F in one reading and 16°F in a second test a few minutes later. The TXV and the system need time to reach a steady state.

10. After checking the superheat, make sure that the evaporator airflow is correct. Reduction in airflow may affect the system operation and your ability to accurately measure the subcooling.
11. Allow the system to operate. In close-coupled units, the amount of time needed is less than with systems that have remote condensing units.
12. Measure the subcooling.
13. If subcooling is not correct, adjust the refrigerant charge by metering liquid into or out of the system. This method will always add or remove the right amount of refrigerant without affecting the vapor/liquid ratios of refrigerants (such as refrigerant mixes that use temperature glide).

Tech Tip

Follow these steps to determine subcooling:

1. Measure the actual liquid-line temperature at the inlet of the TXV with an accurate electronic thermometer.
2. Measure the pressure at the discharge of the compressor—this is the discharge saturation pressure.
3. Use a pressure-temperature chart to cross-reference the refrigerant in the system with the pressure taken at the compressor—this is the discharge saturation temperature of the refrigerant.
4. Subtract the liquid-line temperature from the discharge saturation temperature.

5. The temperature difference is subcooling.

The subcooling of a system will fluctuate for various reasons, such as high load conditions, high outdoor temperature, and outdoor air damper maladjustment. If the system is still pulling down the temperature of a building that has not satisfied the room thermostat, an excessive amount of heat is being pumped from the building. The system may need to stabilize before the correct subcooling is reached.

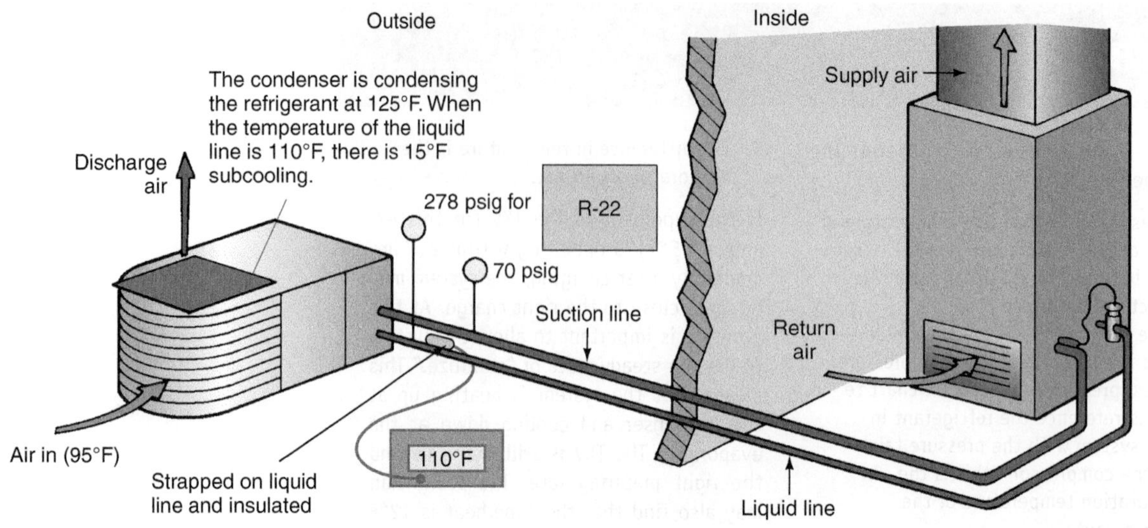

Figure 6–14
A unit with a TXV controls the superheat at the metering valve; monitor by checking the subcooling. With the condenser blocked (enough to simulate a 95°F day by producing a 125°F saturation temperature), the subcooling of a fully charged system should be 10–20°F.

Tech Tip

There are various methods for charging. Manufacturers have different preferred charging methods. TXV systems are charged by the subcooling method. Fixed bore systems are charged to a certain superheat. See *HVACR 201* (page 185) for details on proper charging methods.

Recharging the system requires the technician to know or find the type of refrigerant and the correct amount of charge for the unit. Next, the type of metering device must be determined. With this knowledge the technician is ready to begin the recharge procedure. *Always follow manufacturer's charging specifications.* Recharge procedures are generally as follows:

TXV   With a thermal expansion valve, measure in charge and monitor subcooling and/or adjust charge as necessary. Monitor superheat to ensure the system is not flooding back should the TXV malfunction (see Figure 6–14).

Fixed bore With a capillary tube and metering orifice, measure in charge and monitor the superheat. Adjust the charge as necessary.

Tech Tip

Condenser airflow can be blocked and adjusted by using a plastic sheet drawn against the coil.

Tech Tip

When adding refrigerant to any system with a water/refrigerant heat exchanger (e.g., chillers and systems with water-cooled condensers), water flow should be established first whenever possible and only then should refrigerant vapor be added. Charge the system until system pressure is above saturation pressure for a temperature of 32 degrees Fahrenheit; this will prevent any damage to the heat exchanger that could be caused by rapid freezing and expansion of water in the heat exchanger. For example, R-410a at a pressure below 17.9" Hg boils at a temperature of less than −50°F.

System Isolation and Pumpdown

Refrigerant in most systems can be isolated in the "condensing unit." By isolating the refrigerant in the condensing unit, you can open the low side for repair without the need for recovery. Anytime any part of the system has been opened, it must be evacuated to 500 microns before refrigerant can be released back into the system. The condensing unit includes the compressor and condenser and (in some systems) the receiver. Service valves on the condensing unit allow this part of the system to be closed off from the rest of the system.

The process of isolating refrigerant is called "pumping down." By closing off the liquid service valve that leaves the condensing unit and continuing to operate the compressor, all of the refrigerant vapor in the rest of the system is pumped through the compressor and condensed to a liquid in the condenser (see Figure 6–15). After the compressor pulls enough refrigerant out to drop the pressure to zero gauge or below (as per minimum vacuum levels required by the Environmental Protection Agency), turn off the compressor. This completes the pumpdown and isolates the condensing unit from the rest of the refrigerant system.

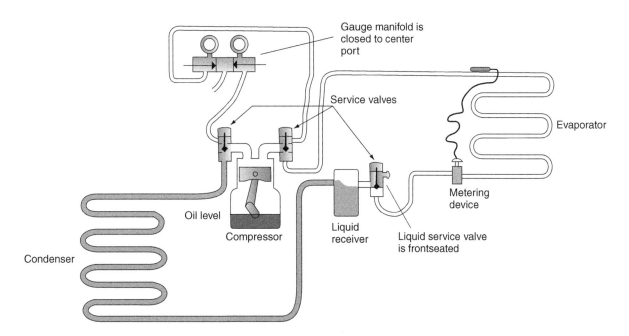

Figure 6–15
Pumping down backs up refrigerant into the condenser and receiver while the compressor pulls vapor from the low side. After the pumpdown is completed, the compressor's high- and low-side service valves are frontseated to isolate the condensing unit from the rest of the system.

Safety Tip

When jumping out safety controls, two things must be considered. The first is that the technician is required to actively monitor the system pressure. Attention to gauge readings is a must! Quick action must be taken to shut down the system if pressures are exceeded. The second consideration is how the control is being jumpered (bypassing the low pressure control by jumpering between the low pressure contacts). The more physical the jumper connection, the safer the connection. Use of a jumper with alligator clips is not as safe as a heavy-gauge wire physically connected between two electrical terminals.

It is also important not to let the compressor run for more than a few minutes in this condition, because the flow of refrigerant (which normally cools the compressor motor) is greatly reduced.

Tech Tip

Scroll compressors should not be operated below certain pressures. Some should not be operated below 5 psig because they will lose their scroll tolerances and then create mechanical problems. When working with scroll compressors, always check with the manufacturer before using the compressor to recover refrigerant.

Tech Tip

Most systems have suction safety controls that prevent the compressor from operating below set pressures. In order to keep the compressor running for pumpdown, it may be necessary to "jump out" (i.e., use jumper connections to bypass) the safety control. Whenever this is done, the technician must pay extremely close attention to system pressures. If there is too much refrigerant in the system (overcharge) there will not be enough room in the receiver to store the refrigerant. This will be evident when discharge pressures rise quickly, at which point the technician must quickly stop the pumpdown process.

Metering Devices

To replace a metering device—a thermostatic expansion valve (TXV), capillary tube, or piston—in an otherwise working system, the system is first pumped down to isolate the refrigerant charge in the condensing unit. Then the system pressure is brought back to zero gauge and the old metering device removed. After the new metering device is installed, the system is leak-checked by pressurizing the system with inert gas (see Figure 6–16).

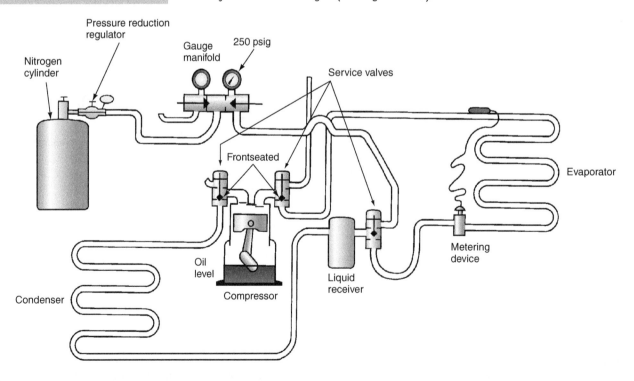

Figure 6–16
Isolating the compressor to pressurize the rest of the system for a leak check. Note that the only time the discharge service valve should be frontseated is for compressor isolation such as compressor removal.

After verifying that the connections are leak proof, the system low side is evacuated and the refrigerant contained in the condenser is allowed back into the system.

COMPRESSOR REPLACEMENT
Why Do Compressors Fail?
There are two classes of compressors; those that are separate from the drive motor (open drive) and those where the drive motor is inside the refrigerant (hermetic). These two classes of compressor fail for slightly different reasons (see Table 6–4). Open-drive compressors are susceptible to one set of problems, but hermetic compressors can have more problems because the electric motor is in the system.

Verification of Failure
When ever a compressor is diagnosed as having failed, the technician wants to be absolutely sure. In order to do that, every failed compressor must be verified as having failed. Table 6–5 lists the verification methods used to ensure that the diagnosis of compressor failure is correct.

Table 6–4 Reasons Compressors Fail

Cause	Effect	Open	Hermetic
Metal fatigue, poor material specifications, and poor production techniques can all lead to valve, piston, and crankshaft failure.	Mechanical component failure	✓	✓
Low refrigerant charge, improperly set or applied metering device, ambient higher than design temperature, evaporator temperature set too low; and refrigeration system used for the wrong application will all lead to greater compression ratio.	Large differences in pressure between the low and high side	✓	✓
Oil leaks with refrigerant, oil not being returned to the compressor, and low oil charge can lead to lack of lubrication.	Loss of oil	✓	✓
Blocked condenser, overcharge of refrigerant, high inside ambient temperature, hot gas bypassing, and improper compressor size will lead to high suction temperature and can result in high head pressure/temperature.	Warped discharge valves	✓	✓
Slugs of refrigerant liquid or slugs of oil returning to the compressor intake valve can enter the cylinder and stop or inhibit compression.	Broken valves, pistons, and connecting rods	✓	✓
Oil breaks down in the presence of moisture and forms acids that erode electrical insulation.	Electrical insulation loss on motor winding		✓
Oil foaming at start-up.	Broken valves, pistons, and connecting rods	✓	✓
Electrical leaks in the winding allow electrical arcing that fuses windings; this can lead to winding separation (open windings).	Arcing		✓
High suction temperatures, loss of refrigerant, and high evaporator loads lead to a reduced ability to cool the compressor motor.	Overheating		✓
Low discharge pressure and high suction pressure.	Inefficient compressor	✓	✓

Table 6–5 Methods of Verifying Compressor Failure

Source of Failure	Procedure
Electrical	• Verify that windings are not shorted to ground by using the ohmmeter at the compressor terminal with the power off and the wires disconnected from the compressor • Verify winding resistance values using the same technique as above • Re-verify the above resistance readings by measuring at the contactor with the wires now connected to the compressor terminals • Allow compressor to cool in order to reset internal overload • Verify the operation of all motor starting and limit controls: relays, capacitors, and external overload devices • Verify that the compressor terminals are receiving voltage by measuring at the contactor (*Caution:* Never read voltage at the compressor terminal, as this is an extremely hazardous task.) • Verify amperage readings • Operate the compressor while bypassing the system starting controls (manual start controls) • Operate the compressor while using alternative starting controls (alternative start controls)
Mechanical	• Test compressor efficiency per manufacturers specifications • Use compressor amperage readings and manufacturer data to determine compressor pumping capacity • Check refrigerant charge • Check oil charge • Verify operation of compressor unloaders (if equipped)

Tech Tip

Some manufacturers recommend that all start components be changed whenever a compressor is changed. Reversing valves and accumulators might need to be changed as well. In addition, liquid-line filter-driers should be changed anytime the system is open for repair.

Manufacturer Warranty Data

Manufacturers typically warrant a compressor for 5–10 years. Additional warranty coverage can be purchased to extend the warranty for an additional period. Compressor failure during the warranty period is normally covered by the manufacturer and/or the OEM (original equipment manufacturer) supplier. Data about failures—such as the type of failure, the required repair or adjustment to the system, and frequency of failure—are difficult to obtain. However, failed compressors that are returned to be remanufactured or rebuilt are often analyzed for cause of failure. In this procedure, technicians must seal the compressor inlet and outlet, place the old compressor in the same box used to ship the new compressor, and supply a small amount of information on the return slip. Remanufacturing shops may receive the faulty compressor, tear it down, and log the reasons for its failure.

Typically half of all compressor warranty claims result from wanting to maintain good customer relations: compressors are replaced at the customer's request and the replacement cost is simply written off. The balance of compressor warranty claims concern four groups of problems:

- Compressor and system application, 20% of failures
- System design, 10%
- Control malfunction, 10%
- Defects, 10%

It should be noted that there are very few manufacturing defects, which account for a mere half of one percent (0.5%, or 1 of every 200) failures. The point is that, if a compressor is actually broken, the cause is almost always some other condition in the system. If this cause is not identified and corrected at the time of replacement, then the new compressor will also likely fail—sometimes within minutes of being placed in service. When installing a replacement compressor it is advisable to check a number of items, which include (but are not limited to) the following:

1. Condition of contactor and compressor starting components
2. Condition of system heat exchangers
3. High pressure control (condition and settings)
4. Low pressure control (condition and settings)
5. Oil pressure safety controls: setpoints and time delays
6. Oil levels
7. Proper functioning of metering devices
8. Refrigerant charge
9. Operating conditions: pressures and temperatures

Keep in mind that if the cause is not found, then the new compressor will soon fail as well.

Another telling piece of information stands out in the failure percentages: many returned compressors are actually functional. If 4% of 5,000 compressors torn down have nothing wrong with them, then 200 compressors were returned unnecessarily. This number becomes more significant when you add in the costs of a new compressor, installation, and customer downtime. If a dollar figure could be assigned to customer satisfaction after experiencing downtime and the added costs of the new compressor, then the cost of this type of technician error would be staggering. Errors of this type can easily be avoided by the technician. Before condemning a compressor, always verify that it is actually broken!

TECHNICIAN SALES

The technician's job is not limited to repairing and servicing equipment. In addition, the technician is the company ambassador. In that role, the technician is expected to be the company's PR (public relations) person: always upholding the company's image and following its mission statement. Because all companies are selling customer comfort and satisfaction, the technician is expected to do likewise.

Technicians sell products and services by discovering what their customers need and want and then informing their customers about the realities of the situation and the available alternatives. Technicians also inform their customers about system performance, longevity, wear, and upgrades. Toward this end, the technician needs "tools" to make the sale: company business cards, stickers, or appointment calendars. The technician is expected to take an active role in promoting and selling products and services. It might be said that the technician's job is not limited to fixing the equipment but is instead "fixing the customer."

SUMMARY

This chapter covered the operation and problem solving of refrigeration systems, beginning at start-up and ending with sales. The ability to rule out potential problems allows the technician to eliminate parts of the system that cannot be the root cause of a service problem. Understanding the cause and effect of various problems can also help a technician pinpoint the problem. During scheduled maintenance, the technician can identify potential causes of problems that have not yet occurred; by knowing what to look for, the technician can correct the problem or prevent one that could later cause a system to fail. Finally, using the correct repair and service techniques can extend the life of a refrigeration system.

REVIEW QUESTIONS

1. Describe why checks are done before a system is started for the first time.
2. Describe the process of ruling out parts of the system.
3. How can cause and effect be used to identify problems in a refrigeration system?
4. What does "scheduled maintenance" mean to a customer?
5. Why is cleanliness so important to refrigeration service?
6. Describe the purpose of and procedures for purging the system with inert gas.
7. What process must occur before recharging can be done? Why?
8. How does pumping down a system isolate its low side?
9. What are two of the largest causes of compressor failures?
10. What is one reason for compressor warranty claims that the technician can eliminate?

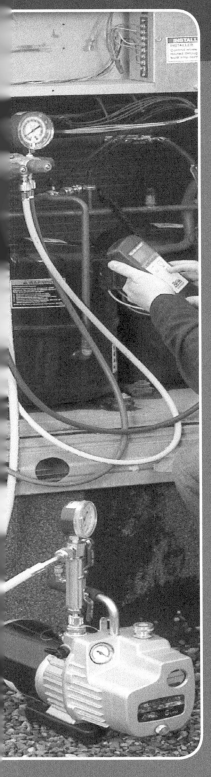

CHAPTER

7

Troubleshooting with the Psychrometric Chart

LEARNING OBJECTIVES

The student will:

- Identify the four basic parts of the psychrometric chart
- Plot a point on the psychrometric chart using two measured conditions of the air
- Recognize the relationship of the conditions of air and the operation of heating and cooling equipment

INTRODUCTION

This chapter applies the properties of air to the use of a psychrometric chart. The psychrometric chart greatly simplifies the science of air applications to comfort heating and cooling. As you read and learn about psychrometrics, it would be wise to follow along by plotting the points given in the text. A point can be found on the psychrometric chart when given two conditions of the air. These conditions can be temperature, humidity, volume, or moisture. Two conditions that can easily be measured and plotted on the psychrometric chart are dry-bulb and wet-bulb temperatures. The psychrometric chart is based on one pound of dry air. As air is heated, cooled, humidified, or dehumidified, the conditions of the air change.

As air moves through comfort conditioning equipment, the air conditions change. When the air's conditions are measured as they enter the return air grill and compared with measurements of air leaving a diffuser, the technician can understand how the conditioning equipment affected or changed the conditions of the air. Understanding those changes can give technicians insight on how effectively the conditioning equipment is working.

Field Problem

The customer is complaining that the air temperature seems to be too high, but more importantly he says that it seems "sticky"—as if there is a high level of moisture in the air. The technician has a suspicion that it is a cooling system problem, but before jumping to conclusions she takes a few readings with an aspirating psychrometer as she continues listening to the customer and asking the standard troubleshooting questions to determine the possible causes. The technician asks: When was the condition first noticed? Was there any other symptom or noise that was noticed? What has been done to try to fix the problem?

The technician took her reading from inside the occupied space, which represents the return air condition. She also needs a sample from the supply duct, so she finishes with the customer and thanks him for his candor. Going to the system, she looks for an outlet as close to the discharge air plenum as possible. Here she takes another sample with her aspirating psychrometer. In her head she can imagine the psychrometric chart as the aspirating psychrometer compares the two readings. Just to be sure, she returns to the truck and pulls out a psychrometric chart, plots the return and supply values, and draws a couple of lines.

Summarizing the conditions for herself, she notes the following.

1. Room temperatures are slightly higher than they should be (73°F when the thermostat is set for 72).
2. Humidity measurements are too high in the occupied space (close to 80% RH as measured by the handheld aspirating psychrometer).
3. Temperature at the closest outlet to the coil is too high (66°F).

Trying to avoid jumping to any conclusions, the technician makes a list of possible causes (*Note:* Write down your own list of possible causes.)

The two measured conditions and the corresponding lines drawn on the psychrometric chart quickly indicate that there is not enough water being taken from the air. This validates the customer's complaint of feeling sticky. This condition could be caused by one of several problems. The best place to start is by conducting an external check of the system. Beginning with the first external check on her list of possible causes, she walked out to the condensing unit and looked at the condenser—her suspicions were confirmed! We will return to this problem at the end of the chapter after discussing psychrometrics and the properties of air that it evaluates.

CHART PARTS

The parts of a psychrometric chart are identified as if you were referring to a foot. In other words, the chart features a toe, a heel, a sole, an instep, and an ankle (see Figure 7–1).

Inside the chart's outline there are more parts, shown in Figures 7–2 through 7–7. These parts consist of lines that may seem confusing at first. After you use

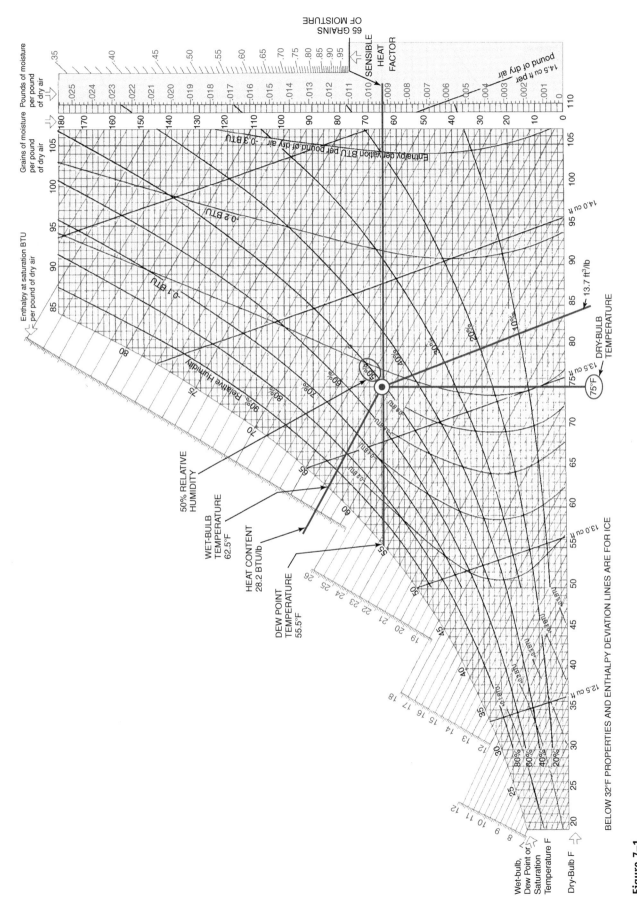

Figure 7-1
A psychrometric chart with the basic parts labeled. The point plotted on the chart shows desired indoor conditions of 75°F with 50% RH. (Courtesy of Carrier Corporation)

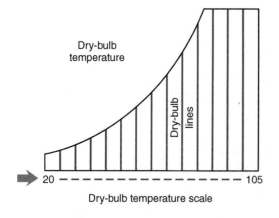

Figure 7–2
The dry-bulb temperature lines on the psychrometric chart.

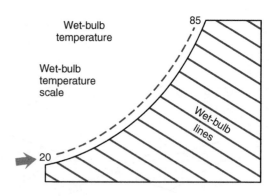

Figure 7–3
The wet-bulb temperature lines on the psychrometric chart.

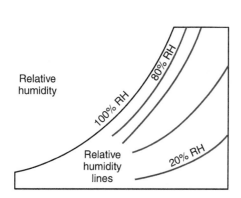

Figure 7–4
The relative humidity lines on the psychrometric chart.

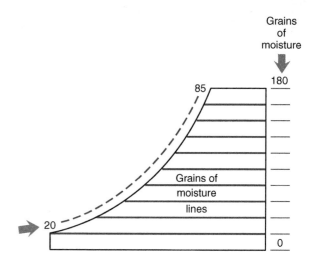

Figure 7–5
The moisture (grains per pound of air) lines on the psychrometric chart.

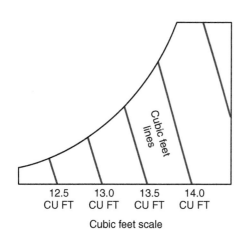

Figure 7–6
The volume (cubic feet) lines on the psychrometric chart.

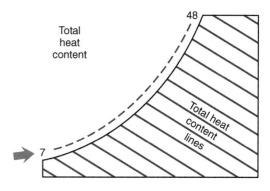

Figure 7–7
The total heat content (BTU) on the psychrometric chart.

each line and understand what each one means, the chart becomes less confusing and more useful. The psychrometric chart is based on one pound of air at a specific elevation. Charts are selected based on the elevation of the service area. This is an important point to remember. Converting measurements to cubic feet (i.e., the amount of space taken by one pound of air at a particular elevation) requires that the conditions be evaluated only by using the correct chart for a given elevation.

Definitions

Here is a list of terms that are frequently used in the measurement of air.

- *Dry bulb* (DB)—air temperature as measured by standard thermometers.
- *Wet bulb* (WB)—the temperature produced by evaporation of a wet wick-covered thermometer.
- *Dew point* (DP)—air temperature at which relative humidity is 100%, which is the point at which moisture will condense out of the air (upper curved line on the psychrometric chart).
- *Relative humidity* (RH)—amount of moisture in the air as a percentage of the amount of moisture the air will hold at a given temperature.
- *Psychrometrics*—the study of the properties of air, as distilled in the psychrometric chart.
- *Saturation*—the temperature at which air can hold no more moisture (upper curved line on the psychrometric chart).
- *Grains of moisture* (Gr/lb of air)—a fine measurement of the amount of moisture in one pound of air; 1 pound of water = 7,000 grains.
- *Volume* (ft^3/lb of air)—the amount of space one pound of air occupies at a given set of conditions.
- *Enthalpy* (BTU/lb of air)—the amount of heat that is present in one pound of air at a given set of conditions.
- *Air*—a combination of water, vapor, and gases.

PLOTTING POINTS OF A SYSTEM

The psychrometric chart can be used to evaluate the conditions of any air sample. If two air samples are compared, then the difference between the two readings will indicate what has happened to the air: if it has been cooled, dehumidified, heated, or humidified. An air sample from the return air and another from the output of an operating heating system will show any increase in the dry-bulb temperature of the air. A horizontal line, often referred to as a *process line,* can be drawn between these two points because the amount of water does not change when air is heated. The reduced relative humidity shown by the process line also indicates that the air can hold more water at the new higher temperature.

Only two conditions of an air sample are required to plot a point on the chart. The conditions most easily measured are the dry-bulb and wet-bulb temperature. It should be noted that, once a point is plotted, all of the air sample's other properties can be determined by using the chart.

Air conditioning systems are much more interesting than heating systems. Air is cooled and water is removed during the cooling process. Removing moisture means that the air has been cooled to the dew-point temperature and that water has condensed out of the air and onto the coil. Moisture removed means that latent heat of condensation has occurred and that 970 BTU of heat per pound of water has been removed. Moisture removal is an important part of the cooling process because it increases comfort conditions by lowering the relative humidity. When an air conditioning system doesn't remove moisture effectively, the air can feel sticky because it is nearer the dew point (closer to 100% RH).

Suppose an air conditioning system's return air temperature is 72 degrees DB and 64.4 degrees WB and that its discharge air temperature is 65 degrees DB and 62 degrees WB. When these two points are plotted, the relationship looks like a heating system in reverse. The air is dropping in temperature, but there is no moisture being removed. Only "sensible" heat is being removed, and the relative humidity is 66.8% in the room. Conditions like this indicate that the evaporator's temperature is too high and it is not condensing enough moisture. The horizontal line goes through each of these two points and crosses the instep of the psychrometric chart at approximately 60 degrees WB, which is also read as the DP (dew-point) temperature. This is the temperature where moisture will condense out of the air for both of these two points. In order to remove moisture (condense moisture on the coil), the temperature must fall below 60 degrees WB. If the coil were visible you could see that it is dry, because the coil temperature is not below 60 degrees WB. A room under typical conditions of 72 degrees DB, 54 degrees WB, and 30% RH would have a DP of approximately 40 degrees. This would mean that the coil was removing latent heat and sensible heat from the room. Most systems operate at or slightly above these conditions once the system and building reach an equilibrium (balance) between the system size and the amount of heat entering the building.

Human comfort conditions are general conditions under which no mechanical cooling or heating is required. The middle of a comfort chart would be the most ideal condition. Under conditions that are closer to the edges of the plotted area, humans begin to feel uncomfortable. Hence, the aim of comfort conditioning systems is to maintain conditions within this acceptable range. Measurement of human comfort conditions should be taken in the occupied zone of a room—in the area of chairs, tables, and work areas and at approximately 3 to 5 feet from the floor. When a room is heated, warm air tends to rise to the ceiling; in the cooling season, cool air drops to the floor. In both cases, diffusers are responsible for directing and mixing air in the room. If good mixing does not occur, then the air will stratify or become stagnant within the occupied zone. On the other hand, if diffuser air is being directed to the occupied area and blowing on a person, the moving air will tend to cool him by forced evaporation of sweat on the skin. In the cooling season this might be a welcome additional source of cooling, but fast-moving air will also cool a person in the heating season, which reduces desired warmth.

CONDITIONED AIR

Conditioned air is air that has been changed in quality or heat content. Air quality includes the amount of moisture suspended in the air as water vapor. Heating systems without humidification deliver heated air to rooms of a building with no change in moisture content. In these systems the psychrometric chart would show a horizontal change in sensible heat from the return air temperature to a higher supply air temperature. With the addition of humidification, a rise in the moisture content would be seen, and the supply air's condition on the psychrometric chart would increase in latent heat. Heated and humidified supply air would be at a higher (vertically) point on the psychrometric chart and correspond to a higher relative humidity condition. An example is when air is returned at 70 degrees DB and 55 degrees WB. The air is heated and delivered to the room at 110 degrees DB and 69 degrees WB. Plotting between these two points on the chart would yield a horizontal line, indicating no change in moisture.

Cooling systems are designed to reduce the temperature of the air and, in the process, reduce sensible and latent heat. When both of these are reduced, the supply air is lowered in temperature and moisture is removed. Supply air is discharged into rooms in the building and mixed with the room air. Mixing absorbs some of the moisture in the room air and reduces the DB temperature. Good indicators that an air conditioning system is functioning are when room temperatures are a minimum of 10 to 20 degrees DB below outside conditions and the air is at 30–50% relative humidity (depending on outdoor ambient conditions and system design). For example, air might be returned at 70 degrees DB and 65 degrees WB. The air is cooled to the dew point of 62.5 degrees WB and then further cooled to 40 degrees WB (the coil temperature, also known as the apparatus dew point). As the air is cooled from 62.5 to 40 degrees, moisture condenses on the surface of the evaporator. If all of the air touched the evaporator and all of its moisture were condensed out (this doesn't happen), then two lines could be drawn (one from 40 WB and one from 62.5 WB) horizontally across to the moisture scale (at the heel of the chart). The difference between these two lines represents the amount of moisture removed from the air (latent heat) in addition to any reduction in sensible heat. In reality, some air is not conditioned as it moves through the coil and mixes with conditioned air. The mixed air condition is warmer and contains more moisture; it also depends on the coil's efficiency. If the delivered air at the diffuser were 55 degrees DB and 49 degrees WB, then the total amount of heat (sensible and latent) could be determined by subtracting the enthalpy reading of the delivered air from that of the return air: approximately 0.116 BTU/lb of air. In this case the change in moisture is approximately +42.66 Gr, or 0.006 pounds of water, per pound of air.

Dehumidification is a process that involves only the removal of moisture. In contrast, an air conditioning system both dehumidifies and also cools the air. There are situations where the temperature or cooling of the air is not of greatest concern, as in southern locations of high humidity. In these locations or situations, only the removal of moisture is important. Reduction of humidity will reduce the possibility of mold growth. In a home, the dehumidification system can be a small, refrigeration-based system for dehumidifying the basement or a larger system that dehumidifies the entire house. Commercial systems reduce humidity by using compression refrigeration to remove moisture or a desiccant wheel (see Figure 7–8) to absorb moisture from the air. In either case, dehumidification systems are designed to remove moisture and not to reduce the air's

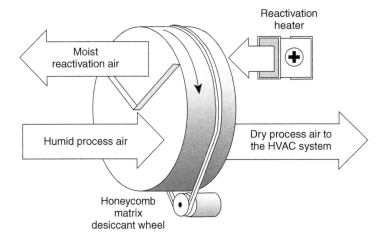

Figure 7–8
The desiccant wheel turns while absorbing moisture from one air stream and giving up moisture to air being sent out of the building. (Courtesy of RSES)

temperature. Air delivered by these systems shows only modest reductions, if any, in temperature. Measured supply air will correspond to a lower (vertically) point on the psychrometric chart than return air, indicating a drop in the amount of moisture (latent heat). For example, suppose air is returned at 70 degrees DB and 64 degrees WB. The moisture in the air is removed and delivered to the room at 70 degrees DB and 52 degrees WB. Plotting these values on the chart will yield a vertical line, showing that only moisture is being removed. Horizontal lines could be drawn from each of these two points to the moisture scale at the heel of the chart. The difference in the two lines represents the amount of moisture removed from the air (latent heat only). For practice, plot these points on the psychrometric chart (Figure 7–9).

Blowers receive return air from the occupied space and deliver it to a comfort conditioning machine. The condition of return air affects the amount of air delivered by the blower. The density of the return air will determine the weight of air in relationship to its cubic volume. The amount of air delivered in cubic feet per minute (cfm) by the blower is directly linked to the density of the air in cubic feet per pound (ft^3/lb) of air. Simply put, if the air is denser, the fan will deliver more pounds of air per minute. More pounds of air delivered to the comfort conditioning machine will require the machine to add or remove heat. If most air returned to a blower is 70 degrees DB, then air density should not be of any concern. Draw a line on the psychrometric chart that runs vertically from 70 DB to 70 WB. Along that line you will see that cubic volume varies from below 13.4 to 13.7 ft^3/lb.

MEASURING SYSTEM OPERATION

Condensate water is produced when moisture is formed on the evaporator and then drained away. An air conditioning system can be evaluated in terms of how much condensate it removes over a period of time. If less condensate is flowing, then the amount of moisture in the air could be low. If tests of the room air show high relative humidity, then we conclude that the system is not working efficiently to remove moisture. A properly operating system will maintain relative humidity between 30% and 60% within the occupied space. The difference between moist and dry air is the difference between sensible and latent heat, and this difference is used to calculate the *sensible heat ratio* (SHR). The psychrometric chart has an SHR scale. The SHR is determined by the system designer and reflects the relationship between the amount of sensible heat removed and total heat (sensible and latent) removed. An SHR of 0.5 means that half of the removed heat was sensible heat. Suppose the system capacity is 10 tons; then an SHR of 0.5 means that 5 tons are being used to reduce the DB temperature and the other half to remove moisture. In this case, a 10-ton system would produce about 7 gallons of condensate per hour: (12,000 BTU/ton × 10 ton × 0.5) ÷ (970 BTU/lb water × 8.33 lb/gal water) = 7 gal/hr.

The SHR depends on regional weather conditions. A system designed for the Southwest will have a lower SHR than a system designed for use along the East Coast or in Florida or Michigan. In the Southwest, outside air is dry. On the eastern seaboard, there is a lot of ambient moisture to remove. If the SHR is known or can be determined, then a technician can measure the condensate to determine whether the system is operating near capacity. Armed with the formula

$$\text{Condensate (gal)} = \frac{12{,}000 \text{ BTU/ton} \times \text{tons} \times \text{SHR}}{8{,}080.1 \text{ BTU/gal water}}$$

(here 8,080.1 is 970 BTU/lb water × 8.33 lb/gal water), the service technician could calculate the amount of condensate per hour (or divide the number of gallons by 60 min/hr and measure the amount of condensate for a minute). If the

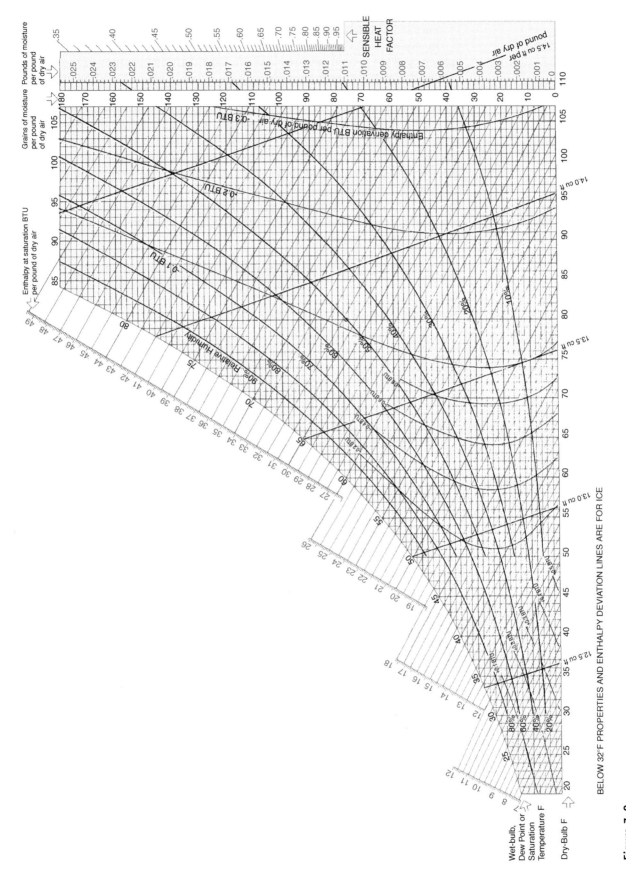

Figure 7–9

A common psychrometric chart used for determining the properties of air. A psychrometric chart with higher or lower temperature ranges is used for heating or refrigeration applications. (Courtesy of Carrier Corporation)

system is operating normally, then the amount of condensate delivered from the system should closely match that calculated as above using the known SHR value.

Supply air temperature is one of the most important measures of system operation, and the second most important is return air temperature. If the air being delivered is not hot, cool, or reduced in moisture content, then conditioning of the space is not possible. Air conditioning systems must supply air that is both cooler and drier than the room conditions. The most accurate measurement of air supply conditions is one that is made 3 to 5 duct diameters from the heat exchanger (refrigerant evaporator). It is at this point that the air has had enough distance to mix; most of the air has been conditioned but some has slipped past the heat exchanger without being affected. Engineers know this and factor this dynamic into their calculations. Service technicians need only keep in mind that the air needs some distance to mix. Accurate measurement of the air delivery conditions (temperature, humidity, pressure, speed, and volume) are critical for determining system problems.

Tech Tip

When taking the discharge air temperature of a fossil-fuel heat exchanger, it is important to be aware that the temperature will read higher than it should if it is taken within the line of sight of the heat exchanger. Line of sight means that the sensing element of the thermometer is "seeing" the surface of the heat exchanger and is measuring not only the heat in the discharge air but also the radiant heat given off by the heat exchanger surface. The thermometer needs to be shielded or moved to a location where radiant energy from the heat exchanger will not affect the reading.

Moisture in the delivered air (supply air) is important when being either humidified or dehumidified. The amount of moisture removed (dehumidification) can be a determining factor in the diagnosis of air conditioning systems. Moisture can also be an important factor in humidification. Buildings spaces that need to maintain specific humidity levels may require that moisture be added to the air delivery system. Some humidification systems allow water to evaporate into the air stream. Commercial systems may use spray or steam injection to add large amounts of moisture. Moisture can be measured by taking any two readings from the supply air. Humidification system capacity can be checked by comparing the return air (room air) to the supply air conditions. Increases in moisture can be seen by a vertical rise in the psychrometric chart and a corresponding increase in the dew point. It should be noted that, as the air absorbs moisture, heat energy is being used to evaporate the water. This means that a certain amount of heat energy is needed to increase the humidity without changing the dry-bulb temperature. If one gallon of water is evaporated per hour to increase the relative humidity in a building, then slightly more than 8,000 BTUH are required to evaporate the water (calculation given below). The amount of water added to the air is also the amount that is needed to offset outside air that is being heated for ventilation. As the relative humidity of outside air decreases, a greater amount of moisture will be needed to bring the supply air up to design humidity levels.

Evaporating water calculation: 1 gallon of water = 8.33 pounds; 8.33 lb/gal $\times$ 970 BTU/lb latent heat of evaporation = 8,080.1 BTU of heat required to evaporate one gallon of water per hour.

Another instance where water is not added is in a typical heating system without humidification. Outside air is brought in both for ventilation and for make-up air. Outside air in the winter might have a relative humidity of 80% at temperatures of 50 degrees DB. The intersection of these two measurements on the psychrometric chart determines a point on the horizontal moisture line (grains per pound of air). To find the new relative humidity, draw a line to the right along this moisture line to the new room temperature. If the outside air is raised to 70 degrees DB then the new RH value would be approximately 40%. As the air temperature increases, the amount of moisture that the air can hold also increases. The increased ability of the air to hold more moisture is a function of the amount it already contains. Therefore, as the air temperature rises, the air's relative humidity drops.

SOLUTION TO THE FIELD PROBLEM

What was the top item on your list of possible causes? By this point you may have determined for yourself what the technician saw when she looked at the condenser. If you did, you have combined your knowledge of the air conditioning system and the system operating conditions with new knowledge of psychrometrics. The possible causes could be listed as follows.

1. A plugged condenser coil
2. An evaporator coil that is plugged with ice resulting from
   a. Dirty filters
   b. Bad fan bearings
   c. Dirty blower wheel
   d. Blockage of return air
      i.   Dirty filter
      ii. Blocked intake grilles
      iii. Collapsed return air ductwork
   e. Some device or some process that is adding more moisture to the air
      (*Note:* This is not typical; however, since the service problem did not specify whether the building was residential or commercial, many external problems could be the cause of this malfunction.)
      i. Too many people taking a shower (too many people using the house)
      ii. A new commercial process that gives off steam and water vapor
3. Overcharge
4. Undercharge
5. Open return air ductwork in the attic or crawlspace that is drawing in warm, humid air
6. Airflow too high for system
7. System not sized correctly
   a. Mismatched evaporator and condensing unit
   b. Building addition or building usage patterns have changed
   c. Failing insulation, windows going bad, or worn doors and gaskets
8. Blower bearings are failing
9. Shrubs, plants, or buildings are obstructing air flow to condenser
10. Outside air temperatures are above design
11. System is oversized for the application
12. Open windows or doors
13. Too many people for the building or too much human activity

If your list included many of the conditions just listed, then you are on the right track to solving the problem. A plugged condenser would not allow enough heat to move from the high-temperature vapor in the condenser. Not very much of the vapor would condense. The little liquid that was formed would be used to cool itself down to suction pressure, and the suction pressure would remain high. The higher than normal suction pressure would be responsible for a higher evaporator temperature at the coil. The coil could not get cold enough to remove moisture, and the conditions in the occupied space would show a higher relative humidity when tested. The higher humidity is what the customer was complaining about when he stated that it felt "sticky." Because he noticed this sticky feeling and made the complaint, the technician was able to fix the problem, reduced stress on the compressor, and saved him hundreds of dollars in future repair costs. When you plot the conditions given in the field problem on a psychrometric chart, you will find that 73°F DB and 80% RH correspond to a point on the comfort chart. Reading to the left, you will find the outlet reading of 66°F DB and 90% RH. This shows that the coil was not removing any moisture, since the two readings are on the same grains of moisture line.

SUMMARY

In this chapter you learned some fundamentals of psychrometrics and the use of the psychrometric chart. We hope that you used a psychrometric chart and plotted each of the examples given in this chapter. With practice you should be able to plot a point on the chart when given any two conditions of air. You also learned how the chart relates to comfort conditioning. Heating systems, by themselves, do not add or remove moisture from the air. The chart shows that the statement "this heating system is a drier heat" is not correct, because no moisture is being removed. You were also introduced to dehumidification and the process of air conditioning, which both dehumidifies and cools the air. In addition to becoming familiar with the psychrometric chart, it is important to understand the concepts of how air is conditioned and how the conditions of air affect comfort conditioning. With a good grasp of air and its properties, you will be better able to troubleshoot problems that involve air.

REVIEW QUESTIONS

1. What can be read at the sole of a psychrometric chart?
2. What can be read at the heel of the psychrometric chart?
3. What can be read at the toe of the psychrometric chart?
4. What can be read at the instep of the psychrometric chart?
5. Describe or demonstrate how to plot the following points on a psychrometric chart: (a) 68°F DB and 80% RH; (b) 73°F DB and 80% RH; (c) 80°F DB & 50% RH; (d) 80°F DB and 20% RH; (e) 68°F DB and 20% RH.
6. Describe or demonstrate how to plot the following points on a psychrometric chart: (a) 68°F DB and 80% RH; (b) 62°F WB and 100% RH. Describe or show the amount of moisture change between the two points.
7. Describe or demonstrate how to plot the following points on a psychrometric chart: (a) 68°F DB and 80% RH; (b) 73°F DB and 80% RH; (c) 80°F DB and 50% RH; (d) 80°F DB and 20% RH; (e) 68°F DB and 20% RH.
8. Describe or plot 56.5°F WB and 80°F DB on a psychrometric chart, and then read the following conditions of air at that point: volume, total heat content, grains of moisture, and relative humidity.
9. Describe what is happening to air that enters the return air grille of a comfort conditioning system at 70°F DB and 58.5°F WB but leaves the diffuser at 55°F DB and 51.5°F WB.
10. Describe what is happening to air that enters the return air grille of a comfort conditioning system at 70°F DB and 50°F WB but leaves the diffuser at 95°F DB and 68°F WB.

CHAPTER 8

Cooling Towers and Evaporative Condensers

LEARNING OBJECTIVES

The student will:

- Identify the four basic parts of a cooling tower
- Describe the function of a cooling tower
- Describe the function of an evaporative condenser
- Recognize the relationship of good maintenance and tower efficiency

INTRODUCTION

A cooling tower is technically a cooling system that relies on the evaporation of water to cool water. As 1 pound of water evaporates, it removes 970 BTU of heat. The evaporation cools the surrounding water as outside air receives the heat and water vapor. Cooled water drops to the bottom of the cooling tower and is used to remove the heat from water-cooled condensers. The water returns to the cooling tower hot, ready to give up heat to the outside air conditions.

Evaporative condensers are essentially cooling towers with the refrigeration system condenser tubing placed inside the tower. They operate in the same fashion by evaporating water. Both the evaporative condenser and cooling tower are used in commercial cooling systems.

In this chapter we will take a closer look at cooling towers and evaporative condensers. The emphasis will be on cooling towers, since they are extensively used. The parts of a cooling tower will be identified, and the function of both the cooling tower and evaporative condenser will be better explained. The chapter will finish with maintenance procedures to keep cooling towers operating properly.

Field Problem

The technician met the maintenance supervisor at the Stroh's Brewery Corporate Headquarters building. The supervisor explained that the chiller would shut down and that they had been resetting the HPC on days of warm weather, and occasionally the HPC would be tripped when they came in the morning.

Symptoms:
- Chiller going off on high pressure control (HPC) and having to be reset to restart the chiller.

Possible Causes:
High head pressure caused from:

- Plugged water-cooled condenser
- No or insufficient water circulating through condenser
- Noncondensibles (air) in system
- Refrigerant overcharge
- Cooling tower not cooling water because of dirty tower, plugged nozzles, or insufficient air across cooling tower

Being familiar with the 100-ton centrifugal chiller and cooling tower on the roof, the technician started with a visual check in the mechanical room and confirmed that water was flowing, but it seemed warm. Going to the cooling tower, he confirmed that water was flowing through the tower, and the atomization of the water coming out of the tower nozzles looked good; the two large fans on the top of this induced draft tower were operating. Before looking into refrigeration problems, the technician thought he better apply his knowledge of cooling towers and psychrometrics.

With his aspirating psychrometer and temperature meter, the technician took some readings and found the following:

- 88 degrees dry bulb
- 80 degrees wet bulb
- 98-degree water out of the cooling tower
- 112-degree water into the cooling tower

The temperature difference (14 degrees) of the water in and out of the tower was not out of line, but both readings seemed to be on the warm side. Wondering why the water temperatures were so high, the technician decided that this was caused by a high building load, the tower not doing its job of cooling the water, or water flow rates lower than the approximate 300 gallons per minute required. Because the water flow through the tower nozzles seemed good and the tower did not appear dirty, the building load should not be excessively high at 88 degrees ambient. The technician remembered that with cooling towers and evaporative condensers, the maximum approach temperature that can achieved is 7 degrees of (or above) the wet-bulb temperature of the ambient air being drawn into to the tower. Going back to the readings, he realized that, rather than something close to a 7-degree approach, this system was operating with an 18-degree approach (98 degrees − 80 degrees = 18 degrees). Because the tower appeared to be clean and the water distribution through the tower appeared to be good, the only remaining possible cause was that insufficient airflow through the tower was not creating enough evaporation to cool the water sufficiently. Climbing to the top of the tower, he felt the airflow over one fan and then the other. One was discharging air (as an induced application should), and the other seemed to be pushing the air. Knowing he was on to something, the technician shut down the system and observed that the second fan was turning the wrong direction. He went back to the fan disconnect and reversed two wires of the three-phase fan, and upon restarting the system, he found that both fans were now inducing air through the tower and not recirculating much of the air through the fans. After allowing the system to operate for awhile, he grabbed his instrumentation and took some more readings. They were:

- 88 degrees dry bulb
- 80 degrees wet bulb

Field Problem (Continued)

- 89-degree water out of the tower
- 106-degree water into the tower (and out of the condenser)

Now that there was a 9-degree approach (the difference between the 89-degree tower water temperature and the 80-degree wet-bulb temperature) and the water temperature from the condenser coming back to the tower was down to 106 degrees, it was evident that the problem was solved, and the system was working to pull the load out of the building. After steady state, the approach temperature would probably come down from the present 9 degrees to 8 or even 7 degrees, proving that the tower was doing all it could do. With the cooler water temperature in the chiller condenser, the head pressure was certainly coming down and approaching a typical 105-degree saturation temperature.

So what was the cause of the problem, and why was one fan suddenly turning the wrong way? After consulting with the maintenance supervisor and checking the records at the company, the technician found that the fan blade on the fan that was turning the wrong direction was replaced in the spring. Apparently, when the fan was reinstalled, the direction of the three-phase motor was not checked. Problem solved.

OVERVIEW OF COMMERCIAL SYSTEMS

A large number of commercial systems use cooling towers to remove heat from condenser cooling water instead of continually using potable (city or well) water that would be wasted down a drain. A few commercial systems use an evaporative condenser to do the same job. The choice of which technique to use is usually based on where the heat will be rejected (given off) and where the components are placed (system configuration).

A cooling tower is a piece of equipment used to force the evaporation of water. By evaporating water, the heat that results from evaporating the water is used to reduce the temperature of the water that was not evaporated. The cooler water is sent to an air conditioning or refrigeration condenser where it picks up heat and returns to the cooling tower to be cooled again. During the process of evaporation, large quantities of water are evaporated. Evaporating water reduces the amount of water and increases the concentrations of minerals in the remaining water. The system corrects for these conditions in two ways: (1) new water is introduced to maintain the water level, and (2) water is bled from the system to reduce the concentration of minerals in the water.

Evaporating water requires 970 BTU of heat for every pound of water evaporated. Water can evaporate only if air is low in moisture content (relative humidity). Higher percentages of relative humidity mean that the air is increasingly saturated with water and thus can hold less additional water. The *approach temperature* indicates how close to the wet-bulb temperature the water temperature can get in a cooling tower. If the wet-bulb temperature of the air is 78°F, then it is possible for the approach temperature to be within 7 degrees, or 85°F. Many manufacturers design their towers to operate between 5 and 7 degrees approach. So even if the air dry-bulb temperature is 95°F, the cooling tower can reduce the water temperature below this ambient temperature, to 85°F, as long as the wet-bulb temperature is 78°F.

Instead of using raw water and sending it down the drain, a cooling tower uses water in a refrigeration-like process to evaporate water from a liquid to a vapor and then transfer heat to the vapor (see Figure 8–1 and Figure 8–2). The amount of water that is evaporated is directly related to the system capacity and the amount of heat that is being removed from a building. If a 10-ton system is operating to its design capacity, then it will evaporate approximately 14 gallons of water an hour. (This figure is calculated as follows: (10 tons $\times$ 12,000 BTUH/ton) $\div$ (970 BTU/lb latent heat $\times$ 8.33 lb/gal water) = 14 gallons of water per hour.) By measuring the make-up water (minus the bleed water), the technician can get a rough estimate of the amount of heat being removed at the cooling tower. An

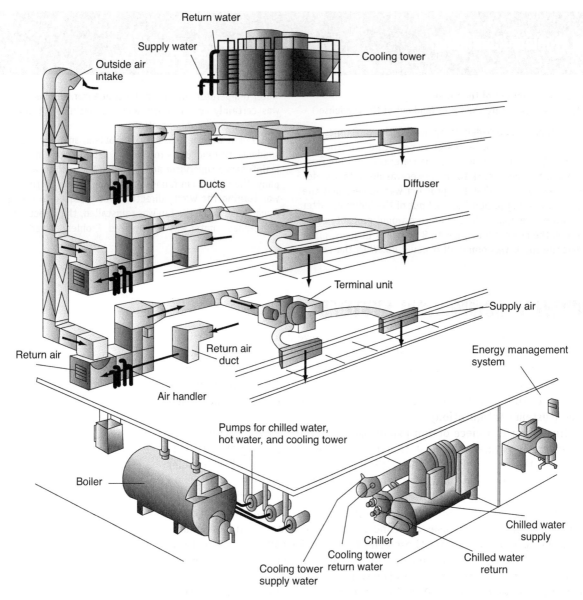

Figure 8–1
A commercial chiller with a cooling tower.

"equivalent ton" is another way of looking at the heat rejection capacity of a cooling tower. This unit equals 15,000 BTUH, or the amount of heat rejected in cooling 1,500 pounds of water (3 gal/min) by 10°F for a properly operating tower.

Tech Tip

Ton of refrigeration effect versus a cooling tower equivalent ton. We know that a ton of refrigeration effect is the amount of heat that is transferred in an hour: 12,000 BTU. You may recall that this number is derived from a ton (2,000 lb) of ice melting in a 24-hour period of time: 144 BTU/lb ice × 2,000 lb = 288,000 BTU per day. Dividing 288,000 BTU by 24 hours/day then yields 12,000 BTUH. The cooling tower equivalent ton is this same amount of heat *plus* the heat energy of the compressor, which is approximately 3,000 BTUH per ton; thus, 12,000 + 3,000 = 15,000 BTUH for a cooling tower equivalent ton.

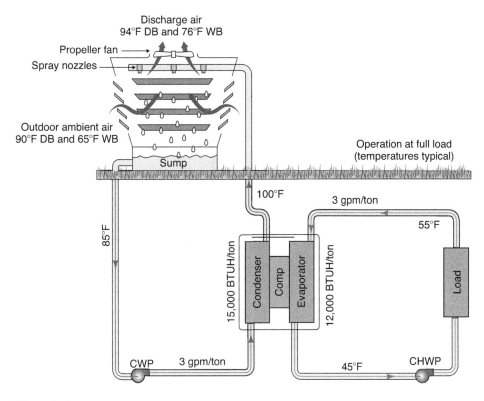

Discharge air
94°F DB and 76°F WB

Propeller fan

Spray nozzles

Outdoor ambient air
90°F DB and 65°F WB

Sump

Operation at full load
(temperatures typical)

100°F

3 gpm/ton

85°F

55°F

15,000 BTUH/ton

Condenser   Comp   Evaporator

12,000 BTUH/ton

Load

CWP   3 gpm/ton

45°F   CHWP

Figure 8–2
An example of a condenser water circuit with a cooling tower.

TOWER FUNCTION AND TYPES

In a cooling tower, water is typically moved down through the tower while air is moved up; this causes a portion of the water to evaporate and carry heat away. Heat is transferred from warmer water to the cooler air through conduction, but even greater amounts of heat are removed from the tower water through the process of evaporation. The cooling effect is similar to the process humans use to get rid of excessive heat during vigorous exercise. Sweat beading on the surface of the skin evaporates as air passes over, cooling the body. There are two basic types of cooling tower design, counterflow (see Figure 8–3) and crossflow (see Figure 8–4).

There are several ways that air can be moved through the cooling tower: by natural draft, induced draft, and forced draft. *Natural* draft uses a chimney style cooling tower that allows warm moist air to rise naturally (see Figure 8–5). *Induced* draft is designed with a fan that pulls air from the tower and discharges it to the atmosphere (Figure 8–6). Induced draft typically has fans on the top. *Forced* draft forces air into the tower (Figure 8–7). Fans for this type of tower are located near the inlet of the cooling tower and blow air into the tower. Forced draft fans are typically positioned at the sides of the tower.

COMPONENTS

Cooling towers (see Figure 8–8) and evaporative condensers (Figure 8–9) are built nearly in the same fashion but with one distinct difference: evaporative condensers cool condensing refrigerant directly because the refrigerant tubes are inside and directly in contact with the water. For ease of discussion, both the

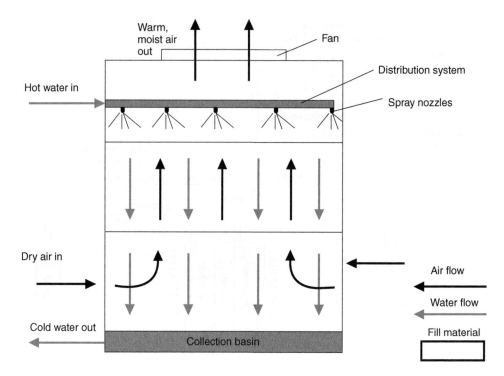

Figure 8–3
A counterflow cooling tower.

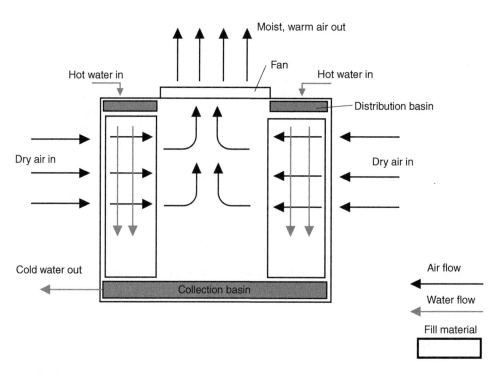

Figure 8–4
A crossflow cooling tower.

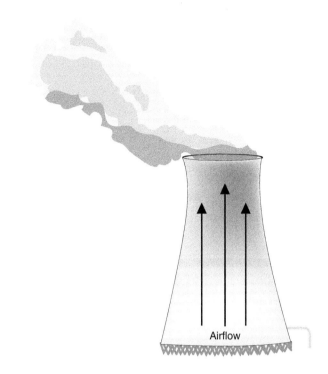

Figure 8–5
Natural draft.

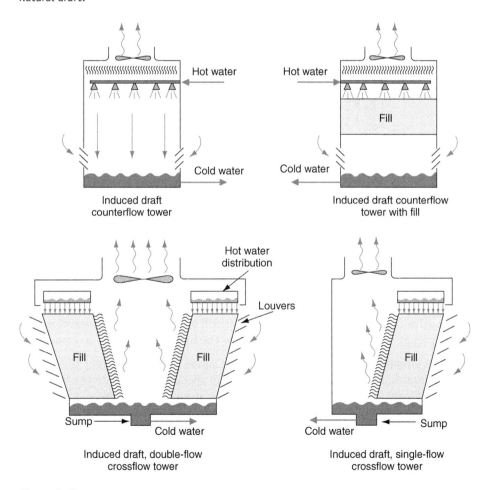

Figure 8–6
Induced draft.

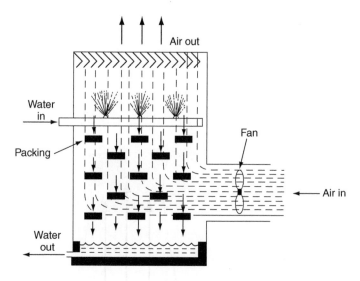

Figure 8–7
Forced draft.

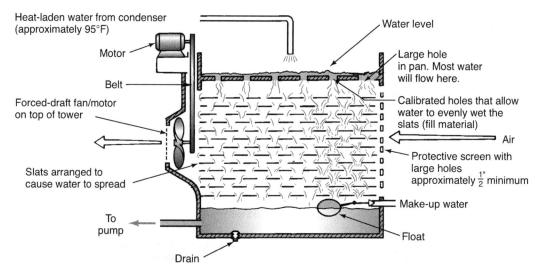

Figure 8–8
Calibration holes for distributing water in the tower are enlarged because of rust. Most of the water is running down the right side of the tower.

evaporative condenser and the cooling tower will be referred to as "towers" unless a distinction is specifically made. Both have the following features in common.

1. A circulating pump that continuously moves water throughout the tower.
2. Nozzles or openings that direct water in specific ways inside the tower.
3. Ambient air is drawn through the tower, usually with a fan, so that the air and water are in constant contact with each other.
4. Either a wetted medium or (with an evaporative condenser) tubing surface increases the surface area of the water.
5. Water is monitored and conditioned to inhibit biological and bacterial growth.
6. Water level is maintained in the basin (or sump) of the tower, most often by using a float valve.

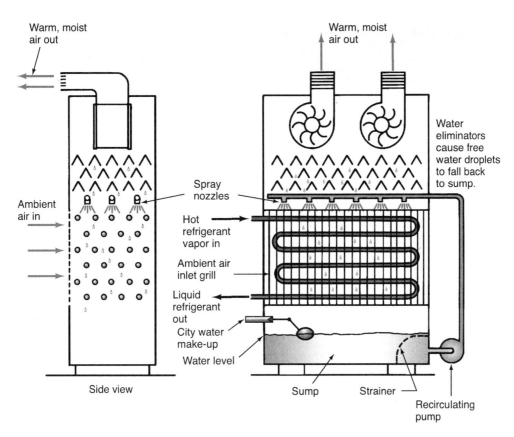

Figure 8–9
Water recirculates in the evaporative condenser. The condenser tubes are in the tower rather than in a condenser shell located in a building.

7. Bleed water is allowed to drain from the system to reduce concentrations of hardness (suspended minerals) in the water.
8. Screens/filters in the water and air streams are used to capture debris.

OPERATION

The normal operation of a cooling tower requires that it be able to evaporate water. This evaporation is not through natural processes, whereby water gently evaporates to the ambient. Instead, the *forced evaporation* process requires that ambient air be forced over the surface of the water, whose area is increased by distributing it over a matrix known as the *fill*. Greater surface area and the associated greater air contact with the water surface increases the amount of evaporation. This is the mechanics of a cooling tower, but there is another factor: humidity. Maximum cooling tower efficiency is based on ambient wet-bulb (WB) temperature. Air must be able to absorb evaporated moisture. When the air's humidity increases (an increase in WB temperature), this decreases the amount of moisture that a pound or cubic foot of air can absorb. That's why cooling towers are sized for the average, "worst case," outside ambient conditions: high temperature and high humidity. Cooling tower rating is accomplished using the concepts of approach and range. *Approach* is the difference between the wet-bulb temperature of ambient air entering the tower and the temperature of the water leaving the cooling tower. *Range* is the temperature difference between the water that enters and the water that leaves the tower.

Case Study

Is it normal? The cooling tower was being checked for operation while being placed into service at the end of April. This particular day seemed less like the beginning of the cooling season and more like the beginning of summer. Using an electronic, handheld aspirating psychrometer, the technician took temperatures and humidities. Ambient air dry-bulb (DB) temperature was 85 degrees Fahrenheit and the relative humidity was 92%. Air leaving the tower was 89°F DB with 62% RH. Water entering the cooling tower was 95°F and leaving was 85°F, as measured by a contact thermometer.

Knowing that the water temperature difference (range) should be between 10 and 15 degrees Fahrenheit, the techni-

cian made the comparison: 95 − 85 = 10°F, within the desired temperature range. The approach needed to be between 7 and 15 degrees. In order to evaluate this, the technician had to determine the WB condition based on the DB and RH of the ambient air. She selected a psychrometric chart for the altitude at her location, on which the WB was plotted as 78°F. The approach was determined by subtracting the WB from the temperature of the water as it left the tower: 85 − 78 = 7°F approach temperature. Rechecking her measurements and calculations, the technician confirmed that the cooling tower was working great.

Make-up Water

Make-up water is typically maintained at a prescribed level in the sump or bottom of the tower, which uses a float valve for this purpose. The water level will drop for two design reasons: (1) water evaporates from the tower, and (2) water is bled from the system to reduce hardness concentrations in the tower water. Make-up water can be taken directly from a well or city water source, but in most cases it should be conditioned before it reaches the tower. Each water source has specific issues that need to be monitored. Water used in a cooling tower must be monitored for conditions that the make-up water might bring to the tower and for conditions that may be added to the water because of contact with air. Water is the prime mover of heat energy from the tower. In order to do this, water must evaporate efficiently. The major reason for water make-up should be to replace water that has evaporated. Water that is lost for any other reason, including bleed water, is of concern to the building owner and to the service technician alike.

Chemical Treatment

Because water needs to come in contact with the air to create forced evaporation, water picks up anything that air carries to the tower. Air can carry *particulate* matter: soot, dust, pollen, leaves, and other material. In addition, air can carry such unseen biological agents as spores, germs, viruses, and so on. One of these agents—Legionella—receives a lot of attention because it was discovered in a cooling tower as the source of "Legionnaires' disease" (so called because many attendees at an American Legion convention became infected from the hotel's building make-up air being contaminated by the air from the tower). Chemicals are used to treat all types of water problems. The most common problem is the growth of algae, a green, stringy, plant that thrives in warm water. Algae also likes sunlight and grows on the most exposed and wet surfaces of the tower. Whatever the problem, it is important to catch it quickly. A few hours or days are all it takes to foul a working tower. Heavy deposits may necessitate cleaning that requires the tower to be disassembled. Continuous chemical treatment is usually needed to prevent possible problems. A tower is similar to a swimming pool in that the water needs to include a certain level of chemicals to ensure that the water is clean (see Chapter 10, "Water Treatment").

Field Problem

The technician was shown the location of the cooling tower after asking the owner some initial questions about the service call. The owner left the technician and returned to the relatively cool interior. "Not enough cooling" was the service complaint, and the technician decided to start with the cooling tower, since the customer took him here, and work his way to the chiller evaporator. This was not his normal routine in cases in which the complaint was that there was not enough cooling, but the cooling tower could be one of the possible causes if the water is not being cooled enough. Other possible causes include fouled heat exchangers, refrigerant system problems, and high load conditions.

The cooling tower had a strange odor, and inspection revealed that the surfaces of the baffles were wet and coated with slimy brown material. Taking caution not to touch the slime, the technician put on goggles, a pair of gloves, and a face mask to avoid breathing the vapor. Looking around the tower, the technician saw large masses of growths in the basin of the cooling tower and on the float valve. The largest of these completely covered the cooling tower outlet. Assuming that the strainer would have to be pulled, he decided that he should not do any maintenance until he was sure of what he was dealing with. He had never seen anything like this, so the technician knew he would need some help.

On the phone with his supervisor, the technician described what he was seeing. His supervisor suggested that he tell the customer to contact the office for an explanation of the need to call in an expert on cooling towers and water conditioning. The supervisor informed the customer that the cooling tower was in need of a complete cleaning and that a chemical analysis was needed to determine the type of biological growth in the tower before they could resolve the problem.

CLEANING AND MAINTENANCE

Cooling towers require a high level of maintenance. The evaporation of water in a mechanical system will accelerate wear and corrosion, requiring the replacement of parts. Motors and fan blades will need maintenance. Sensors, valves, and connectors will often have to be changed out. In addition to the normal wear and tear of any mechanical system, cooling towers and evaporative coolers will need to be manually cleaned. Most often this will involve the use of a power sprayer. Power washing will start at the top of the tower and end with bailing out or shoveling out the sump. The debris may need to be handled as if it were hazardous, depending on the chemical residue captured in the debris.

Poor tower performance can be caused by scale, clogged spray nozzles, reduced airflow, and/or hampered pump performance. The U.S. Department of Energy estimates that, for a poorly operating tower, a 1-degree increase in water temperature corresponds to a 2.5–3.5% increase in energy costs. Manufacturer's operating and maintenance instructions should always be followed to maintain efficiency and prolong length of service. Warranty on a tower may depend on following the manufacturer's instructions. Table 8–1 lists some basic suggested maintenance activities for a typical tower.

Safety Tip

Be aware of apparent biological growth around cooling towers, since this could cause illness and disease in the presence of open wounds or even normal breathing.

ADJUSTING FLOW

Environmental Changes

Cooling towers and evaporative condensers are typically installed outside, so they experience all of the changes in the environment. Everything from cottonwood and pine pollen in the spring to leaves in the fall is brought into the tower. As the temperatures change from hot dry summers to rain and eventually snow in the winters, towers use varying amounts of water. In colder climates, towers are shut down and drained to prevent freezing and then started up again when the weather warms. Wind speed can affect the amount of water that is blown away

Table 8–1 Tower Maintenance

Annual	Periodic	Weekly
• Remove debris with a power sprayer.	• Check and lubricate all mechanical systems.	• Check and clean the water strainer in the basin.
• Look for leaks and corrosion in the water system; repair them with material that prohibits corrosion.	• Operate flow control valves manually and check for proper setting.	• Operate the make-up water float manually to ensure proper functioning.
• Seal leaks in the airflow that would reduce proper airflow patterns.	• Ensure that spray nozzles provide even water distribution.	• Check the mechanical drive system (motors, shafts, pulleys, belts, etc.).
• Inspect and replace baffles as needed.	• Check for leaks.	• Check for excessive vibration in the mechanical system.
• All periodic and weekly checks from this chart.	• Check for sediment levels.	• Check for leaks in the water system and gear boxes.
	• Remove debris from the basin.	• Check for structural problems: loose fasteners, bent louvers, and broken or cracked panels.
	• Check the baffles and louvers for scale buildup.	• Test water for dissolved solids (calcium, etc.) and adjust bleed water flow.
	• Look for damage to the tower parts.	• Measure and ensure that water treatment chemicals are present.
	• Inspect for excessive wear or cracks in motor supports, fan blades, housings, and belt covers.	• Adjust the water treatment chemical system to recommended levels.
	• All weekly checks from this chart.	• Check for ice buildup in freezing weather; ice should be within acceptable limits, and winterizing equipment should be working.
		• Check gearbox lubrication levels.

and hence doesn't contribute to the cooling effect (this is often referred to as *drift*). Environmental changes also affect the amount of airflow actually required to effect the desired amount of cooling. All of these conditions require monitoring and modulation. Before modern automatic control, the cooling tower was controlled manually. Today, cooled water from the tower is regulated by flow control valves. In addition to valves that regulate the flow automatically, motor speed control (electronically commutated motors and frequency drives) can be used to vary the water and airflow through the tower. Regulating the flow of bleed water is still a manual operation on many systems, but some automatic systems monitor the level of dissolved solids (such as calcium) and then adjust bleed water amounts to reflect that level. Whether manually or automatically, towers must be monitored and adjusted for airflow, water flow, and bleed water. Reductions in the amount of air, cooling water, and bleed water can lead to large savings in the operation of a tower over the course of a year.

Water Flow Adjustment

Water flow is adjusted for two reasons, to control the spray nozzles and to control water being drained from the sump (bleed water). Pumping pressure (pump speed) and regulation of water volume are two ways that water is adjusted for spray nozzle supply (also called fill water). Water is controlled so that spray patterns are optimized.

The bleed water flow should be adjusted as weather conditions change. Water is consumed (evaporated) during the normal operation of the tower. As the water evaporates, it leaves behind the suspended minerals or solids that were in the water. The concentration of minerals will cause these solids to stick to surfaces in the tower. Although the solids that stick to surfaces will tend to increase the surface area of the water being evaporated, they also build up and restrict the flow of air through the tower. Small amounts of buildup can be tolerated, but it is increasingly more difficult to remove the solids as they continue to build up. Allowing water with a high concentration of minerals to bleed (drain from the tower) reduces the buildup of solids on the tower surfaces. In a process known as *cycles of concentration,* concentrated minerals in the water are compared to the original source water. When the concentration is twice the level of the source water, there are "2 cycles of concentration." Bleed rate is adjusted to maintain concentration levels in the tower basin (sump).

Air-speed Adjustment

Airflow is also adjusted as the weather changes. Dampers are one means of controlling airflow to the tower. Dampers open or close depending on the total amount of cooling that is needed. Sometimes there is enough natural cooling that fans can be cycled off until system requirements and weather conditions call for them to be turned on again by the temperature monitoring controls. Systems with multiple fans will operate fans as needed by turning off one or more as the mechanical system load decreases. System load can be monitored by measuring pressure or temperature on the condensing side. Full airflow is required at the upper design limits of the tower, but at other times it is possible to save energy by cycling or turning off fans. Fan cycling can be accomplished by temperature or pressure controls monitoring head pressure or water temperature. Table 8–2 describes several ways in which a two-fan, induced draft tower can control the amount of air.

Tech Tip

The typical maximum water temperature for cooling towers is 100°F (80°C) because most modern towers use a plastic material as the fill, and the plastic material is highly susceptible to damage from temperatures above 100°F. Always check with the manufacturer for the design specifications and operating characteristics of the cooling tower.

Table 8–2 Fan Cycling for Two-fan Induced Draft Tower

Tower Fan Control	Fan 1	Fan 2
Cycling Fans		
Scenario 1	Fan 1 remains on continuously	Fan 2 cycles based on leaving water temperature from the tower
Scenario 2	Fan 1 cycles off only at a predetermined low air temperature	Fan 2 cycles based on leaving water temperature from the tower
Scenario 3	Fan 1 cycles based on leaving water temperature from the tower	Fan 2 cycles based on leaving water temperature from the tower
Variable-speed Fans		
Scenario 1	Fan 1 remains on continuously	Fan 2 speed varies based on leaving water temperature from the tower
Scenario 2	Fan 1 speed varies only at air temperatures below a predetermined value	Fan 2 speed varies based on leaving water temperature from the tower
Scenario 3	Fan 1 varies based on leaving water temperature from the tower	Fan 2 speed varies based on leaving water temperature from the tower

Dry and Wet Operation

Some cooling towers and evaporative coolers may be designed to be operated both "wet" and "dry." Operating a tower wet is the normal condition, where water is being force-evaporated to cool the remaining water. In contrast, for dry operation water is not being sprayed into the tower. In this case, water is pumped to the basin or sump and is naturally cooled by the ambient conditions. Whether it is possible to employ dry tower operation depends on the cooling system load, the weather conditions, and the amount of cooling that can be obtained before spraying is needed to increase the amount of evaporation.

EVAPORATIVE CONDENSERS

An evaporative condenser is nearly the same as a cooling tower. Both use the evaporation of water to cool condensing refrigerant. Both pump water to a spray header and force air and water to mix; the water evaporates as it mixes with the air. Both components require monitoring of their water to maintain quality and design performance. What distinguishes the evaporative condenser is that the refrigerant heat exchanger (condenser) is cooled directly via contact with the evaporating water inside the tower. Condenser tubing is positioned inside the tower, so there is no need to pump water to and from the tower. The surface of the condenser is exposed both to air and to water. Water coming in contact with the condenser tubing surface directly cools the refrigerant as the water evaporates.

Most evaporative condensers are positioned outside, but they can also be installed within a mechanical room and ducted to the outside, as depicted in Figure 8–9. During cold weather it is easier to control the amount of outside air that enters the evaporative condenser by operating intake and exhaust dampers.

Field Problem

The technician was preparing to inspect the system on an "insufficient cooling" service call. Because this was the first time he had been to this building, he was unfamiliar with the component layout of the chiller system. The complaint indicated that there was not enough cooling for the building, so his first suspicion was that something was not getting cool on the hot side of the system—namely, the condenser.

The technician found the evaporator and compressor easily enough, but the condenser's location was not obvious. Looking for a water cooled condenser, he traced the discharge refrigerant line from the compressor to a large vertical device located on an exterior wall in the mechanical room. This device seemed to be a cooling tower ducted through the wall of the mechanical room. Nearby, water was running at a drain. The technician

had never seen one before, but he had read about evaporative condensers—and figured that this must be one. Removing an inspection door at the top of the unit revealed several spray nozzles, half of which were not functioning. Using what he already knew about cooling towers, the technician began a complete inspection of the evaporative condenser and discovered that the unit had not been maintained for a long time and needed a complete cleaning. The condenser tubing bundle was covered with mineral deposits and sections of it were dry from lack of water spray, causing high condenser temperatures and pressures and a loss of chiller capacity.

After cleaning the unit and replacing the plugged nozzles, the evaporative condenser was back in working condition and cooling the building.

SUMMARY

In this chapter you learned about cooling towers that evaporate water in order to cool the remaining water, which is then used to cool a refrigeration water-cooled condenser. Water can be cooled to below the ambient dry-bulb temperature if the relative humidity is low. The tower "approach" describes how close the water temperature can get to the wet-bulb temperature of the air. Instead of using raw water (city or well water) and then sending it down the drain, a cooling tower can use the same water continuously and replace only the water that is evaporated. As water evaporates, it leaves behind minerals that were suspended in the water. This concentrated mineral content is bled off before replenishing the tower with make-up water.

There are several different types of cooling towers and configurations, but all of them force the evaporation of water to generate a cooling effect. The structure of the cooling tower is generally the same in all configurations and includes a pump, nozzles, a wetted surface (that can be made from one of several different materials), a float or level control, a bleed water valve, and screens or filters. All of these components and the tower itself require regular maintenance. The water requires the addition of chemicals to hold minerals in suspension and to prevent algae and bacterial growth.

A learning technician would be wise to pay attention to the requirements of a cooling tower. In most cases, problems can be prevented by regular scheduled maintenance and cleaning of the tower's evaporating condenser along with proper water chemistry control.

REVIEW QUESTIONS

1. Describe the function of the wetted surface of the cooling tower.
2. Describe the function of the spray nozzles in a cooling tower.
3. Describe the function of a cooling tower fan.
4. Describe the function of the water-level control.
5. Describe the function of the bleed water control or valve.
6. Describe how water is monitored for biological growth.
7. Describe the function of a screen and what should be inspected.
8. Describe the function of a cooling tower.
9. Describe the structural difference between a cooling tower and an evaporative condenser.
10. Describe the benefits of good cooling tower maintenance.
11. A cooling tower is operating, and the outside air temperature is 90°F. How could water leaving the cooling tower measure 85°F?

CHAPTER

9

Commercial Air Conditioning and Refrigeration Systems

LEARNING OBJECTIVES

The student will:

- Describe the operation of centrifugal compressors
- Describe the operation of screw compressors
- Describe the benefits of central plant systems
- Describe basic troubleshooting methods for any of the following: metering devices, compressors, heat exchangers, solenoids, reversing valves, capacity control, sensors, filter-driers, level controls
- Describe the benefits of service and maintenance records

INTRODUCTION

Commercial refrigeration and air conditioning systems come in many and varied applications, styles, capacities, and piping arrangements. This chapter looks at those refrigeration systems that condition the air for comfort cooling applications.

Chillers are systems that cool water. Cooled (chilled) water is circulated to buildings and rooms, where a fan coil moves the room air over the chilled coil. Chillers can be small, reciprocating (piston) compressor systems, but systems with the largest capacities are centrifugal and screw compressor systems. We will spend some time with centrifugal and screw compressors and then revisit many of the same system components used on other air conditioning and refrigeration systems.

Chilled water systems are used in large commercial and industrial buildings because water piping takes up far less space than ductwork. In a chilled water system, the air handling unit and air distribution ductwork is located in or near the conditioned space, and the conditioning equipment is located remotely. Imagine trying to run ductwork from a central station air handler in the basement to all of the floors of a 10-story building. The first couple of floors would have no room for anything other than the ductwork going up. Chilled water piping from the basement would take up an acceptable amount of space, and the air handling equipment would be in close proximity to the conditioned space.

Field Problem

After the chiller condensing tubes were cleaned and the system was placed back into service, the chiller would operate for a few minutes and then shut off. The operator explained that the microprocessor control indicated a low pressure safety trip and that suction and head pressures were both low.

The technician determined through conversations with the operator and the maintenance person that the system had been shut down for a routine cleaning. One of the maintenance policies was to place the equipment into the same condition as shipped, so all of the valves were placed in the original shipping position as when received. The condenser was drained of water, cleaned, and then placed back into service. Water pumps were working, and the pressure drop through the condenser was within the manufacturer's specifications.

Thinking about the safety trip conditions and the valve placement, the technician had a hunch. When systems are shipped, all valves are placed in the frontseated position in order to minimize system leaks. If the valves were placed in this position during maintenance and not backseated afterward, this could explain the low pressure trip. A quick check of the receiver valve positions indicated that one was not backseated. After repositioning, the system was back in operation.

CHILLER EVAPORATOR STYLES

Evaporators used with most chillers are either direct expansion or flooded. With *direct expansion* chillers, the refrigerant is allowed to expand and turn into a vapor as it passes through the evaporator tubes (heat exchanger). Refrigerant enters one end of the evaporator as a liquid/vapor mix and leaves the other end as a vapor, having absorbed the heat from the water that surrounds the tubes. Direct expansion evaporators are associated with reciprocating, scroll, and some screw compressor applications (see Figure 9–1).

Flooded evaporators allow the refrigerant to boil much as a pan of water boils on a stove. Large quantities of liquid refrigerant cover tubes that are filled with water to be chilled. Refrigerant changes from a liquid to a vapor and then boils to the top of the liquid level, which is maintained by a liquid-level control that can be as simple as a float valve. Flooded evaporator designs are generally associated with centrifugal compressors and some high-capacity screw compressor designs.

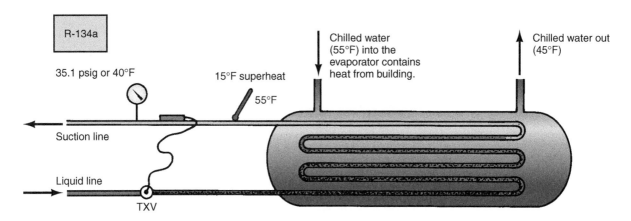

Figure 9–1
A direct expansion evaporator in which water being chilled surrounds the evaporator coil. Baffling (not shown) is used to direct the water flow efficiently over the entire evaporator coil.

CENTRIFUGAL CHILLERS

Centrifugal chillers (see Figure 9–2) constitute a class of refrigeration system that is identified by the type of compressor used. The primary purpose of centrifugal chillers is for large-tonnage applications, in which chilled water can be pumped to remote air handlers or to machinery for process cooling in large or multiple buildings, reducing the amount of refrigerant piping and the overall amount of refrigerant needed for the application and increasing efficiency. One chiller can be used to air condition one or more buildings. The compressor is designed to move a high volume of refrigerant vapor. Depending on the refrigerant application, most centrifugal chillers (with a few exceptions) are low-pressure systems. Typically, the low side operates in a vacuum, and the high side operates under 10 psi.

Centrifugal systems are high-volume systems, and they require a compressor that can move many pounds of vaporous refrigerant in a short period of time. These compressors are similar in operation to a radial fan. Refrigerant is moved into the center of a blowerlike impeller, is slung to the outside by centrifugal force, and is captured in a scroll. These non–positive-displacement compressors can move vast amounts of refrigerant—while boosting the pressure from the suction to the discharge side by only a few psig—from the evaporator to the condenser, a distance of only a few feet.

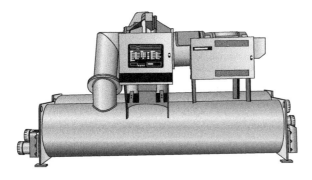

Figure 9–2
A centrifugal system, showing compressor, evaporator, condenser, and control box.
(Courtesy of York International)

Tech Tip

Positive-displacement versus non–positive-displacement compressors. The term "positive displacement" means that if the suction were blocked, the compressor would continue going into a vacuum. If the discharge were blocked, the compressor would continue to build pressure to dangerous levels. Reciprocating compressors are positive-displacement devices, but centrifugal chillers are non–positive-displacement devices. A certain amount of refrigerant is allowed to slip past the impeller clearance from the discharge to the suction side of the impeller. This is normal by design. When the suction of a centrifugal chiller is blocked or any time the pressure difference between the high and low sides becomes greater than normal, then excessive amounts of refrigerant will slip past the impeller clearances. This excessive slip will result in large fluctuations of refrigerant flow through the compressor and noisy operation, sometimes referred to as "surging." The amperage of a non–positive-displacement compressor will be reduced when the refrigerant flow is restricted or reduced. The amperage of the centrifugal compressor will drop when the refrigerant flow rate is reduced for capacity control.

The compressor is similar in shape to a large water-circulating pump or a squirrel cage fan. A motor drives an impeller that pulls vapor into the center of a blower wheel (see Figure 9–3). Vanes on the blower wheel (impeller) throw the refrigerant vapor to the outside of the impeller, where the compressor housing (volute) captures and defines the stream of vapor, discharging it to the condenser. The centrifugal motion of the refrigerant gives the compressor its name. If the compressor has two stages (two impellers), then one stage will discharge into the other and

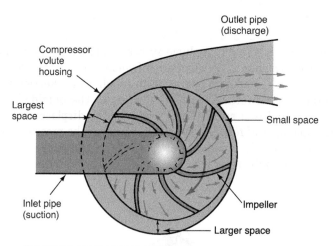

The turning impeller imparts centrifugal force on the refrigerant, forcing the refrigerant to the outside of the impeller and into the compressor volute housing. Because the volute increases in size as it nears the discharge, the refrigerant builds up pressure and is guided toward the discharge. The refrigerant moving to the outside creates a low pressure in the center of the impeller where the inlet is connected.

Figure 9–3
The centrifugal action of the refrigerant is induced by the impeller of a centrifugal compressor.

Tech Tip

Ice storage systems. Some utility companies will discount the purchase of electrical energy during the night or at times when there is lower demand (also called "off-peak" demand). During this period of low energy demand, ice can be produced and then used to cool buildings when it melts. This system of using ice during the day that was produced at night is called *thermal storage*. Thermal storage systems are designed to reduce energy consumption during periods of peak energy demand. The size of the system depends on the amount of heat that needs to be stored.

there will be two or more volutes visible from the outside. Capacity control is accomplished with a set of intake louvers or vanes that open or close to allow more or less refrigerant into the center of the impeller. Bearings for centrifugal compressors are sometimes different from standard bearings. Some systems use a magnet bearing that holds the shaft in suspension to create a "no contact" (levitation) bearing. Bearings and capacity control are both important, because centrifugal compressors operate at very high rpm and are not cycled (turned off and on). The compressor continually operates until maintenance is needed; they are not cycled because they are operated by very large motors. So the compressor continues to operate, and capacity control is achieved by closing or opening a set of vanes (sometimes referred to as pre-rotational vanes). Closing the vanes reduces the amount of refrigerant that the compressor moves, which reduces capacity but maintains rpm. The direction of rotation of the impeller is critical to prevent loosening of the impeller nut, which could occur if it is run in the wrong direction.

When you look at a centrifugal system, the compressor is easy to recognize. The motor is either external (using mechanical seals) or internal. If the discharge is followed, it leads to the condenser. Both the condenser and evaporator may be the same size and shape. From the compressor suction (center of the scroll) the pipe can be followed, in reverse of refrigerant flow, back to the evaporator. Many times the condenser is positioned slightly higher than the evaporator, but if you want to make sure, always follow the connections from the compressor to the evaporator and condenser. Between the evaporator and condenser is the metering device, through which the liquid refrigerant flows from the condenser to the evaporator. These parts are the same as on any compression refrigeration system: compressor, condenser, metering device, and evaporator.

A device not common to previously discussed refrigeration systems is the purge unit (see Figure 9–4), which is unique to centrifugal systems. It is designed to automatically remove air and any other noncondensables that enter the system on the suction side. Because the suction side is in a vacuum, any leak will allow air to leak into the refrigeration system. Air in this refrigeration system has the same effect as noncondensables in any other refrigeration system: it increases the condensing pressure. Noncondensables tend to gather at the top of a condenser, and the purge system gathers vaporous refrigerant from the top of the condenser. Both refrigerant and noncondensable vapors move into the purge. The refrigerant vapor is condensed and returned to the evaporator during the purge cycle, and the noncondensable vapor is pushed out of the system. If the purge is running, there is a leak. If the purge is running more often, then the leak is getting bigger. Most purge systems have an hour meter or software that tracks the purge run time; thus, information can be logged to detect trends that indicate air leaks in the system.

There are many different configurations of purge units. The most popular are systems that cool and condense the vapor to separate the refrigerant from

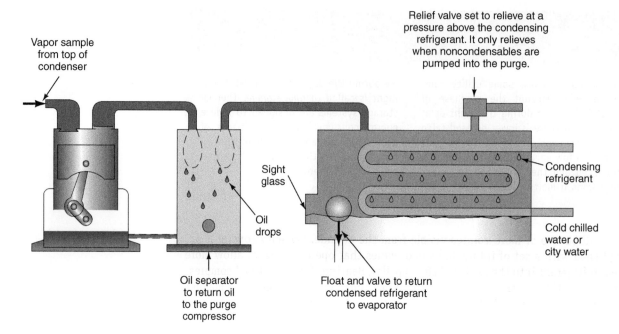

Figure 9–4
A simplified purge unit in operation.

noncondensables. The purge unit can be part of the centrifugal system (and use the system refrigerant for cooling) or a separate refrigerating device. Purge units generally operate as follows:

1. A sample of vapor is drawn from the condenser.
2. The sample is cooled and condensed.
3. When enough liquid refrigerant is in the cooling chamber, a level detector initiates the purge cycle.
4. Noncondensable vapor is expelled from the purge unit, and liquid refrigerant is returned to the system.

There are still some centrifugal chillers that use CFC-11. This refrigerant is slowly being replaced as systems undergo major repair. Now HCFCs are being used to replace CFCs in new centrifugals. Chillers use many pounds of refrigerant, so leaks in a mechanical room could be significant. Because the mechanical room could become saturated with refrigerant, ASHRAE Standard 15 requires them to have alarm systems that activate when the concentration of refrigerant is too high. Short-term exposure to large concentrations of refrigerant can have deadly effects. The technician should also be aware that the EPA (U.S. Environmental Protection Agency) has a program called SNAP (Significant New Alternatives Policy) that lists approved substitute refrigerants. This list is continually being updated. Always go to SNAP or to the OEM (original equipment manufacturer) for information on centrifugal refrigerants and replacements.

Recovery, leak detection, repair, and recharge techniques are a little different on low-pressure systems. To recover refrigerant, there is a valve at the lowest part of the condenser that may, during cooler ambient temperatures, allow the refrigerant to be recovered in liquid form and contained in a low pressure drum. Residual vapor is removed using a recovery machine. Water must be continually circulated (through evaporator and condenser) to keep the water from freezing and to add heat for the recovery process. Leak detection must be done at pressures no higher than 10 psi, as higher pressures may cause the rupture disc to fail. Increasing the pressure is usually accomplished by using a built-in heater or other heat source to heat the water being circulated through the chiller (evaporator) and condenser. Leak repair typically involves replacing gaskets and O-rings or resealing threaded fittings. The system must be evacuated to a pressure of 25 mm Hg absolute (29 inches Hg) before being recharged. Recharge involves using vapor to bring up the pressure from vacuum to a saturation temperature above 32 degrees Fahrenheit for the particular refrigerant used and then continuing the charge using liquid.

Tech Tip

Removing the vapor too quickly could cause any water trapped in the condenser or evaporator to freeze. The use of high-volume vacuum pumps must be closely monitored. Whenever refrigerant is added to or removed from a refrigeration system containing a water heat exchanger (water-cooled condenser or chiller), special attention should be paid to prevent the water from freezing during these operations.

SCREW CHILLERS

Screw chillers are also classified by the type of compressor used. The rest of the system is similar to other compression refrigeration chillers. Screw compressors are unique in that they have two rotors that mesh together and thereby create a long interlocked compression chamber. Also known as a "rotary screw" (see Figure 9–5 and Figure 9–6), this compressor can better tolerate liquid returning in the vapor line than can piston compressors. Screw compressors may have refrigerant cooled motors that are sealed to the atmosphere or may be of the open-drive type, requiring a mechanical seal. On the open-drive type, the motor drive connection is a flexible coupling that must be aligned to very tight

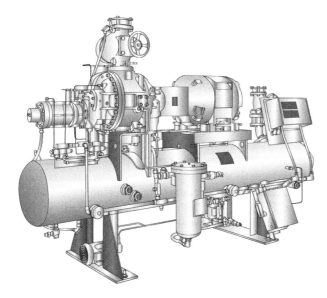

Figure 9–5
A rotary screw compressor, showing condenser and control system.

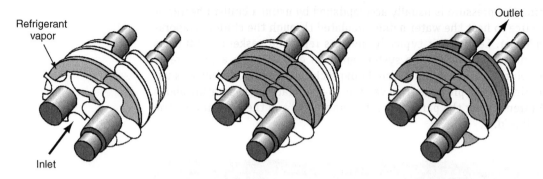

Refrigerant
vapor

Inlet

Outlet

One stage of compression as refrigerant moves through the screw compressor

Figure 9–6
The compression of refrigerant vapor in a screw compressor.

tolerances, because the compressor operates at high rpm. The direction of rotation is critical with screw compressor designs as well (direction of refrigerant flow and bearing design).

All other operating characteristics of a screw chiller are the same as for reciprocating chillers. Valves, controls, and safeties are identical with few exceptions. One difference is that the screw compressor can have built-in capacity control. The control is achieved by sliding the compression chamber or screw to cover or uncover more of the compression area of the screw. Capacity control is continuous (not in discrete stages) and allows the compressor to match, within a wide range, any capacity required or encountered.

Another difference is that screw compressors tend to have much higher levels of oil carryover, and many have oil separators on the discharge of the compressor. This high amount of carryover is by design. The oil is used not only to provide lubrication but also to increase compressor efficiency by sealing spaces between the rotors and between the rotor tips and the compressor walls. Some screw compressors have no oil pump, relying instead on the pressure difference between high side and low side to provide lubricating oil flow. Most other operating and design characteristics of screw chiller systems are the same as for reciprocating compressor chiller designs.

CENTRAL PLANT SYSTEMS

Central plant systems are used to cool a large building or more than one building in a complex. Central plants tend to reduce maintenance costs by consolidating operations in one central cooling system. The costs of a central plant chiller are less per ton of refrigerating effect, as larger systems tend to be more efficient. System capacity can sometimes be expanded with the same machine.

Central plant systems can be upgraded and additional systems added as needed (see Figure 9–7 and Figure 9–8). The additional equipment acts as a redundant system when one machine requires maintenance or is down. Redundant or backup systems ensure fail-safe condition when problems occur. Central plant systems are used for college and university campuses, large industrial complexes, and some municipalities. A central plant system may service all or a group of buildings that are structurally disconnected and at some distance from one another. As we discussed in the introduction to this chapter, water piping takes up much less space than air ductwork.

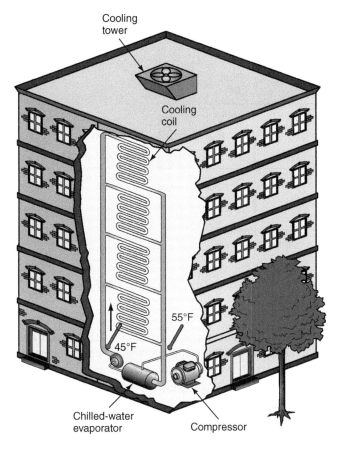

Figure 9–7
A central plant used to cool individual rooms within a building via cooling coils mounted in air handlers on each floor (air handlers are not shown). Water is chilled and sent to each of the cooling coils; the cooling tower for condenser water is mounted on the roof.

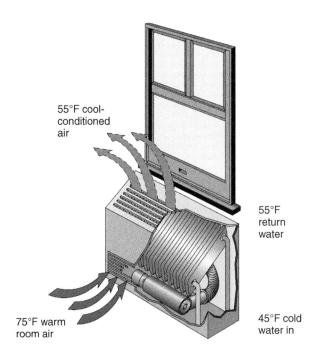

Figure 9–8
Chilled water is sent to each room, where a fan-coil system moves air over the chilled water coil to condition the room.

TROUBLESHOOTING AND ISOLATING SYSTEM COMPONENTS

System components for commercial refrigeration systems should be located such that they can be isolated and removed. Larger systems have more valves, making the job easier. Smaller systems have fewer valves. Compressors may have service valves on both sides, but sometimes only on the discharge side. Condensing units have valving that allows the system to be disconnected and the condensing unit removed. All systems should have enough valving to enable pumping the system down, isolating the refrigerant charge in the condenser and receiver.

Metering Devices

The most commonly used metering device for high-pressure refrigerants (reciprocating, screw, and scroll) is the TXV or TEV. Centrifugal chillers, which are low pressure, typically use an orifice plate or float valve system.

The evaporator float system monitors a "flooded evaporator" condition (see Figure 9–9). An evaporator is *flooded* when it is filled to a certain amount (typically half full) of liquid refrigerant. The float (sometimes called a low-side float) allows more refrigerant into the evaporator as the liquid level drops from

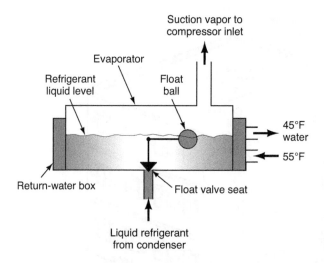

Figure 9–9
A simplified schematic of a flooded evaporator float valve (also called a low-side float).

evaporation. The float control configuration can be a direct float that opens and closes a float valve, much as a cooling tower float; it could also be a pilot-operated valve (a small valve that operates a larger valve), a setup sometimes referred to as "master and slave": the master valve is small and operates the slave valve, which is much larger.

Tech Tip

Float-style metering devices usually have sight glasses. There is either a "high and low" sight glass or a "column" sight glass that allows the operator to see and re- cord the level. Look for one of these two indicators on the evaporator or reservoir where the float is located.

A condenser float system monitors the level of liquid refrigerant in the condenser (see Figure 9–10). The float (sometimes called a high-side float) allows liquid into the evaporator only when there is enough to fill the condenser or another float chamber. As the float rises, the liquid is allowed to flow to the evaporator.

Another type of metering device that is gaining popularity is the electronic expansion valve (EXV or EEV). The EXV uses an electronic stepper motor to open or close a valve (see Figure 9–11). It is easily mistaken for a solenoid, because it resembles a solenoid in shape. A single valve can be used for a wide range of capacities and for many different applications. It can be used in place of a float (or level) control in flooded evaporator applications. Another benefit of the EXV is that it can be controlled precisely using solid-state controls. An EXV may have 1,500 steps in only a quarter inch of travel between fully closed and fully open.

Here are some tips for troubleshooting metering devices.

1. *Check for the correct superheat with the system operating in a steady state*—compare readings with manufacturer specifications.
2. *Check the distributor*—each evaporator run should have an equal amount of "sweating" or frost before it reconnects to the suction manifold at the evaporator.

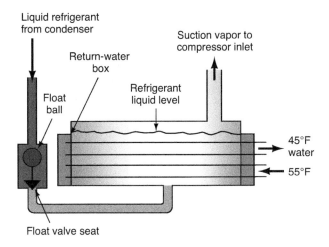

Figure 9–10
A simplified schematic of a condenser float system (also called a high-side float).

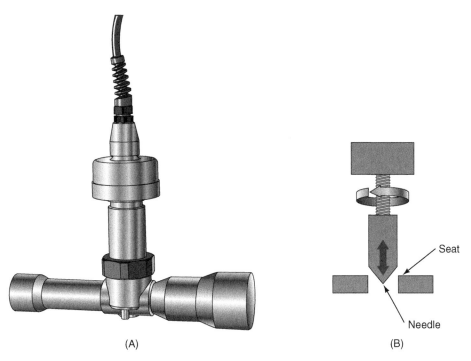

(A)　　　　　　　　　　　　　　　　　　　(B)

Figure 9–11
(A) The electronic expansion valve shown with stepper motor attached. (B) The motor turns a screw that opens or closes the needle to the seat of the valve.

3. *Look for the frost line*—the suction line leaving the evaporator should sweat or frost evenly. If the load is steady, the line should not move very much. If it moves, it may be searching for the load.
4. *Check the thermal bulb connection*—TXV thermal sensing bulbs should be clean, tight, and insulated from the ambient. Externally equalized TXVs should have the bulb positioned between the external equalizer line and the evaporator.

Compressors

Compressor service valves are normally supplied on the suction and discharge (see Figure 9–12). These valves will allow the compressor to be changed without

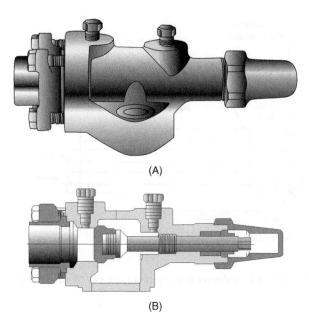

Figure 9-12
(A) A service valve used on a large refrigeration system. (B) Cutaway view of the service valve.

Safety Tip

No matter what type or style of compressor, at no time should it be left operating with the valves closed (front-seated). Always be sure that the service valves are open before placing a compressor back in operation! The only exception may be on a reciprocating compressor when performing a pumpdown test for a short period of time. Failing to open the service valves may ruin the compressor and cause a dangerous, explosive condition for the technician.

Tech Tip

Metering devices generally operate in a superheat range from 5 to 15 degrees Fahrenheit, with most running in the middle at about 10 degrees. When checking a metering device for operation, a superheat reading of approximately 10 degrees usually assures the technician that the metering device is working. However, always check with the manufacturer for specific operating characteristics.

having to recover refrigerant from the system. Depending on the size of the system, the valve connections may vary from brazed, welded, or flanged (bolted) connections to all flange fittings.

Larger compressors start with the compressor unloaded and use several ways to start. Starting a large motor involves a lot of amperage. To reduce the amount of amperage on start-up, the compressor is started in stages or ramped up to speed. A *part winding start* compressor uses half of its windings to start the compressor; when it reaches a certain rpm, all of the windings are used to bring it to full speed. An *auto transformer* start uses reduced voltage to the motor until it reaches a certain rpm; then the power is routed around the transformer to give the motor full voltage. In the *Wye-Delta* start, contactors electrically configure the motor as a Wye-wound motor to start and then reconfigure it as a Delta to run. A starting configuration that is gaining popularity is the *frequency*, or *electronic* start. These start systems start the motor smoothly by using frequency or voltage to ramp the motor to full speed. Because of their size and the extra heat generated during their starting process, these motors are limited in the number of starts they can incur within a given timeframe. Capacity control allows the motors to operate continually.

Listed below are some checks that will help you determine whether a compressor is operating properly.

1. Take amperage and voltage readings to compare with manufacturer specifications.
2. Listen to the compressor for abnormal noises.
3. Measure the compressor for vibration.
4. Measure the temperatures and pressures of the compressor. The compressor should be receiving relatively cool refrigerant on the suction side and discharging appropriate temperature for the condition.

All of these readings can be used to discuss the operating characteristics with manufacturer representatives.

Heat Exchangers

Commercial refrigeration heat exchangers are shell-and-tube, tube-and-tube, fin tube, or welded plate evaporators (which are increasingly popular on small-tonnage chillers). Heat exchangers use water to cool the refrigerant (condenser) and refrigerant to cool water (evaporator), which is pumped to the point of use (see Figure 9–13). Flow characteristics of both refrigerant and water are determining factors in how well the exchangers are working.

Refrigerant flow is determined by metering devices and flow controls. Water pressure drop is the measurement used to determine water flow. Flow curves are provided by the manufacturer and are used to evaluate the flow of water in gallons per minute through the exchanger (see Figure 9–14).

Tech Tip

On systems that use a welded plate type evaporator, there is no way to disassemble the evaporator for cleaning; only chemical-type cleaning can be done. It is therefore important that the heat exchanger be protected from chilled water system contaminants that may foul or plug this type of evaporator. During installation, it is crucial to follow the manufacturer's advice regarding strainers or filters.

Maintenance involving water chemistry is the most important first step in maintaining a chilled water system. If pressure drops are too great, this calls for additional maintenance. Cleaning the water tubes of a shell-and-tube heat exchanger involves opening the exchanger end bells and mechanically removing

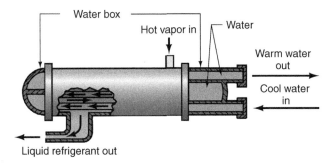

Figure 9–13
Four-pass shell-and-tube condenser, showing the water box and end bell that is used to direct water flow in this heat exchanger.

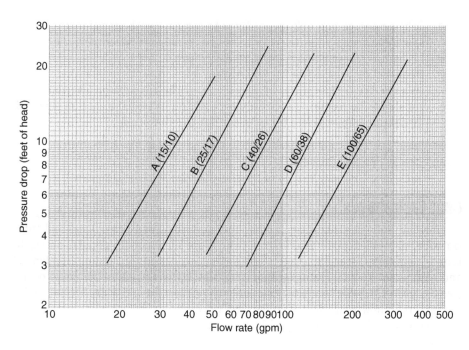

Figure 9–14
A pressure drop chart for five models of heat exchangers.

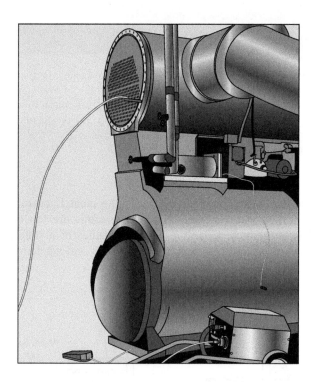

Figure 9–15
Using a power brush to clean the tubes of a shell-and-tube condenser.

buildup (see Figure 9–15). After the tubes are cleaned they are inspected for pits, cracks at welded connections, and other conditions that could lead to failure. In many cases, special tools (depending on system design) may be used to cut out, replace, and re-roll damaged tubes. Sometimes special plugs can be used to plug some fraction of the tubes without significantly affecting system operation.

Tech Tip

Feet of head is calculated as ΔP in psig $\div$ 2.31 $= \Delta P$ in feet of head.

Pressure drop charts such as Figure 9–14 are useful for system design and verification of proper flow at initial system start-up. They may also be useful for troubleshooting but must be used with caution. For example, a heat exchanger with partially blocked water tubes will exhibit a higher pressure drop as indicated on the pressure drop chart, indicating higher flow, when actually the reverse is true. When in doubt, the tubes must be visually inspected for cleanliness.

Tech Tip

Refrigerant in water. If a water-cooled condenser of a high-pressure refrigerant system is leaking, then refrigerant is being put into the water, which will produce bubbles. To check the condenser, capture water being discharged and direct it to a pail or bucket. As the water exits the discharge, bubbles may be seen. To check for very small leaks, use an electronic leak detector and hold the detector close to the rising water level (do not allow the tip to contact the water). Refrigerant is heavier than air and will be detected at the surface of the water. Alternatively, the water could be drained and an electronic leak detector used to check for the presence of refrigerant. On low-pressure refrigerant systems, the system pressure would need to be raised using heat from the system—but not to more than 10 psig. Higher pressures might cause the rupture disk to blow.

The troubleshooting processes for heat exchangers are as follows.

1. *Check water flow*—determine if the (water) pressure drop through the heat exchanger is within the manufacturer's specifications.
2. *Check refrigerant flow*—based on superheat (evaporators) or subcooling (condensers).
3. *Leak check*—check for refrigerant vapor in the water; check purge unit run times.

Solenoid Valves

Solenoid valves (see Figure 9–16) are used for many different purposes in a refrigeration system. Here are a few uses for solenoids:

- Pumpdown
- Bypass
- Flow control
- Oil control
- Capacity control
- Purge unit suction
- Discharge control

The operation of a solenoid is either to open or to close the flow of fluid through the valve. In some cases a double-acting solenoid will open one port and close another when the coil is energized. Solenoid valves are referred to as normally open, normally closed, or double-acting. *Normally open* (NO) means that when the solenoid is deenergized (as if lying on the shelf), the valve is open and thus will allow flow. The opposite is true for the *normally closed* (NC) valve (see Figure 9–17). When this valve is deenergized, it is closed and thus will *not* allow

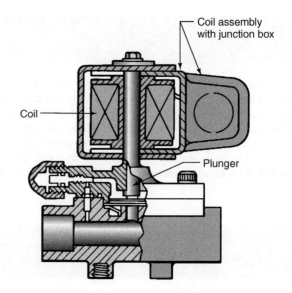

Figure 9–16
Cutaway view of a solenoid.

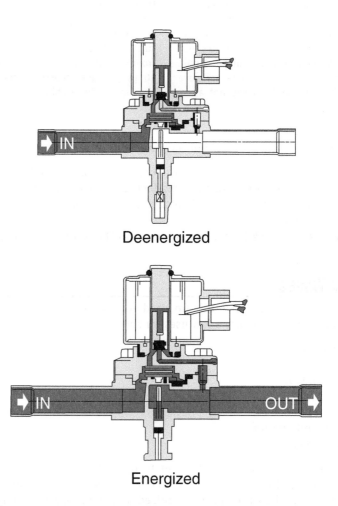

Figure 9–17
A pilot-operated NC solenoid valve in the normally closed position (deenergized) and in the energized position. The pilot valve is operated by the solenoid to move a piston in the main valve and open it for full flow.

flow through the valve. Double-acting valves have one port designated as NO and another as NC for the deenergized position.

There are two steps to determine if a solenoid valve is functioning properly. First, *check to see that it is operating.* A small screwdriver held close to the electromagnet will buzz when it encounters the magnetism induced by 60-Hz power. If the valve has a DC operator rather than an AC, there won't be a buzz but there will be a magnetic attraction.

Second, *check for flow through the valve.* Checking operation of valves can be done through a combination of different methods, which may include checking temperatures, pressures, sound, sight glasses, and system response.

Reversing and Diverting Valves

The purpose of a reversing or a diverting valve is to change the flow through components and heat exchangers. One example of this is the process of hot gas defrost, in which hot gas is used to defrost an evaporator, similar to what a heat pump does. In most cases, reversing valves (see Figure 9–18) are a type of solenoid, but some reversing valves are motorized. The same principles of operation for a solenoid (and the same troubleshooting steps) can be used for a solenoid-style reversing valve (see Figure 9–19). Many reversing valves are pilot operated; a small solenoid opens or closes to actuate the movement of a larger valve or piston.

Motorized reversing or diverting valves can also be a ball-style valve that rotates to change ports (see Figure 9–20). Motorized valves can be used for any number of applications. The motor can be operated, in some cases, to completely divert flow or to mix fluids of two different temperatures (see Figure 9–21). Such mixing of fluids allows the system to adjust in response to small changes in ambient conditions.

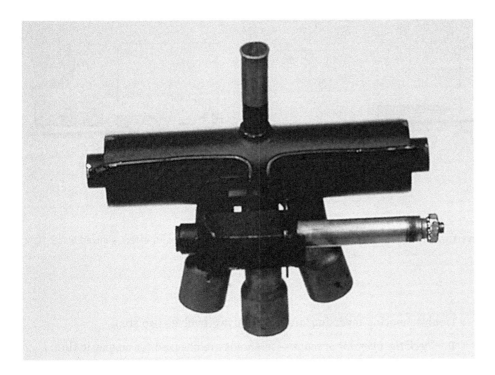

Figure 9–18
A four-way reversing valve. (Photo by Bill Johnson)

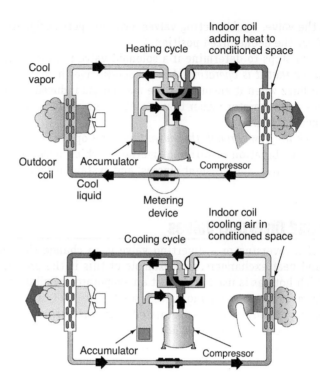

Figure 9–19
The heat pump refrigeration cycle shows the direction of the refrigerant vapor flow.
(Courtesy of Carrier Corporation)

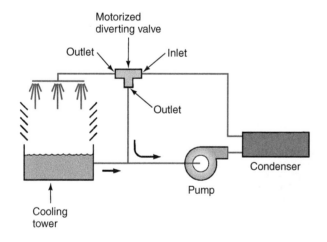

Figure 9–20
Motorized diverting valve being used to divert water flow
around a cooling tower.

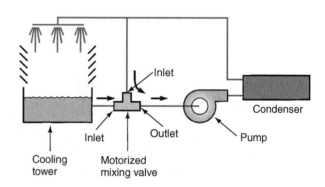

Figure 9–21
Motorized mixing valve being used to effect a mixed water flow
from a cooling tower.

Troubleshooting reversing/diverting valves involves two steps:

1. *Check the operator or motor*—solenoids are checked for magnetic flux;
   motors are checked for their position.
2. *Check for flow*—in the same way as checking a solenoid for flow,
   check via temperature and/or pressure readings or by inspecting a
   sight glass.

Capacity Control

Capacity control can be accomplished in many different ways, as outlined below.

- *Reducing refrigerant flow through the compressor:*
  - *Unloading valves* (electrical solenoid or pneumatic) that hold open one or more compressor intake valves in a multi-cylinder compressor.
  - *Operating more or fewer compressors*—that is, increasing or decreasing the number of operating compressors; the compressor bank may also have compressors of several different sizes that are operated to match capacity needs.
  - *Reducing suction vapor flow*—limiting or regulating the flow of suction vapor to the compressor using vanes or other means to partially obstruct vapor flow (see Figure 9–22).
- *Hot vapor bypass*—bypassing the compressor discharge vapor around the condenser directly to the inlet of the evaporator to raise low-side pressure and artificially load the system.
- *Liquid bypass*—moving liquid around the evaporator to decrease flow to the evaporator and thus maintain liquid flow to the expansion valve.
- *Limiting flow through the evaporator:*
  - Shutting off evaporator passes. Figure 9–23 shows a screw compressor with a slide valve, which is controlled by a load-limiting device.
  - Using multiple coils.
  - Employing suction pressure control, also known as evaporator pressure regulation; an EPR (evaporator pressure regulator) can be used for this purpose.

The EPR is set by monitoring the evaporator pressure. Adjusting the EPR will change evaporator pressure and temperature. The temperature is determined by the saturation pressure, as determined by the P-T chart for the refrigerant being used.

The CPR is set by monitoring the compressor pressure and/or the motor amperage. Adjusting the CPR will change the crankcase pressure (i.e., the compressor suction). The CPR is most often used to safeguard the compressor by limiting its crankcase pressure during system pulldown, which would otherwise be excessively high because of high load conditions. For this reason, the CPR is sometimes viewed as a refrigeration safety device.

Figure 9–22
The vanes inside the suction opening of a centrifugal compressor can be opened or closed to provide variable capacity control.

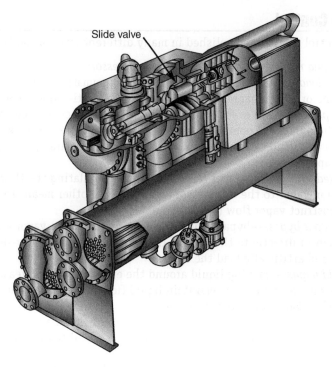

Slide valve

Figure 9–23
Screw compressor with slide valve.

Troubleshooting capacity control involves the following steps.

1. *Determine whether the capacity control is functioning properly*—this may involve taking electrical, temperature, pressure, and/or flow measurements. The control must be energized in order to engage and reduce capacity.

2. *Check for reduced capacity*—the temperature of a bypass system indicates how well it is operating as compared to manufacturer data. External piping and multiple compressors can be checked by measuring their temperatures. Unloading valves in the compressor can be checked by measuring compressor amperage.

Tech Tip

CPR or EPR? An EPR (evaporator pressure regulator) and a CPR (crankcase pressure regulator) look and work in nearly the same way. Both valves are located between the evaporator outlet and the compressor. The difference is that the EPR monitors evaporator pressure and regulates by preventing the evaporator pressure from getting too low, whereas a CPR monitors compressor crankcase pressure and regulates by preventing the crankcase pressure from getting too high and overloading the compressor. Another way of looking at these two valves is that the EPR is sensitive to inlet pressure and the CPR is sensitive to outlet pressure.

Temperature and Pressure Sensors

Mechanical sensors operate bellows, dials, springs, and other like components (see Figure 9–24). Electronic sensors measure changes in resistance or amperage of an electronic device mounted in or on the system (see Figure 9–25 and Figure 9–26).

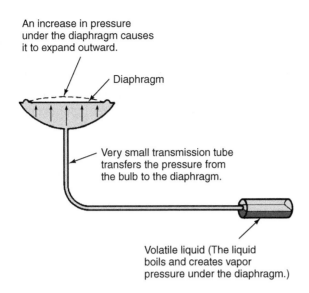

Figure 9–24
The liquid-filled bulb transmits temperature information to a diaphragm that presses on a set of mechanical contacts. A tube connects the bulb directly to the system, whose pressure moves the diaphragm.

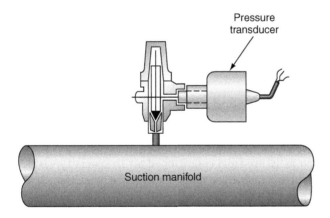

Figure 9–25
A pressure transducer measures pressure and then sends the information electronically to a controller.

Figure 9–26
An electronic temperature sensor (thermistor). (Photo by Bill Johnson)

Mechanical and electronic sensors can either respond to some event (e.g., turn on or turn off at a set temperature or pressure) or modulate (i.e., send pressure or temperature information to a controller that will determine operation).

Troubleshooting temperature and pressure sensors requires the following steps.

1. Determine whether the sensor is electronic or mechanical.
2. Check *electronic sensors* by confirming proper input (amperage or voltage) at the controller. Check *mechanical contacts* by:
   a. Voltage check with electrical power on
   b. Ohm check with electrical power off
3. Check for operation by changing pressure or temperature.

Tech Tip

Do not mechanically force contacts closed with screwdrivers or other tools to check pressure or temperature controls. Doing so will bend, warp, or distort the delicate mechanical functioning of the control. Always change temperature or pressure to check sensor operation.

Filter-driers

Commercial systems use many types of filter-driers (see Figure 9–27) and sometimes use several different types on one system. Filter-driers are used for:

- Removing moisture and debris from the liquid side of a refrigeration system before liquid moves into the metering device
- Removing moisture and debris from the suction side of a refrigeration system before vapor moves into the compressor
- Removing moisture and debris from the oil in a refrigeration system before the oil moves back into the compressor
- Removing moisture from the air of a pneumatic compressor used for pneumatically controlled air
- Removing moisture from conditioned air in specialty systems (test chambers), where air inside the chamber must be drier than the refrigeration system can control

Figure 9–27
Suction filter-drier with access to the removable core. Cores can be changed without disconnecting the filter-drier housing.

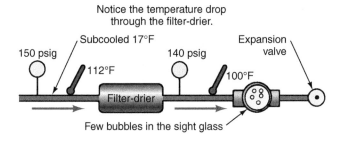

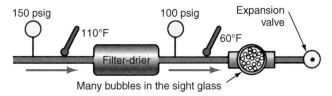

Figure 9–28
Filter-driers that have large pressure drops in the liquid line may exhibit a temperature drop and generate bubbles in a sight glass.

Special filter-driers containing activated charcoal can be used to help reduce acid in a system after compressor burnout.

Troubleshooting filter-driers involves the following steps.

1. *Checking moisture indicators* mounted in the system downstream from the filter-drier.
2. *Checking pressure drop across the filter-drier*—pressure drops are prescribed by the manufacturer, but generally there should be no more than a 2-pound pressure drop across a filter-drier (see Figure 9–28). If the filter-drier is in a liquid line, then excessive pressure drop will cause a temperature change and bubbles may show in a sight glass downstream of the filter-drier (see Figure 9–28 and Figure 9–29).
3. *Checking for moisture* if no indicators are installed (see Figure 9–30). This check may require the technician to test the refrigerant for moisture content. This test involves the use of a sampling tool to take a sample of the vapor oil and/or refrigerant and then determining the moisture content based on its color. Sampling kits can be obtained from a wholesaler.

Oil and Refrigerant Levels

Oil and refrigerant levels are generally easy to check. Sight glasses are placed at the level where liquid should be seen (Figure 9–31). If there is a single sight glass, liquid levels are read to the center of the sight glass. If there are marks on the sight glass (in with a tube sight glass) the level is read to the indicating marks. If two sight glasses are placed one above the other, then the level of liquid should be maintained between these two sight glasses (Figure 9–32). Some sight glasses have moisture indicators (Figure 9–30), which are used to determine if moisture is present in the system.

Mechanical and electrical level controls are also used (see Figure 9–33). Mechanical float valves and switches can be used to monitor liquid levels for refrigerant (Figure 9–34), oil, and water systems.

Figure 9–29
Suction-line filter-drier showing access ports for checking pressure drop and taking refrigerant samples. (Photo by Bill Johnson)

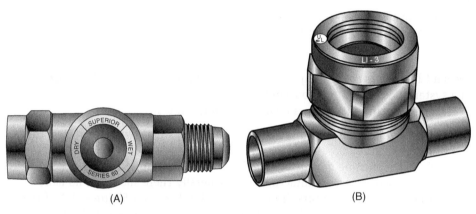

(A)   (B)

Figure 9–30
Sight glasses for liquid refrigerant: (A) the moisture-indicating element in the center changes color when encountering moisture; (B) a sight glass used to determine liquid/vapor content.

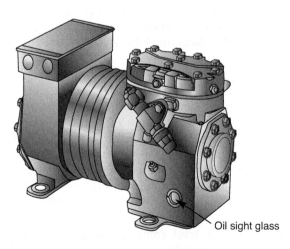

Oil sight glass

Figure 9–32
Oil reservoir with two sight glasses for checking the oil level.

Figure 9–31
A semihermetic compressor; the oil sight glass is in the lower right corner.

Tech Tip

Liquid-line filter-driers should have no measurable temperature drop during operation. The temperature drop indicates a partially restricted or possibly under-sized filter-drier that should be replaced and/or checked for proper sizing.

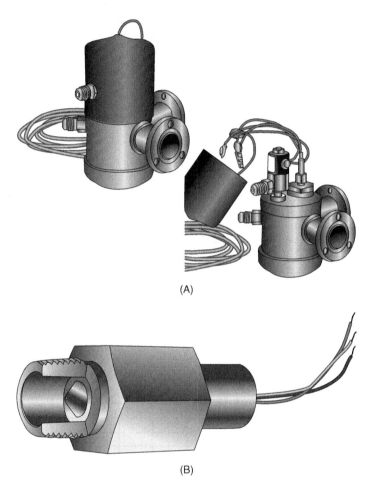

(A)

(B)

Figure 9–33
(A) An electromechanical oil-level regulator; (B) an electronic oil-level detector.

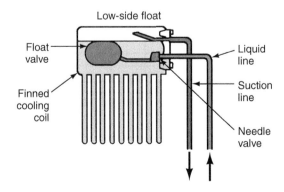

Figure 9–34
A float valve used in a flooded evaporator to control liquid refrigerant levels.

SERVICING AND MAINTENANCE RECORDS

Records of service and maintenance are used to help identify potential problems with the system. Most systems have daily maintenance and operations logs (see Figure 9–35). These logs are continually reviewed by the operator, who looks for changes in system operation. As logged conditions start to slowly change over a period of time, the operator can spot the change and plan to conduct maintenance operations to correct the problem. An example would be documenting a change in water temperature. The condenser water approach temperature will begin to spread (see Tech Tip on approach). An operator documenting and watching conditions will note this change and take action to correct the problem. Correction may include altering the water chemistry and/or cleaning the condenser tubes.

ROTARY SCREW LIQUID CHILLER LOG SHEET											CHILLER LOCATION _____ SYSTEM NO. _____				
Date															
Time															
Hour Meter Reading															
O.A. Temperature D.B./W.B.		/	/	/	/	/	/	/	/	/	/				
Motor	Volts														
Motor	Amps														
Cooler / Refrig.	Suction Pressure														
Cooler / Liquid	Inlet Temperature														
Cooler / Liquid	Inlet Pressure														
Cooler / Liquid	Outlet Temperature														
Cooler / Liquid	Outlet Pressure														
Cooler / Liquid	Flow Rate — GPM														
Condenser / Refrig.	Discharge Pressure														
Condenser / Refrig.	Corresponding Temperature														
Condenser / Refrig.	High Pressure Liquid Temperature														
Condenser / Refrig.	System Air — Degrees														
Condenser / Water	Inlet Temperature														
Condenser / Water	Inlet Pressure														
Condenser / Water	Outlet Temperature														
Condenser / Water	Outlet Pressure														
Condenser / Water	Flow Rate — GPM														
Compressor	Oil Level														
Compressor	Oil Pressure														
Compressor	Oil Temperature														
Compressor	Suction Temperature														
Compressor	Discharge Temperature														
Compressor	Filter PSID														
Compressor	Slide Valve Position %														
Compressor	Oil Added (gallons)														

Figure 9–35
Manufacturer's operation log sheet.

Tech Tip

Approach. When talking about heat exchange processes, we sometimes use the term "approach," and we often ask for the value when communicating with equipment manufacturers. Remembering the laws of thermodynamics, we know that heat always travels from hot to cold, and therefore in a perfect heat exchange process, one fluid would give up heat to the other until the temperatures of the two fluids were equal. This would be an approach of 0°F. The cost and size of heat exchangers would be prohibitive in practical applications; therefore they are designed with a limited design approach temperature, typically 10°F. In other words, approach is simply the measure of how close the temperature of the fluid being cooled approaches the temperature of the fluid absorbing the heat.

For example, consider a refrigerant system with a water-cooled condenser (see Figure 9–36). In this example, the "condenser approach" is found by comparing the refrigerant condensing saturation temperature (105°F) with the tower water outlet temperature (95°F): 105°F − 95°F = 10°F condenser approach.

The evaporator is also a heat exchanger (see Figure 9–1). In this example, we would compare the chilled water outlet temperature (45°F) with the evaporator refrigerant saturation temperature (40°F) to get a result of 5°F "evaporator approach."

Approach is also used to evaluate operation of cooling towers, although theirs is not a pure heat exchange process because the fluids (air and water) come into direct contact with each other and some of the water leaves with the air as a result of evaporation. "Tower approach" typically refers to the comparison of the temperature of the water leaving the tower and the wet-bulb (WB) temperature of the air entering the tower (see Figure 8–2). In this example, the tower approach would be 20°F: 85°F tower water outlet temperature − 65°F WB temperature = 20°F tower approach temperature.

Approach values on system heat exchangers can be a very useful tool for evaluating or troubleshooting system performance issues, especially when combined and compared with other current and historical system data. Factors that affect approach include system load conditions, system refrigerant charge, heat exchange surface cleanliness, fluid temperatures, and fluid flow rates. For example, consider a chiller with a water-cooled condenser exhibiting signs of high refrigerant head pressure. The condenser is not rejecting enough heat. Possible causes of poor heat transfer include:

1. Higher-than-normal tower water temperature into the condenser
2. Low water flow
3. Fouled or dirty heat exchange surfaces

If the tower water inlet temperature was determined to be normal, the approach temperature could be used to determine which of the two possible causes should be checked next. Clean tubes with a low water flow (pumping system problem) would exhibit signs of a lower-than-normal condenser approach, while normal flow and fouled heat exchanger surfaces would exhibit signs of a higher-than-normal approach temperature.

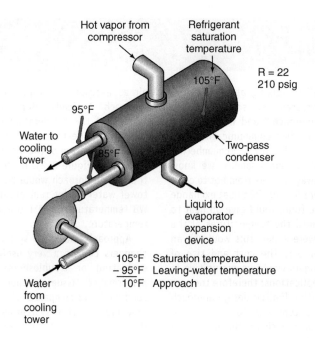

Figure 9–36
Approach temperature is shown for a two-pass condenser. Water leaving the condenser and going to the cooling tower is compared to the refrigerant saturation temperature.

SUMMARY

Chillers make up a large majority of the commercial refrigeration systems being used. There are three types of compressors: centrifugal, screw, and reciprocating. This chapter spent more time on centrifugal and screw compressors. Both of these systems are large and operate the compressor continuously. Large motors use large amperages. Various means are used to start compressor motors slowly, bringing them partially up to speed before full operation in order to reduce starting amperage needs.

Chillers are generally designed for central plant configurations, where one large chiller cools many floors of a large building or several buildings in a complex. The larger the central plant, the less (per ton) it costs to operate. Central plants can be very efficient in their operation and involve lower maintenance costs.

There are standard parts to all commercial refrigeration systems. Most of these can be isolated and removed from the system. For each component, there are some basic troubleshooting checks that need to be done. Most of these checks can be performed using the senses of touch, sight, hearing, and listening.

Service and maintenance records are critical for large commercial systems. Records can pinpoint problems that slowly affect the operation of the system. This gives the operator and the maintenance supervisor time to plan for corrective action.

REVIEW QUESTIONS

1. Explain why centrifugal chillers are different from other chillers.
2. Explain what distinguishes a screw chiller from other types.
3. Explain the benefits of a central plant system.
4. Describe one troubleshooting check for metering devices.
5. Describe one troubleshooting check for compressors.
6. Describe one troubleshooting check for heat exchangers.
7. Describe one troubleshooting check for solenoid valves.
8. Describe one troubleshooting check for reversing/diverting valves.
9. Describe one troubleshooting check for capacity controls.
10. Describe one troubleshooting check for sensors.
11. Describe one troubleshooting check for filter-driers.
12. Describe one troubleshooting check for liquid-level controls.
13. Explain why maintenance records are important for service and maintenance of equipment.

CHAPTER

10

Water Treatment

The student will:

- Describe the physical conditions of water
- Describe the treatments used to condition water
- Describe the reason for blow-down
- Describe the difference between manual and automatic treatment
- Describe when scheduled maintenance is performed
- Describe safety requirements related to water treatment
- Describe one troubleshooting technique for cooling water

INTRODUCTION

Cooling towers, chillers, and boilers all use water as a heat transfer medium (see Figure 10–1). Since water is not pure, it requires treatment. The treatment of water is a science, requiring knowledge of water, substances in the water, processes used to condition water, and chemicals used to treat water. *Water chemistry* is the term used to describe all of the conditions of a sample of cooling water. Water chemistry experts are people who specialize in the treatment of water.

Water used in an HVAC system is contained in either a closed-loop (sealed system) or an open-loop (open to the atmosphere) system. An example of a closed-loop system is the boiler, where water is contained under pressure; boiler water remains in the boiler as long as there are no leaks or repairs. An example of an open system is a cooling tower where water is sprayed into air to evaporate and cool the remaining water (see Figure 10–2). Open loops are open to the atmo-

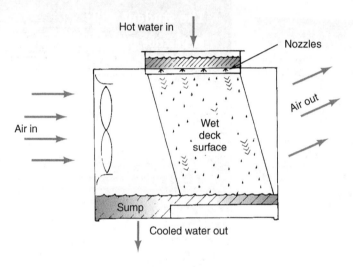

Figure 10–1
Cooling water comes in contact with ambient air to force the evaporation of water, cooling the water to be returned to the system.

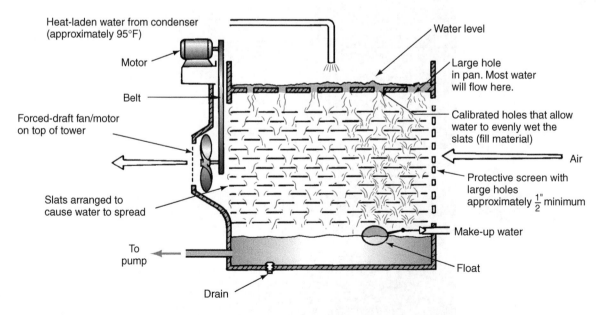

Figure 10–2
A typical cooling tower, showing water flow, make-up water, sump, and drain plug.

sphere, whereas closed systems are contained and generally under some pressure. Water in either of these two systems may be treated for any of the following conditions.

1. Lime, iron, and other mineral scale and deposits
2. Micro-biological growth, such as algae, bacteria, fungus and molds
3. Suspended solids accumulations, such as dissolved mineral deposits left from evaporated water, airborne dirt, and debris that is washed into the cooling tower water

Treatments are also available to minimize dissolved oxygen and control for pH, which will reduce corrosion of system materials.

This chapter explains the conditions of water and some of the common treatment that is applied to water for cooling.

Tech Tip

Hardness. A simple explanation of water hardness is the amount of minerals held in suspension or dissolved in a water sample. As water evaporates, it leaves behind deposits of minerals in the form of crusty, hard, material that builds up on surfaces (see Figure 10–3).

These deposits interfere with the operation of various components. Water with high concentrations of minerals is either treated or eliminated from the system, depending on the amount of minerals in the water.

Figure 10–3
Hardness collecting on the inside of a pipe.

Field Problem

A production plant was using a cooling tower to remove heat from a process. The tower was working effectively but seemed to be using too much water. A review of the heat rejection process showed that approximately 80% of the water was being evaporated—a rate consistent with the engineering specification. The rest of the consumed water was being used to blow-down the hardness accumulating in the water. The technician was asked to increase the cycle from the designed 4 cycles/day to 6 cycles/day. *Cycles* is short for "cycles of concentration," or the amount of hardness building up in the water during evaporation. Cycles of concentration summarizes the comparison between the levels of solids being recirculated and the levels in a sample of make-up water. If the recirculated water has 4 times the concentration of solids as make-up water, then we say it has 4 cycles of concentration. By increasing the cycles of concentration to 6, it was expected that a substantial reduction could be made in the amount of make-up water being used.

After testing a sample of water, the technician was able to determine that pH could be held constant but that the amount of solids was going to be a problem. Scale inhibitors to suspend the hardness could be added, but the concentration of inhibitor would have to be continually adjusted as the amount of hardness increased. Manual chemical treatment would require someone on duty 24 hours a day to monitor the water condition. Instead, automatic controls were prescribed when a sales representative was consulted. The controls were installed after engineers determined the correct fit. The new controls yielded 6 cycles of concentration, which produced savings for the company in make-up water.

Tech Tip

Cycle of concentration. This is a way of referring to the amount of solids or minerals being held in circulating water as compared with new (make-up) water. If the level of minerals is 4 times greater than in the make-up water, then the circulating water is said to have 4 cycles of concentration. If the water can be treated so that it holds greater amounts of minerals and solids, with less being deposited on tower parts, then it is reasonable to assume that less blow-down (concentrated) water will need to be made up by new water. The cycles of concentration are normally determined by the cooling tower representatives or water treatment personnel.

REASONS AND PROCESSES

Water is an abundant resource in some areas, but in other areas water is a precious commodity. Aside from its availability, water is not the same everywhere. Water carries particles; has dissolved solids; can be acidic or alkaline; and, when exposed to air, will absorb what's in the air. Water is necessary for all life, and this includes bacteria, mold, fungus, and algae. Because of what water may carry and the types of life it can support, water usually needs to be treated.

The process of treatment can be as simple as filtering or as complicated as marshalling chemicals and automatic systems to eliminate various problems. In many cases, water treatment requires the attention of a knowledgeable maintenance person. This person is charged with the responsibility of monitoring the water and adding chemicals as necessary. Sometimes this person is familiar with only one type of chemical or test, and the HVAC technician is often called when other problems occur. If the water condition continuously occurs (as with, e.g., hard water), then an automatic system can be used that will dispense the appropriate chemicals in the right proportion and on a regular basis.

CONDITIONS OF WATER

It is rare when water does not require treating, and water may have more than one condition to treat. The following sections explain typical treatments applied to control various water characteristics.

Hardness

The *hardness* of water refers to the amount of dissolved minerals (e.g., calcium and magnesium) that it picks up while being filtered through the ground; see Table 10-1. In the United States, 89% of all homes and businesses have hard water. The hardest waters are in the Southwest and the softest are in the Northeast. Exact measurements of mineral content or hardness are made using a *wet titration* process in which chemicals are added to the water until the solution changes color. The color is then compared to a scale that assesses hardness. A simple test is the *soap test,* where dish soap is added (a few drops at a time to a glass of water) to water and shaken. The amount of bubbles produced indicates the level of hardness. The more soap that is needed to produce bubbles, the harder the water. More bubbles (or less soap) means softer water and therefore fewer dissolved solids.

If the water is too hard, water softening may be required. Sometimes dissolved solids can be precipitated out of solution (dropped out) by using a chemical. Sump water is then blown off periodically to remove the concentration of solids. For these types of systems, the automatic bleed and feed is generally controlled by conductivity meters, which are calibrated on a scale that is based on the amount of total dissolved solids allowed in the system. Other cases may require the use of salt-based or RO (reverse osmosis) systems.

pH

Water that is neither acidic nor alkaline has a pH of 7 and is sometimes referred to as "neutral" water. The unit "pH" stands for "potenz of Hydrogen." Potenz simply means potential, so pH is the potential concentration of hydrogen ions in any solution. When the number of hydrogen (H) ions equals the level of hydroxide (OH) ions, the solution is neutral and has a pH value of 7.

Neutral water is usually preferable because it is neither alkaline nor acidic. Although a certain level of acidity does prevent certain types of growth, water that is too acidic will eat away at tubing and metallic components of the system. Acidic water can be neutralized by adding an alkaline, and vice versa. Swimming pool water is a good example. If the water test is alkaline, then hydrochloric acid (muriatic acid) is added to bring it back to neutral; if the water test is acid, then sodium hypochlorite is added to bring it back to neutral. This is the same process

Safety Tip

Use acids and alkaline (caustic) solutions with care. Both acidic and alkaline chemicals can cause severe burns. Always use and apply these chemicals using the PPE (personal protective equipment) listed on the MSDS (material safety data sheet). Typically, this means goggles, rubber gloves, and a face shield.

Table 10-1 Hardness Scale Corresponding to Readings from a Hardness Test Kit

Hardness	Measurement of Calcium		
Soft	0–20 mg/L	or 0–0.0027 oz/gal	or 0–0.04 grains/gal
Somewhat soft	20–40 mg/L	or 0.0027–0.0053 oz/gal	or 0.04–0.08 grains/gal
Slightly hard	40–60 mg/L	or 0.0053–0.0080 oz/gal	or 0.08–0.12 grains/gal
Somewhat hard	60–80 mg/L	or 0.0080–0.0107 oz/gal	or 0.12–0.17 grains/gal
Hard	80–120 mg/L	or 0.0107–0.0160 oz/gal	or 0.17–0.25 grains/gal
Very hard	>120 mg/L	or >0.0160 oz/gal	or >0.25 grains/gal

Table 10–2 pH Levels

Fluid Type	pH
Stomach acid	1.5
Cola	2.5
Beer	4.5
Urine	6
Neutral (pure water)	7
Sea water	8
Hand soap	9
Household ammonia	11
Chlorine bleach	12.5

More acidic
Neutral
More alkaline

Tech Tip

Alkalinity. Water with a pH value higher than 7 is considered alkaline (the opposite of acidic). Alkalinity reflects the amount of dissolved carbonates or bicarbonates in the water that can produce CO_2. Other substances that can cause alkalinity include borate, hydroxide, phosphate, silicate, nitrate, and dissolved ammonia. Be sure to listen to how the term "alkalinity" is used. Other terms for alkaline solutions are "bases" and "caustics," which may not be technically correct. Table 10–2 is a practical reference for pH values.

that is used to bring any water to a desired pH level with the chemicals prescribed by manufacturers and water experts. Always read and understand instructions for all chemicals and application methods.

Chlorination

Most city water is chlorinated in order to kill bacteria before the water is dispensed in municipal systems. If this water is being used for HVAC systems, it can control some types of slime without additional treatment. However, chlorine is an oxidizer and thus can react with some system components. Chlorine can be used in conjunction with other chemicals to disinfect water, but the proportions of these other chemicals must be monitored and adjusted so that system components are protected. Chlorine is corrosive, so high concentrations of chlorine may require the addition of corrosion inhibitors.

TYPES OF TREATMENT

Water treatment is a science that is continually evolving with the development of new technologies, products, and procedures. Some of these may become widely adopted. This section will discuss some of the popular products and procedures being used. It should be noted that all of the treatments discussed here are applied only after the water has already been treated for pH and hardness. Once the water is prepared for those two conditions, it is ready for further treatment.

Biocides and Legionella

A biocide is a chemical that kills bacteria. One biological agent that is of particular concern is *Legionella pneumophila,* the causative agent in Legionnaires' disease. Legionella can grow in stagnant water that ranges in temperature from 69 to 120 degrees Fahrenheit. It was first discovered when many people attending the 1976 Legionnaires' convention in Philadelphia became sick with flu-like symptoms. More than 30 people died. A cooling tower was found to be responsible: its vapors were being pulled into fresh air intakes.

Tech Tip

Biofilm. Biofilm is a slippery coating that forms on the surfaces of cooling towers and may also be seen as a film on top of the water in a cooling tower basin. If it is allowed to grow, it colonizes into mushroom shapes and pillars, which may clog pipes and foul equipment. If biofilm is felt or seen, the water chemistry expert should be notified immediately.

Safety Tip

Biofilm can be harmful to human health. It is involved in 65% of all human bacterial infections, and it plays a role in such medical conditions as ear infections, urinary tract infections, kidney stones, and gum disease. Always wear gloves and goggles in the presence of biofilm, and be sure to wash your hands after contact with any water used in cooling equipment.

There are two classifications of biocides: oxidizing and nonoxidizing. Oxidizing biocides destroy bacteria by stripping hydrogen ions from cell walls. Nonoxidizing biocides are organic compounds that are toxic to organisms, including human beings and animals.

Common oxidizers include:

- Chlorine products
- Bromine products
- Mixed oxidants of chlorine and bromine products
- Ozone
- Peroxide

Common nonoxidizers include:

- Trisnitro (hydroxymethyl nitro)
- Methylene bisthyiocyanate
- Quats
- Glutaraldehyde
- Isothiazolin
- DBNPA (dibromo-nitrilo-propionamide)

Biological organisms will build up immunity to one biocide at a time, so alternating biocides is recommended for water treatment programs. Additionally, since the amount of biocide in the water must be maintained at the designated amount for a 3–4-hour period, the biocide should be added during off-peak hours so that the bleed water can be shut down without causing a spike in its hardness or conductivity.

Safety Tip

Biocides are extremely dangerous. They can harm the person handling them in addition to plants and animal life not intended for extermination. Do not open a container or use application equipment unless you are completely knowledgeable about the use of the specific biocide. Special certification is recommended (and may be necessary) before handling these materials.

Tech Tip

Certification. The use and application of biocides may require that the individual applying them be certified. It may also be necessary to obtain certification of approval from state and local departments of environmental quality. Before opening a biocide container, dispensing any of the product, or applying it to any HVAC system, be sure that you have obtained the necessary training, certification, and authority to use the product.

Foaming

Water, by itself, does not foam; it is agents in the water that produce foam. As the water becomes more neutral in pH, soaplike solutions will tend to foam when the water is pumped or sprayed in a cooling process. Soaplike solutions can form in many ways:

- Incorrect pH (high alkalinity)
- A system leak into the water
- Soap residue from a cleaning process
- Cellular residue from destroyed bacteria (the result of using biocides)
- Interaction with or response to the use of algaecides
- Too many chemicals being added to the water
- Excessive suspended solids (e.g., dust from chemical or agricultural sources and dirt from fields, roads, and mining operations)

Foaming can damage equipment, ruin paint, and provide an inhalation point for bacteria (such as Legionella). Foaming can be reduced or eliminated by adding antifoaming agents or simply by blowing down the cooling water system to eliminate concentrations in the water.

Field Problem

Dispatched to a cooling tower that was foaming, the technician could see the problem as he pulled into the employee parking lot next to the building that had the cooling tower on top. Foam was blowing from the tower and descending like snow, covering the tops and hoods of all of the vehicles in the lot. This was the worst case of foaming he had ever seen.

Talking with the owner, the technician suggested that the employees be told to wash their cars to remove the foam—the foam could be caustic to the paint. Furthermore, he advised that the parking lot be vacated until the foaming condition could be brought under control.

The technician made a mental list of the conditions that could cause foaming:

- Alkalinity is too high
- Surfactants (chemicals) are at too great a concentration
- The chemical feeder is overfeeding
- Too many cycles of concentration and/or too many solids
- Microbacterial growth or biofilms
- Contamination from another process

He checked the chemical feeder and found that it was still full, ruling out this possibility as the problem's cause. He then looked for biofilms but did not see anything growing in the sump or on the fill material in the tower; hence, he also ruled this out as a possibility. Conversations with the owner and maintenance personnel indicated that no one had added any chemicals to the water, so he ruled out surfactants for the time being. If the alkalinity or the cycles of concentration were too high or if there were contamination, then he could reduce the foaming by blowing down the system, although this would not be the same as finding and correcting the problem. To find the problem would require a water sample. He took the sample in a clean container and then blew down the sump. The foaming stopped. The technician took the sample back to the shop, where his supervisor arranged for it to be delivered to a water analysis expert. The technician and the building owner would have to wait until the test results came back, but in the meantime the tower was being watched for signs of foaming.

Scaling and Corrosion

Scaling occurs when the water has too much mineral content (high mineral concentrations or hardness). Water with higher levels of hardness will form more scale and deposit scale faster on surfaces. A standard water heater is a good example of a device that is subjected to large amounts of raw water and has trouble with scale and corrosion. A sacrificial rod (an anode) is installed in all water heaters to help minimize corrosion of the functional tank components. As scale deposits grow in the bottom of a water heater, they block heat transfer. For this reason, water heaters need to have the scale buildup removed periodically.

An *anode* is a positive connection. The anode (negative terminal) of a battery collects electrons from the cathode (positive terminal). In the same way, an anode will allow electrons to be stripped, thereby neutralizing molecules that might interact with other metals in a water-based system. The electrons are in the form of ions, which will bond with molecules that have a net positive ion charge. The anode must be electrically more active than the metal that is being protected. For example, zinc is used as an anode to protect iron when dissolved oxygen in the water might attack iron piping. Magnesium or aluminum alloys are used to protect metals that are in contact with sea water. Outdrives of some boats will have an alloy fin close to the propeller—a sacrificial anode that protects the outdrive. In order for the anode to work, there must be an electrical connection between the anode and the protected metal. Direct contact, a wire, or a ground strap must connect the anode and the metal. If the electrical connection is lost (e.g., because of corrosion), then the formerly protected metal will be attacked.

The sacrificial rod (anode) is being "reduced" when electrons are stripped and the metal moves into solution to bond with positively charged molecules and form new compounds. When this reduction occurs at the anode, the protected metal (cathode) tends not to oxidize (rust or corrode). Dissolved oxygen carries a positive charge and wants to bond with another negative molecule. If a zinc anode is present, the oxygen will bond with the zinc and strip some of the metal away. Zinc is galvanically more active than iron. Without the rod, the oxygen will bond with iron and form rust. Allowing the zinc to reduce saves the other metals from rusting.

Tech Tip

Dissolved oxygen. New water added to a system contains a large amount of dissolved oxygen. If left in a system, the oxygen will start to corrode or rust system components. So, for example, when new water is added to a boiler, the dissolved oxygen must be driven out of solution by operating the boiler.

Whenever water undergoes temperature changes, more of the dissolved minerals precipitate out (fall out) of the solution to form sludge and scale. The higher the temperature difference, the more scale forms unless water is conditioned properly. Scale on the surfaces of heat exchangers impedes heat transfer and reduces efficiency. If raw water enters a system without being treated then, depending on the amount of water and degree of hardness, scale will form. Often make-up water is delivered to the system from water softening equipment designed to lower the hardness and thus minimize scale. If scale does build up, it can be costly to clean. The best, long-term solution is to condition the water before introducing it to an HVAC system.

Tech Tip

Corrosion. Corrosive conditions are the opposite of those that produce scaling. Corrosive water contains dissolved oxygen, is highly acidic or highly alkaline, and causes metallic parts of a system to be attacked. Corrosion is the oxidation or chemical reaction of metals. Metal rusting is a form of corrosion caused by oxidation. High concentrations of chloride (alkaline) and acids can pit and eat through metal surfaces. Dissolved oxygen can cause galvanic (reduction) reactions.

Algaecide

Algae (see Figure 10–4) is a green, slimy, stringy growth that can occur on exposed parts of a cooling tower. It is made up of a diverse set of plants that use photosynthesis (sunlight to produce sugars) to grow. Transported as spores, algae can be brought into cooling tower water in many ways. Airborne spores typically enter the water where it is exposed to the atmosphere.

Algae can be controlled two ways. It can be controlled with the use of chemicals, in particular biocides (called algaecides in this context). Algae can also be controlled by reducing one thing that algae needs to grow: sunlight. Upper decks and sides of cooling towers that are exposed to sunlight provide an environment that is conducive to the growth of algae. Enclosing these parts of a cooling tower helps to reduce or eliminate algae growth.

Magnetic Treatment

Magnetic water treatment has been the subject of much discussion. Some experts claim that it works; others claim that it does not. What we know for sure is that chemicals do work. Here we present both sides of the debate on magnetic water treatment.

Proponents: The claim is that water-borne minerals in solution can be affected by multi-pole magnets whose polarity may change during operation. This magnetic action either holds the mineral in solution or precipitates it out of solution at the magnetic source (depending on the manufacturer's design).

Opponents: Magnets have long been thought to affect mineral suspension and scale prevention; one mythical account describes lodestones (natural magnets) being used as weights in cast-iron pots of soup in the 1800s in order to reduce the hardness. However, tests of magnet water treatment have not confirmed any such effect. Several agencies have conducted tests of magnet water treatment systems and found little or no difference between the treated and untreated water.

Figure 10–4
Algae in a glass of water.

Ozone Treatment

From a chemical standpoint, ozone consists of three atoms of oxygen bonded together (O_3). Recall that oxygen normally bonds as O_2. Ozone can result naturally from electrical discharge (lightning) and from the sun's ultraviolet rays. Ozone is an oxidizer, so it bonds with hydrogen ions in the cellular walls of bacteria and other living organisms. The reaction is similar to that of other oxidizers.

Ozone generators can be electrical or ultraviolet. Air is pumped through the generator to create the ozone that is sent to the system, injected as bubbles. The bubbles of ozone react with the molecules of water, releasing oxygen and allowing it to combine with hydrogen ions.

Safety Tip

The limit for ozone exposure is 0.1 ppm (parts per million) averaged over an 8-hour day. Ozone levels as low as 0.01 ppm can be detected by smell, which is described as a sharp sensation. Sufficient concentrations of ozone irritate the eyes, dry the throat, and may induce coughing. Too much ozone exposure can be fatal.

Freeze Protection and Glycol

Antifreeze is a treatment solution for systems that must continue to operate during the winter. Under these conditions, many systems can operate without the operating fans normally used for system cooling. Whenever water or piping containing water is exposed to temperatures below 32°F (0°C) there is a potential for freezing.

There are many different applications of glycols for freeze protection in HVAC systems (although freeze protection does not require that glycols be used; see Figure 10-5). There is also a distinction made between freeze protection and burst protection, where the difference concerns the particular fluid and whether it needs to be pumped. If the system is turned off and the cooling water solution is stagnant and more viscous, then less glycol needs to be added to the system. In this case, the owner needs burst protection to prevent pipes from bursting during cold weather. If the system is not turned off (because it must continue to operate), then more glycol is necessary.

Propylene is one of two different types of glycol that can be used for cooling systems (see Table 10-3). If the system is self-contained and does not drain or need to be drained periodically, then ethylene glycol can be used (Table 10-4). If the system is open and is drained or blown down, propylene glycol is preferred. The differences between the two glycol types can be summarized as follows.

- *Propylene* glycol is a chemical of relatively low toxicity that is manufactured in many varieties for many applications, including human ingestion. Purchased specifically for HVAC purposes, it is marketed as an environmentally safe antifreeze product. This product can be added to open cooling systems and drained to municipal waste systems.

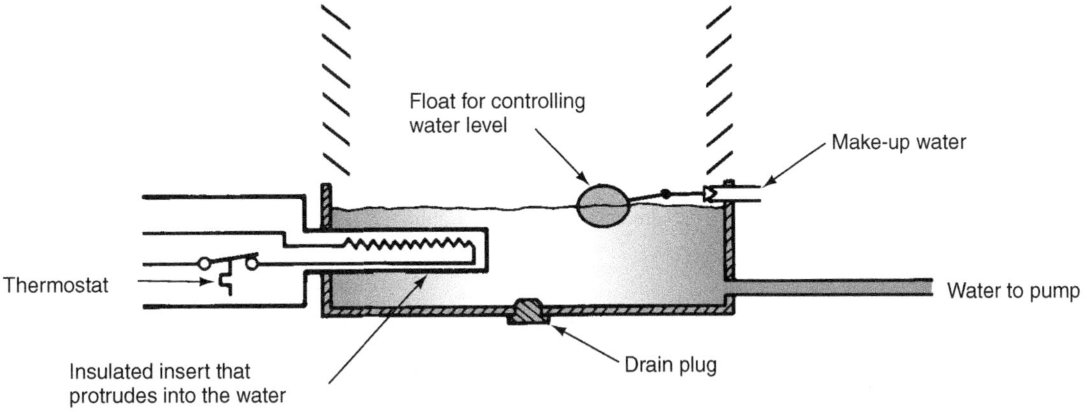

Figure 10-5
Alternative freeze protection using an electric heater instead of glycol solutions.

Table 10–3 Chart of Generic *Propylene* Glycol Concentration in Water

Temperature	Burst Protection	Freeze Protection
20°F (−7°C)	12%	18%
10°F (−10°C)	19%	27%
0°F (−18°C)	24%	35%
−10°F (−23°C)	29%	42%
−20°F (−29°C)	31%	47%
−30°F (−34°C)	34%	50%
−40°F (−40°C)	36%	52%
−50°F (−46°C)	37%	54%
−60°F (−51°C)	38%	57%

Note: Always check the specifications for the propylene glycol that you are using.

Table 10–4 Chart of Generic *Ethylene* Glycol Concentration in Water

Temperature	Burst Protection	Freeze Protection
20°F (−7°C)	12%	17%
10°F (−10°C)	18%	26%
0°F (−18°C)	23%	35%
−10°F (−23°C)	27%	41%
−20°F (−29°C)	31%	46%
−30°F (−34°C)	31%	50%
−40°F (−40°C)	31%	55%
−50°F (−46°C)	31%	59%
−60°F (−51°C)	31%	63%

Note: Always check the specifications for the ethylene glycol that you are using.

- *Ethylene* glycol is the version used in automotive applications (i.e., radiators). It is toxic and so cannot be ingested. The danger is that it has a sweet taste, which attracts animals and young children. Ethylene glycol cannot be drained to municipal waste systems and must be treated as a hazardous chemical; this requires that it be taken to a recycler.

Tech Tip

It is important to know what type of glycol (ethylene or propylene glycol) is in a system prior to adding more if needed to restore concentration level. When in doubt, a sample should be taken and analyzed by a water testing laboratory. It is also good practice to post signage at system fill points that indicate which type of glycol is in the system.

Here is a convenient formula for estimating the heat transfer effect of water:

$$\text{BTUH} = 500 \times \text{Flow (gpm)} \times \text{Change in temperature (°F)}.$$

The fluid factor (500) comes from the weight of the water multiplied by the specific heat of the water multiplied by 60 minutes in an hour: the weight of a gallon of water at 8.33 lbs × 1 (the number of BTUs required to raise 1 lb of water 1°F) × 60 minutes = 499.8, which rounds to 500. By adding glycol to water, we decrease

Tech Tip

Whenever glycols are added to pure water, the characteristics of the solution change: there are changes in heat transfer, fluid viscosity, and specific weight. All of these may require that the system be reevaluated with respect to the solution's characteristics to determine whether the new solution will work in an existing system. Do not add any glycol to a system unless you first check with an authority on solution mixes. Additives in water generally decrease the ability of water to transfer heat. Adding glycol to water changes the amount of water needed to transfer the design load.

the specific heat of that mixture and increase its viscosity, which decreases flow rate. It is important to understand that glycol solutions cannot transfer the same amount of heat as water can without an increase in flow (see Figure 10–6). All things being equal, a glycol solution will result in a decrease in heat transfer capacity over water, and glycol should be added only after a complete analysis of the system's capacities and application has been conducted.

Softening

Water can be softened (i.e., have its hardness removed) by using one of two processes. The oldest and most common is salt and zeolite. Zeolite beads act like a filter, removing the hardness as the water flows through it. When the beads become plugged, the zeolite is expanded with a salt solution, which allows the hardness to be washed out. After washing, the zeolite is rinsed and placed back into service.

The second method involves the use of a membrane. In this process, which is known as *reverse osmosis,* water is forced through the membrane to separate the minerals from the water (see Figure 10–7). When the membrane becomes plugged, water is forced in the opposite direction to dislodge the minerals, cleaning the membrane.

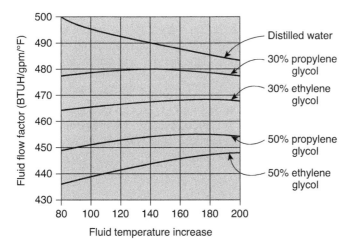

Figure 10–6
At any given temperature, the fluid flow factor is lower for glycol solutions than for water. This means that more fluid must be moved through the system in order to maintain the same amount of heat transfer in BTUH.

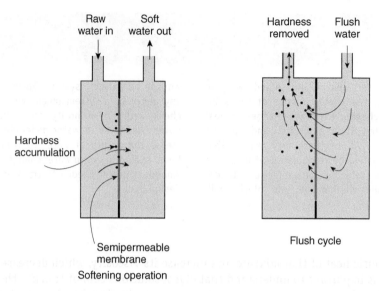

Figure 10–7
Reverse osmosis uses a membrane to remove hardness from the water.

BLOW-DOWN

Blow-down, drain-down, and bleed water are terms generally used to describe the draining of a portion of the water in a cooling system when concentrations of minerals, debris, and other materials become too high. *Turbidity* describes the extent to which water contains suspended material. Water that is turbid is murky and cannot be seen through easily. In this case, blowing down water with the highest concentration of suspended solids is necessary (see Figure 10–8).

Blow-down can create environmental issues (when the water contains biocides and other chemicals) and will increase costs associated with water usage and disposal. Water that is blown down to a municipal waste system must conform to the municipal specifications. If the water does not conform, it must be contained and shipped to a recycler capable of removing the objectionable substances before the water can be sent to the municipal waste system. This adds to the blow-down cost.

The water that is required to replace the blow-down water must be treated, entailing additional costs and time (see Figure 10–9). If raw water is allowed to enter the system, there must be enough chemicals to treat the new water and maintain concentrations for existing water. If the water must be conditioned before entering the system, additional water softening and treating systems may be required. Raw water added to the system will increase the cost of system operation.

For each system, a balance must be found between the need to blow-down system water and the need to contain and treat the water chemically. There is no magic formula that will work for every system. Each system condition must be evaluated for use of necessary chemicals and the manual or automatic processes necessary to keep cooling water clean.

MANUAL AND AUTOMATIC CONTROL

Manual control generally means that chemical levels are checked periodically and adjusted manually. This requires that someone go to the cooling water loca-

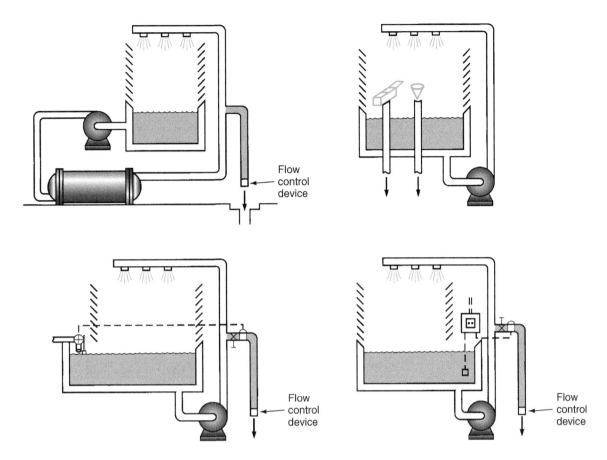

Figure 10–8
Four methods that are used to blow down cooling water systems. (Courtesy of Nu-Calgon Wholesaler, Inc.)

tion and physically check the water chemistry. After the chemistry is checked, sufficient chemical is added to bring the water back to specifications for the application.

Manual control can also mean control that can be achieved through remote digital communication with the chemical metering equipment (see Figure 10–10). In "manual" mode, the operator can add specific amounts of a chemical by sending a digital signal to the metering pump.

Automatic control generally means that some type of timing system is used to meter a small amount of chemical on a regular basis. Automatic control is also associated with a new class of digital monitoring and control systems. Sensors are able to detect many different water conditions and feed this information back to the controller. The controller will activate a metering pump to dispense chemicals whenever the water conditions dictate. Automatic control sensors can detect:

- pH
- Hardness
- Bioactivity
- Conductivity
- Oxidation/reduction potential
- Corrosion
- Turbidity (suspended solids)
- Biofilm

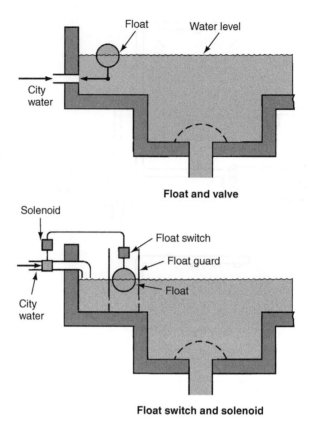

Float and valve

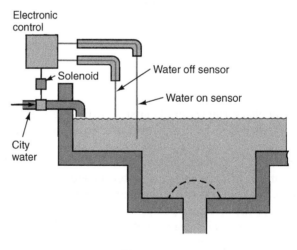

Float switch and solenoid

Electronic water level control

Figure 10–9
Three types of make-up water control used to balance blow-down water.

SERVICE TECHNIQUES

When working on cooling water systems, the technician should observe all aspects of equipment operation, system condition, and the quality of the cooling water. A problem detected early on can save a great deal of time and money. If a problem is caught early enough, there may be time to plan a repair, modification, or water treatment. Caught too late and the problem becomes a crisis—diverting attention, time, and money from other important operations because cooling equipment is off-line.

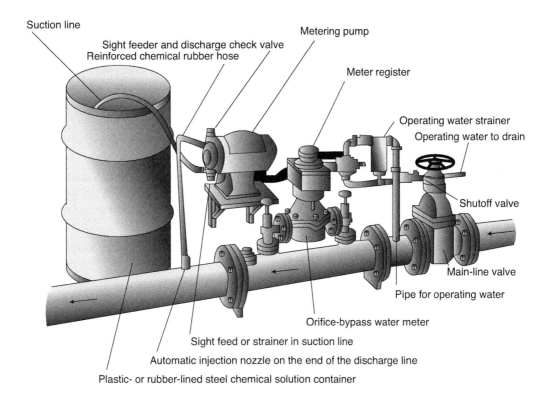

Figure 10–10
Chemical metering system that adds chemicals to cooling water. Metering of the chemicals is based on the flow rate.

If you're on a service call for a system that uses cooling water, take time to inspect the system and its condition; this should tell you if the cooling water needs to be tested. Table 10–5 lists what a technician can do while working on a system that uses cooling water.

The listing in the table is not exhaustive. The most important thing to remember about service techniques is that there are people ready to help. Manufacturer representatives are prepared to advise, assist, and suggest the best service technique for a particular cooling water system. If you are unsure, the best policy is to consult with another expert. That expert could be in your service company, at the wholesaler, or with the manufacturer. Your job is to do the job correctly the first time. Don't be afraid to ask for help.

SCHEDULED MAINTENANCE

Scheduled maintenance for systems with cooling water is performed at several intervals: weekly, monthly, every 3 months, semi-annually, and yearly. See Table 10–6 for details.

SAFETY AND TROUBLESHOOTING

All of the chemicals used to treat water are cause for safety concern. The safe use and dispensing of these chemicals is required at all times. Accurate records of the dates of application and the amounts are necessary for data purposes. If there is any question about air or water quality, these records will be important to answer questions about the system and the cooling water condition. Be sure to read and understand the material safety data sheet (MSDS) for each chemical.

Table 10–5 Inspecting Systems with Cooling Water

System Condition	Look for	Do This
pH is too high	Look for pH to be neutral (7 pH) or close to neutral; the system's pH level should be checked on a regular basis, which is considered a best practice	Use standard litmus paper to check for pH; if the pH level is incorrect, recommend that a pH titration be conducted and the pH adjusted
Pipes are showing buildup of hardness	Look for the chemical pump and determine whether it is working	If the pump is not working or you are unsure, contact the pump or tower manufacturer
The wet surfaces of the system look rusted or corroded	Look for evidence that corrosion inhibitors are being used to treat the condition	If no corrosion inhibitors are being used, discuss your observations with the owner; suggest that a water test and a system surface evaluation be conducted
The water looks murky; high turbidity	Look for the blow-down system and determine whether it is working	Operate the blow-down system to clear the water
Water level is too low	Look for the water level control	Operate the water level control to add water; then check the control for operation
Algae is present	Look for gaps or missing shields, demisters, and louvers	Discuss with owner and recommend replacement parts or that shades be installed to cover the water exposed to sunlight
Leaks are observed	Look for other rust or corrosion points and missing, rusted, or corroded fasteners	Record the location of the leaks and potential leaks; recommend that the system be drained, leaks repaired, and system components have protective coatings applied
Water or vapor escaping the control points (splashing, foaming, misting, etc.)	Look for plugged nozzles, missing baffles, missing or broken fill, and incorrectly installed parts	Unplug or replace nozzles; order and install missing baffles or fill; and correctly install parts

Table 10–6 Scheduled Maintenance

Weekly	The system is inspected for growth. Visual inspections for algae and bacteria are conducted. Water pH may also be checked.
Monthly	The system is visually checked for water flow. Clogging or dry spots where the flow is not covering an area are noted. Buildup of dirt or sediment in the sump or pan is recorded. Debris is removed and the system may be blown-down if turbidity is high. Operation of the water make-up (float) valve is observed. Screens (Figure 10–11) are inspected for obstructions and are cleaned.
Every 3 months	Pumps, bearings, fans, and any rotating equipment are checked for lubrication and operation. Water flow may be measured to determine if there is any flow reduction.
Semi-annually	The system may be cleaned and disinfected. Samples may be cultured to test for Legionella. All filters and strainers are cleaned, and the media may be pulled and cleaned or replaced as needed (see Figure 10–12).
Yearly	The system is inspected for potential leaks, corrosion, missing or rusted fasteners, and damaged system parts. The system is drained and all screens are cleaned. If the system is to be shut down, drain plugs are pulled and left out of the system. The system is cleaned and the protective coating is inspected. When the system is clean and dry, a new protective coating can be applied to the entire system or spot repair may be conducted.

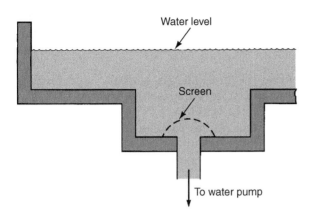

Figure 10–11
Location of screen in the basin of a cooling water system.

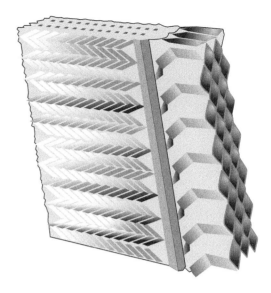

Figure 10–12
Tower fill material used to increase the surface exposure of the water.

Wear all required personal protective equipment (PPE), which may include rubber gloves, goggles, mask, and respirator. The MSDS sheet will specify the PPE that is necessary.

When troubleshooting cooling water, the following basic tests should be used. These tests are not the only ones, nor will they be conclusive in all cases. When in doubt, ask for a water expert to analyze the water.

1. *Always check the pH of the water first.* The pH must be close to neutral (7) in order for the other chemicals to function properly.
2. *Check for scale or foreign substances* circulating in the water (turbidity). This could mean one of two things:
   a. There is insufficient blow-down.
   b. There is a deficiency in the chemical treatment.
3. *Check for hardness.* This test may confirm that the problem involves turbidity, indicating that the system needs to be blown down.
4. *Check for biofilm.* This will be an indicator that biocides are not working or that the dosage rate needs to be increased.
5. *Check for algae growth.* Growth could be a result of:
   a. Missing sun shields, baffling, or demisters that are exposing the water to sunlight
   b. A biocide application problem

Safety Tip

Bacteria. Remember that cooling water may contain bacteria and sources of maladies such as Legionnaires' disease. Do not touch your mouth or eyes after touching the water, since bacterial and chemical agents can easily enter the body at these points. Do not breathe the vapors of evaporating water. The vapors can carry biological agents.

SUMMARY

Water treatment is a science. It involves the understanding of chemicals, the behavior of water, and the condition of water as it goes through a season of cooling. It also requires an understanding of the types of weather conditions and seasonal factors that affect water. Cooling water in an open loop is subjected to all of the conditions weather brings. Anything that the wind can carry will end up in the cooling water.

Water is kept in balance by using chemicals and mechanical procedures that maintain the water's effectiveness and protect the cooling water system. The balance is best maintained by keeping the water in the system as long as possible and blowing down water as infrequently as possible. The cost of treating new (raw) water is one of the factors in maintaining a balance. New water must be conditioned to the same level as existing water, which will usually require additional treatment or chemicals.

Water conditioning also requires the use of safe practices. Safety is necessary for the technician, the equipment, the environment, and the plants and animals that may encounter the cooling water. The technician must avoid personal injury as well as the environmental degradation that would result from spills and misapplication of chemicals.

REVIEW QUESTIONS

1. Describe the physical state of water that creates an acidic condition.
2. Under what condition is calcium high? How is this detected?
3. What is Legionella, and why is it a problem?
4. What is biofilm? How can it be detected?
5. Describe the treatment of water for high pH.
6. A cooling water system needs to be blown down. Describe two conditions that might have made this procedure necessary.
7. What is the difference between manual and automatic water treatment?
8. According to the maintenance schedule, how often should the system be drained and refilled?
9. Why is safety important for the environment?
10. When troubleshooting a cooling water system, what is the first check? Why?

CHAPTER 11

Indoor Air Quality (IAQ)

INTRODUCTION

Indoor air quality, or IAQ, is concerned with the temperature, moisture content, cleanliness, oxygen concentration, and movement of air. It becomes even more important as the cost of energy increases and building construction becomes tighter. Everyone is trying to conserve energy, so outside air is being conditioned at minimum levels. Buildings are not leaking as much, because construction methods are attempting to reduce the amount of infiltration. Working in concert, these two factors have the effect of creating indoor environments with less air exchange with the outside and that allow indoor gases to build up to levels of concern.

There is usually no single cause of an IAQ issue, and several forms of pollution can combine to create poor air quality (see Figure 11–1). In this chapter you will become aware of the issues concerning IAQ and some of the things you can do as a technician to reduce and eliminate those problems.

1. CHIMNEY
- MOISTURE INTRUSION (IMPROPER FLASHING OR POOR MASONRY)
- DOWN-DRAFTING OF COMBUSTION GASES
- IMPROPER COMBUSTION

2. STORAGE/CLEANING ROOM
- CLEANING PRODUCTS (VOCs)
- PESTICIDES
- HERBICIDES
- FUNGICIDES
- AIR FRESHENERS
- STANDING WATER

3. INFANT/CHILD ROOM
- HUMIDIFIER CAUSING EXCESS HUMIDITY & MICROBIAL GROWTH
- HUMIDIFIER FEVER
- PET ALLERGENS
- DUST MITES IN BEDDING

4. STRUCTURAL AREAS
- ICE DAMS
- GUTTER OBSTRUCTION
- EXTERIOR CRACKS
- POOR CONSTRUCTION
- IMPROPERLY SEALED CONCRETE
- IMPROPER EXTERIOR GRADING

5. HOME OFFICE
- VOCs (FURNISHING)
- COMPUTERS (VOCs/OZONE/ PARTICULATES)
- ELECTROMAGNETIC FIELDS/PARTICLES
- IMPROPER VENTILATION

22. PAINTS/STAINS
- VOCs
- LEAD (PAINT PRE-1978)
- CONTAIN PRESERVATIVES AND FUNGICIDES
- FLAKING/AEROSOLIZATION

6. BATHROOM
- AIR FRESHENERS (VOCs)
- CLEANING PRODUCTS & PERSONAL CARE ITEMS (VOCs)
- SEWER GAS
- WATER LEAKS (MOLD/ BACTERIA/VIRUSES)

21. ATTIC
- ASBESTOS
- FIBERGLASS
- VERMICULITE
- MOLD
- BACTERIA
- DUST/DEBRIS

7. KITCHEN
- COOKING ODORS/OVEN FUMES
- TOBACCO SMOKE
- WATER LEAKS (MOLD/BACTERIA)
- ALLERGENS (COCKROACH/ MOUSE/RAT)

20. PEOPLE
- PERFUME/COLOGNE
- BODY ODOR/SMOKING
- SKIN FRAGMENTS
- TRANSMITTABLE DISEASE (TUBERCULOSIS/COMMON COLD/ETC.)

8. INTERIOR WATER SOURCES
- DISHWASHER
- CLOTHES WASHER
- TOILET
- BATHTUB/SHOWER
- WATER HEATER
- WATER PURIFIER

19. DUCTWORK/VENTILATION
- IMPROPER EXCHANGE RATE
- FIBERGLASS INSULATION
- CONTAMINATED DUCTWORK (DUST/POLLEN/FUNGI/BACTERIA)
- TEMPERATURE EXTREMES
- IMPROPER FILTRATION

9. BIRDS
- ODORS
- ALLERGENS
- HISTOPLASMOSIS
- CRYPTOCOCCOSIS
- PSITTICOSIS

18. CARPETING/FLOORING
- POTENTIAL MICROBIAL & CHEMICAL RESERVOIR
- VOCs
- ASBESTOS FLOOR TILES
- IMPROPERLY SEALED CONCRETE
- NON-HEPA VACUUMING

10. PESTICIDES/HERBICIDES
- IMPROPER STORAGE
- EXTERIOR & INTERIOR APPLICATIONS
- INHALATION & TRACKING OF CHEMICALS ONTO FLOORING MATERIALS

17. MOISTURE INTRUSION
- LEAKS
- WICKING OF MOISTURE (SLAB/WALLS/ROOF)
- MOLD, MVOCs & MYCOTOXINS
- BACTERIA & ENDOTOXINS

11. HOME MAINTENANCE
- NON-HEPA VACUUM
- VOCs FROM CLEANING ITEMS
- MACHINERY EXHAUST
- PESTICIDES
- HERBICIDES
- FUNGICIDES

16. PLANTS
- POLLEN/ALLERGENS
- FUNGI
- INSECTS
- INSECTICIDES/ FUNGICIDES/ HERBICIDES
- STANDING WATER

15. FURNACE ROOM
- IMPROPER COMBUSTION (PARTICULATES/GASES)
- IMPROPER VENTILATION
- BYPASSING PROPER FILTRATION
- ASBESTOS INSULATION
- FIBERGLASS INSULATION
- *LEGIONELLA* FROM IMPROPERLY HEATED WATER TANK

14. GARAGE
- EXHAUST INFILTRATION
- VOCs FROM CLEANING PRODUCTS
- STORED PESTICIDES/ HERBICIDES
- GASES FROM STORED FUEL
- GARBAGE CONTAINERS

13. BASEMENT
- EXHAUST/COMBUSTION FUMES
- CARBON MONOXIDE
- MOISTURE INTRUSION
- CHEMICAL STORAGE/ SPILLS
- RADON RISK
- PARTICULATES/ WOODWORKING

12. TRASH CONTAINERS
- ODORS
- BACTERIA
- FUNGI
- RODENT/INSECT/AVIAN ALLERGENS
- PARTICULATES

Figure 11–1
The sources and causes of indoor air pollution. (Courtesy of Aerotech Laboratories, Inc.)

COMFORT AND HEALTH

The entire HVACR industry is concerned with the comfort and health of the occupants of homes and businesses. Supplying air that is filtered, heated, cooled, and properly humidified is the first priority. The next priority is ensuring that indoor air is free of bacteria, fungus, and mold; is low in CO_2 and NO_x; and is free of CO and noxious vapors.

Sick Building Syndrome

The *sick building syndrome* arises not because a building is sick but instead to describe the tendency of people who work in certain types of building environments to exhibit illnesses. The building environment contains both natural and man-made toxins that become concentrated to the extent that they become unhealthful or objectionable to humans. The first signs of sick building syndrome occur when fresh air exchange is limited. The exchange is limited because buildings are being built tighter, energy costs are higher, and design considerations limit the amount of fresh air that can be brought into a building. All of these factors contribute to an environment in which indoor air contains less fresh air. Less fresh air means that relative toxin levels will increase, which can eventually affect human health.

Toxin is a general term encompassing airborne contaminants that affect health. Higher levels of humidity, carbon dioxide, carbon monoxide, and other gases can accumulate as air becomes stagnant. The accumulation of any one or more of these conditions is usually related to the type of activity occurring within the indoor space. It may also result from the concentration of toxins in "fresh" air being brought into a building. If exhaust air from an industrial process is too close to a fresh air intake, then contaminated air can be brought in with fresh air. Table 11–1 lists some forms of indoor air pollutants that can cause sick building syndrome.

Figure 11–2
A room in which humidity was too high for too long, promoting the growth of mold. (Courtesy of Aerotech Laboratories, Inc.)

Table 11–1 Indoor Air Pollutants

Contaminant	Description
Toxin	A general term used to describe poisons that can cause health issues
Carbon dioxide	Usually referred to as stale air, it is caused by oxygen being converted to carbon dioxide through human respiration
Carbon monoxide	Caused by open flames (primarily from cooking over a gas range) but can also come from other fuel appliances used in the home or business
Mold	A mold is a type of fungus; there are many types of molds that can grow in humid environments
Allergens	Substances that cause the human immune system to react and sometimes overreact
Fungus	There are many types of fungus; of these, stachybotrys is of most concern because of its strong effect on human health
Industrial pollutants (airborne)	Depending on the concentrations, this can be a chemical soup of dust and gases that causes smog in many cities
Auto/diesel exhaust	Generally comes from attached garages or parking areas where gas and diesel vehicles are operated
Chemical oxides	Nitrogen and sulfurous oxides (NO_x and SO_x) from combustion processes
Soaps and perfumes	Chemical fragrances; they are synthetic but smell similar to natural odors
VOC (volatile organic compounds)	Chemicals and agents: insecticides, disinfectants, lubricants, and a host of other commercial products used in homes and buildings; this category includes virtually anything that "out-gases" (i.e., gives off gases as it ages)
Radon	A colorless, naturally occurring radioactive gas, radon (Rn) can leak from the ground into basements, crawl spaces, and buildings
Fabrics and furnishings	Resinous chemicals are given off as vapors through the outgassing process
Construction materials	Some man-made and natural materials out-gas their resins; one natural material (cedar) out-gases and is used for closet lining material

Table 11–1 helps us develop a general understanding of the types of indoor pollutants that can cause IAQ problems leading to sick building syndrome. But there are other, more everyday things that contribute to problems in buildings:

- Smoking
- Copy machines
- Cooking
- Washing
- Aerosol sprays
- Correction fluid (e.g., White-Out)
- Perfume
- Inoperative exhaust fans
- Cleaning chemicals
- Negative airflow
- Exhaust from combustion
- New furniture, carpeting, and other floor covering materials (adhesives)
- Too low or no ventilation air
- Dust and pet hair
- Mold
- Pests and pest control products

For pest control, capture and disposal is preferable to poisoning, which may allow rodents to crawl into inaccessible areas and die—their decomposing carcasses creating foul odors and contaminants that cause health problems as well as attracting other pests and insects. Finally, the dust, fumes, and other problems associated with remodeling projects need to be controlled during the renovations.

Safety Tip

Chemicals in the home or workplace can cause IAQ problems for the occupants and the service technician. If strong odors are encountered, use common sense while investigating them. Open windows and provide adequate ventilation while the source is being determined. Chemicals can be corrosive or explosive. Do not use open flames in the presence of most chemicals.

Field Problem

Tight house. During a routine annual maintenance visit, the technician noticed that the furnace was down-drafting, spilling flue gases into the living space. A visual check of the basement area, where the furnace was located, revealed that the rim joist had been foamed in around the perimeter of the house. If this had been done, then perhaps other types building changes had occurred during the past year. The home owner confirmed that a whole host of weatherization improvements had been made and that she was enjoying lower heating bills as a result. However, she had a question about condensation on the windows during colder weather—this had never happened before. The technician then turned on all the exhaust fans in the house, went to the furnace, and observed a down draft in the flue. He suggested scheduling an appointment with the sales manager, explaining to the customer that several possibilities should be considered. One approach would be to install a heat recovery ventilation system that would ensure sufficient ventilation air and reduce the amount of indoor moisture she was experiencing. Another possibility was updating the furnace to a direct vent model and addressing any other potential atmospheric venting issues with other existing appliances.

ASHRAE Standards

The American Society for Heating, Refrigeration, and Air-Conditioning Engineers (ASHRAE) has conducted a large number of studies and produced a number of publications regarding sick building syndrome. One of their standards to improve IAQ sets the requirement for the minimum amount of ventilation. Standard 62 makes recommendations for acceptable indoor air quality and is continuously updated. One version of this standard recommends that homes have 0.35 air exchanges per hour and not less than 15 cfm per person. To determine minimum cfm, multiply the number of bedrooms by 15 (this assumes that there is at least one person occupying each bedroom). If the exact number of occupants is known, multiply that number by 15 cfm to determine the minimum. The same cfm requirement is generally accepted as the minimum both for residential and commercial structures. In some commercial structures the cfm per person may be as high as 60, but most commonly the figure ranges between 15 and 35 cfm per person.

Standard 62 also provides guidelines in terms of square footage: offices, 0.4 cfm/ft^2; kindergartens, 0.5 cfm/ft^2; hallways, 0.7 cfm/ft^2; and so forth. In a residence, fresh air is usually obtained from cracks around doors, windows, and throughout small gaps in the building construction. A design engineer who feels that there will not be enough air—because the building components and construction are too tight—can intentionally bring air from the outside by providing mechanical ventilation. Mechanical ventilation strategies should always be applied to commercial buildings.

DOE Requirements

The U.S. Environmental Protection (EPA), a division of the Department of Energy (DOE), has a website devoted to IAQ. Included in the information that can be found there are strategies to reduce IAQ problems. The EPA takes a three-step approach to controlling indoor air pollution:

1. *Source control*—eliminate the sources of indoor air pollution.
2. *Ventilation improvements*—lower the concentration of indoor air pollution by introducing more ventilation air and exhausting the stale air.
3. *Air cleaners*—use air cleaners to remove particulates from the air. Air particulates can include dust, pollen, and animal hair.

NIOSH (National Institute for Occupational Safety and Health) partners with the EPA to produce information to help reduce IAQ problems in buildings.

Tech Tip

Using a boiler or furnace that has been underwater because of flooding. In terms of safety, HVAC equipment that has been underwater because of flooding must be completely dried and then evaluated for proper operation. This entails a detailed inspection before applying power, further inspection after power is applied, and a final inspection during the combustion process. If the equipment is currently installed, remove it for improved access to all areas of the equipment before conducting the evaluation. Ask the original equipment manufacturer for instructions or checklists in this regard. Inspection checklists for boilers and furnaces may also be found in the *ACCA 4 Quality Maintenance (QM) Standard*. With both checklists, you should be able to assess the general and specific condition of the equipment and, when deemed safe, the operation of the equipment.

Customer Relations

Most IAQ investigations start with a complaint. Interaction with the customer is a crucial component of diagnosing the IAQ problem. Additionally, the technician should be knowledgeable about those things that may contribute to IAQ problems for some people. Different groups of people may have complaints about air quality. The *building occupants* are those who live, work, or visit a building on a regular basis. *Visitors* are those people who frequent a building environment on an irregular basis yet still may be sensitive to a particular IAQ problem. *Customers* are those who come to the building for products and services. *Building owners* and *workers* spend a great deal of time in a building. All of these people could be sensitive to IAQ problems and may take more or less time to notice the affects.

Individuals with certain health conditions may be susceptible to IAQ problems:

- *Allergies*—reactive to indoor gases or products
- *Asthma*—reactive to plants and molds
- *Respiratory problems*—reactive to gases, plants, and dust
- *Immune deficiencies*—reactive to molds and bacteria
- *Contact lenses*—reactive to gases and dust particles
- *Heart patients*—reactive to lower levels of CO
- *Children and adults*—reactive to tobacco smoke, dusts, and gases

People often exhibit symptoms that are unrelated to a building's IAQ. Even so, the following affects or symptoms can be exhibited by people exposed to buildings with IAQ problems:

- Sneezing and coughing
- Headache
- Fatigue sinus congestion
- Eye irritation
- Skin rashes
- Dizziness
- Throat irritation
- Nausea
- Shortness of breath
- Nose irritation
- Sinus congestion

Customer feedback is used to determine if response to a particular symptom is warranted. Listening, being empathetic, showing concern, and documenting the symptom and complaint are the first steps. Inquire about any changes in the environment, system, building, or type and level of activity. The next step is following up on the complaint and making sure you don't add to the problem. In this latter regard, make sure that your clothing does not bring dust and pollutants into the environment.

- Wear shoe covers, and dust off your uniform, hat, and gloves.
- Clean your tools and equipment before entering an interior space.
- Do not vent refrigerant (it is against the law) or other gases into an occupied space.
- Ventilate if there has been a gas or VOC release.

Make recommendations for further testing and to reduce pollutants—for example, suggest that open bottles of chemical products be capped.

Follow up with the customer. Report what you have found and provide recommendations to reduce any IAQ problems discovered. Document the complaint as well as what was found or observed, and provide the customer with a copy.

> **Safety Tip**
>
> Reduce the contamination that you bring into occupied spaces. Be sure that you are not introducing additional problems such as dust, chemicals, or biological agents on your clothing, shoes, or tools. Always reduce or eliminate problems by ensuring that you are clean, wearing covers for your shoes, and providing adequate ventilation when conducting service work. At the end of the service call, clean the work area. The work area should be cleaner than when you arrived.

VENTILATION

Ventilation is the process of adding outside air to a building or a particular room (see Figure 11-3). The reason for ventilation is to remove or dilute CO_2 and other gases that may pollute the air. Depending on the building or the use of a particular room, there may be recommended or required ventilation levels. Ventilation is usually discussed in terms of air change volume—specifically, the number of room air changes per hour. To determine the cfm (cubic feet per minute), the room volume needs to be calculated. For instance, a room that measures (in feet) $20 \times 20 \times 10$ has a room volume of 4,000 cubic feet. If five air changes per hour are required then five volumes of new air, each 4,000 cubic feet large, would have

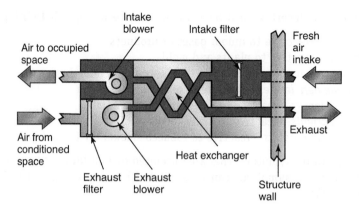

Figure 11–3
An air ventilation system that recovers exhaust heat energy and uses it to heat outside fresh air before delivering it to the living space.

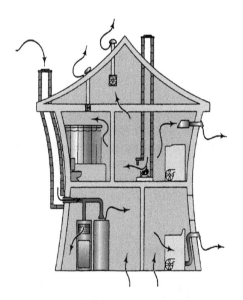

Figure 11–4
Air that is mechanically exhausted from a building creates a negative pressure. Unconditioned, infiltrated air is drawn in from outdoors through openings and natural draft vents. A tight building can cause indoor air pollution problems, as shown in this figure. (Courtesy of Bacharach, Inc.)

to be sent into the room. The total volume of replacement air would thus be 5 × 4,000 = 20,000 cubic feet in 60 minutes (1 hour), or 333.33 cubic feet per minute. When air changes are specified, this is how the amount of air is established in units of cfm. The formula is (Length × Width × Height × Air changes) ÷ Minutes per hour = Desired cfm, or (L × W × H × Air changes)/60 = Desired cfm.

Ventilation is also required to make up air that is removed by exhaust vents, range hoods, vented appliances (furnaces), fireplaces, and any other naturally or power vented devices. Without provision to bring in ventilation air, some appliances that are naturally vented might vent into the home or building if a powered vent is operated. The powered vent might easily overcome a natural vent. Exhausted air from power vents may create a negative pressure within the house. In this case, outside air may enter the house through naturally vented chimneys, pulling the products of combustion into the house and creating a health hazard (see Figure 11–4).

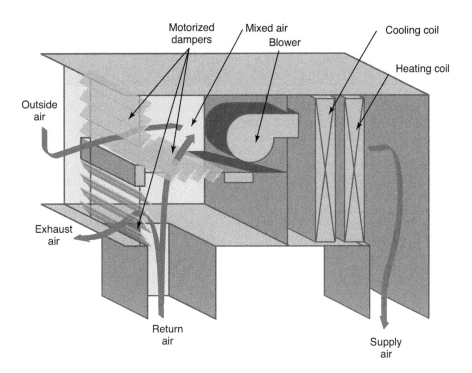

Motorized dampers | Mixed air | Blower | Cooling coil | Heating coil | Outside air | Exhaust air | Return air | Supply air

Figure 11–5
Economizer for an air system, showing the motorized dampers that open and close to allow (respectively) more or less fresh air into the building.

Mechanical Ventilation

Mechanical ventilation can be accomplished by using ERVs (energy recovery ventilators) and HRVs (heat recovery ventilators). The term *energy recovery* usually refers to the recovery of both heat energy and moisture, whereas *heat recovery* is concerned only with the transfer of heat. ERVs should not be used in climates where moisture is subject to freezing. Both ventilator systems move heat from one air stream to the other—inside air to outside air or outside air to inside air, depending on the season.

Mechanical ventilation can also be accomplished by using an economizer, whose dampers allow more or less fresh air into the building depending on the outdoor temperature (see Figure 11–5). If the outdoor conditions are right, the mechanical cooling system can be turned off to allow the outside air to cool the building directly. An economizer is a device that adjusts the amount of fresh air delivered to the building when the outside air is at these appropriate conditions, with a set minimum position to allow for ventilation regardless of the outdoor conditions. Suppose, for example, that the outside air condition is 50°F and low in humidity; then it would be more economical to use the outside air to cool the building than to use electricity to create a cooling effect. In this case, economizer louvers open fully to allow 100% ventilation air into the building and exhaust dampers open to exhaust 100% of the return air from the building. The economizer works between the minimum fresh air setting for the building to full open position for 100% outside air.

Rates for Applications

For purposes of energy conservation, ASHRAE states that "outside air for ventilation purposes should be introduced at the lowest volume necessary to maintain adequate indoor air quality." Different structures require different amounts of ventilation, and residential structures are different from commercial ones. Some

Field Problem

Economizer controls. The system had an economizer with basic enthalpy controls. The complaint was that building employees sometimes complained about the temperature and humidity at the start of the day. During the morning the problem seemed to dissipate. The technician suspected the enthalpy control or the control setting. That morning, the same conditions were evident, and some of the same complaints had been registered with the building owner. Measuring the conditions inside with an aspirating psychrometer, the technician noted the following conditions: 68 degrees DB and 65 degrees WB; outside the measured conditions were 66 degrees DB and 64 degrees WB. Checking the economizer control, the technician noted that the light was on, indicating that free cooling was available. It seemed that the damper was not working. A quick check of the outside air damper position showed that the economizer had not opened the damper. Checking the wiring connection between the changeover control and the damper motor, the technician found it was loose. Tightening the connection started the damper motor. Testing the system while waiting an appropriate amount of time for the system to operate, the technician noted the system problem was corrected, confirming that the damper motor was working. (*Note:* Plot the indoor and outdoor conditions; then compare the points plotted to see if 50% relative humidity could be maintained in the return-air or mixed-air conditions.)

Tech Tip

Leaky dampers. The technician must bear in mind that outdoor dampers can become leaky. Damper leakage will waste energy, and leaky dampers must be repaired so that they seal when closed. If light can be seen through the damper, then its seals need to be repaired or replaced. The ASHRAE Standard 90.1 requires dampers that are integral to the building envelope to have a maximum leakage rate of 4 cfm per square foot of damper area at 1" w.c. pressure difference.

In many cases it is recommended that the damper leakage rate not exceed 0.5% when the damper is closed and working against a 4" w.c. static pressure. The problem is that damper leakage is a function of pressure: the pressure of the return mechanism used to close the damper and the pressure *difference* between the outside air and the pressure on the other side of the damper in the return duct. If airflow were a factor and a 12" × 12" damper were used, then given a 1% leakage specification, a system operating at 2,000 fpm (feet per minute) could leak 20 cfm, whereas a system operating at 500 fpm could leak at only 5 cfm.

A technician measuring a damper leakage might use the following mixed-air formula:

% outdoor air leakage (OA) = (Return air (RA) − Mixed air (MA)) ÷ (RA − OA) × 100

If the size of the damper is 12" × 12", the velocity is 2,000 fpm, and the leakage specification is 1%, as stated above, with the damper closed, the technician would need to measure the OA, RA, and MA temperatures. If the technician measures the outside temperature as 50°F, the return air as 70°F, and the mixed air as 66°F, he would insert the temperatures into the formula as follows:

(70 − 66) ÷ (70 − 50) × 100 = 20%

The technician in this example would need to have an extremely good thermometer that measures accurately to tenths of a degree. If the damper leakage in this example was doubled to 2%, the resulting MA temperature would be 69.6°F—a difference of 0.2°F. This amount if difference in temperature is difficult to measure accurately.

commercial structures (such as hospitals) have specific requirements for each type of room depending on its use. Residential structures may require different rates of ventilation for each room. Table 11–2 summarizes ventilation rates for rooms in a residential structure.

Table 11–2 Rates of Ventilation (Residential)

Room	Requirement/Specification	Recommended Ventilation Rate
Entire house	Based on square footage of the house floor area and the number of bedrooms (e.g., a house with 1,501–3,000 square feet and 4–5 bedrooms)	75 cfm
Kitchen	Vented range hood is required if exhaust fan rate is less than 5 kitchen air changes per hour	100 cfm
Bathroom	If forced ventilation is not provided, the bathroom should have windows that are not less than 4% of the floor area or 1.5 ft^2, whichever is greater	50 cfm

Tech Tip

Economizers have been discussed as a way of bringing in fresh air to provide dilution or ventilation air. However, fresh air can be detrimental if it is contaminated. Technicians should be alert to conditions such as standing water on roofs, which can harbor biological growth, attract birds and other wildlife (droppings), and add significant contamination to the air being drawn into the building. Building owners must be made aware of these conditions, and steps should be taken to minimize them—for example, by reworking roof coverings to provide proper pitch and drainage and/or using ultra-sonic repelling devices to keep wildlife away from these areas. In some cases, pest control specialists may need to be consulted. Air intakes should be securely protected to keep birds out of the ventilation system. Another condition to be alert for is when the air outside is more heavily contaminated than the air inside. For example, you don't want economizers drawing in air near locations (such as drive-up windows) where vehicles are idling or where nearby plumbing system vents could allow sewer gases to enter the air intake under certain weather conditions.

Dilution

Dilution is the process of mixing new air with stale air to maintain a concentration of gases or to prevent certain gases from concentrating. If, for instance, we were concerned only about CO_2, then concentrations of this gas would be monitored. When a certain high concentration was reached, new air with lower CO_2 concentrations would be introduced and mixed, diluting the level of CO_2.

In fact, CO_2 monitors are installed in some buildings to do just that. When CO_2 concentrations meet the setting of the monitor, ventilation fans or dampers are activated to introduce outside air into the air distribution system, where it mixes with return air.

Dilution can also be accomplished by exhausting stale air. Exhausted air will bring outside air (new air) into a building through inlet air dampers and through cracks around doors and windows. Typically, negative air pressure environments are avoided. Negative air pressure brings unfiltered air into a building or room, and unfiltered air carries unwanted particulates. However, there are instances where smells, vapors, and other gases need to be contained within a room. To keep them contained, negative pressure is used to make air leak into the room, which prevents the gases from escaping.

For air that people are breathing, ASHRAE Standard 62 recommends a maximum CO_2 concentration of 1,000 ppm (parts per million). Note that the outdoor concentration of CO_2 at ground level varies greatly and must be checked locally.

Tech Tip

Measuring pressure difference. A simple measurement of pressure difference is to use test smoke as a visual indicator of whether air is moving from one area to another. If air is moving from the outside, through the window gasketing, and to the inside of a room, this will be evident when test smoke is held close to the window gasket. The reverse is true if the air is moving from the room to the outside—test smoke will be pulled through the gasket to the outside. Measuring with a Magnehelic® gauge is done by holding the gauge in the room and having the sensing hose outside. Both negative pressure or positive pressure can be measured relative to the room.

Particulates

Solids that are suspended in air are called *particulates*. If a sample of air is captured and kept in a container, all of the particles suspended in the air would eventually "precipitate out of solution" (fall to the bottom of the container, like rain falls from the sky). We know that dust can be picked up by the wind and carried great distances. When the wind slows down, dust falls out and deposits on the surface of buildings, cars, and the ground. In a building, the air distribution system can keep small dust particles moving in the system, delivering them to every room. Good filtration systems help to reduce the amount of particulates in building air.

Field Problem

Particulate problems. A preseason furnace check was routine until the technician observed rodent fecal deposits. Concerned for the occupants' health, the technician spoke with the owner. The owner had not been aware of the problem and became concerned. The technician suggested some measures to rid the home of the pests, including an extermination company. The technician also suggested installing a better air filtration system, explaining the comfort levels that could be attained with each type of system. The customer was interested not only in getting rid of the rodent problem but also in increasing the occupants' comfort, stating that several members of the family had allergy-like symptoms.

Room Air Changes

Air change in a room is one way to talk about the amount of air necessary to condition the room in one hour. Conditioning includes: heating, cooling, changing moisture levels, filtration, and ventilation. Some specifications call for room air changes. If ventilation is not mentioned, then *room air change* means mixing conditioned air with room air. If ventilation is mentioned, some professionals and allergists suggest a minimum of 2 ACH (air changes per hour).

If a specification calls for 3 ACH in a small room (say, 800 cubic feet, or 10 × 10 × 8), then the amount of air change would be 800 × 3 or 2,400 cubic feet in one hour. In the conversion to cfm, 2,400 is divided by 60 min/hr to determine that 40 cubic feet per minute are needed. For this room, 3 ACH is equal to 40 cfm. At the end of one hour, would 100% of the air in the room be changed? No, but a significant amount of mixing would have occurred—enough to affect the temperature, humidity, cleanliness, and ventilation and to satisfy the specification of 3 ACH while maintaining the comfort of the occupants. You can use this equation: 1 air change in cfm = (L × W × H)/60. Length, width, and height are measured in feet for this calculation.

If 3 ACH of ventilation is needed, the air for ventilation might come from an economizer or from an outside air damper. Air pulled from the outside would be controlled to some minimal amount by the outside air damper. The minimum about would be the total required ventilation for the entire structure. The 800-cubic-foot-room example above would represent a small percentage of the total ventilation air brought through the economizer or outside air damper. Therefore, if 200 cfm of ventilation were required for the entire structure, and the outside air damper was set to allow this amount of air to enter the system, 40 cfm would be allocated for the 800-cubic-foot room to provide 3 ACH.

Humidity Control

The maximum amount of moisture that air can hold at a certain temperature is considered to be 100% relative humidity. Comfort indexes indicate that humans prefer relative humidity between 30% and 50%. The objective of humidity control is to add or subtract moisture from air to maintain an average relative humidity at a given setting. Removing moisture is typically done at the same time air is cooled. Air conditioning systems typically dehumidify the air, reducing moisture and removing latent heat. The drier, conditioned air is distributed to a building or room, where it mixes with air that is more humid. Mixing the drier air with moist air tends to reduce the room's (or the building's) total relative humidity.

The reverse of dehumidifying is humidification. Typically done with hot air in the heating season, water is evaporated and absorbed by heated air. There are several types of *humidifiers* (see Figure 11–6), which can add moisture to a room or building. Evaporative blankets, rolls, and other media allow air to change from a liquid to a vapor as air rushes over the media. Water can be sprayed into the airstream in such a way that will allow the water to vaporize but not deposit on the ductwork. Steam can also be used to inject vapor directly into the airstream. Air that has increased moisture content can be distributed to a building or room, where it mixes with drier air to increase the inside ambient overall relative humidity.

To see how relative humidity is affected by adding water, on a psychrometric chart, plot a return-air condition of 75 degrees DB at 50% RH. Read the grains/pound of air for that point (69 gr/lb). The return air is heated to 105 degrees DB. Plot this new point on the same grains/pound line to show the new RH of the heated air with no moisture added (19% RH). The heated air is moved across a humidifier to pick up moisture. For this example, let's say that the hot air picks up 30 grains/pound (69 + 30 = 99). The relative humidity of the heated air is now approximately 30% RH. As this new moist, hot air is sent out to the occupied space, it cools to the return temperature of 75 degrees DB (and if it never mixed with the other room air, the RH would be higher—approximately 75% RH). If the hot, moist air mixes with the room air (depending on the volume of air mixing) the room RH will continue to rise as new moisture is added. In reality, the dynamics of humidification include ventilation air brought in from outside, air lost through cracks in doors and windows, and the amount of air delivered to a room. All of these things will affect the room's relative humidity.

FILTRATION

Cleanliness of the air is a concern for all people, but particularly for people with allergies. Small particles of dust, mold spores, pet hair, and more can cause

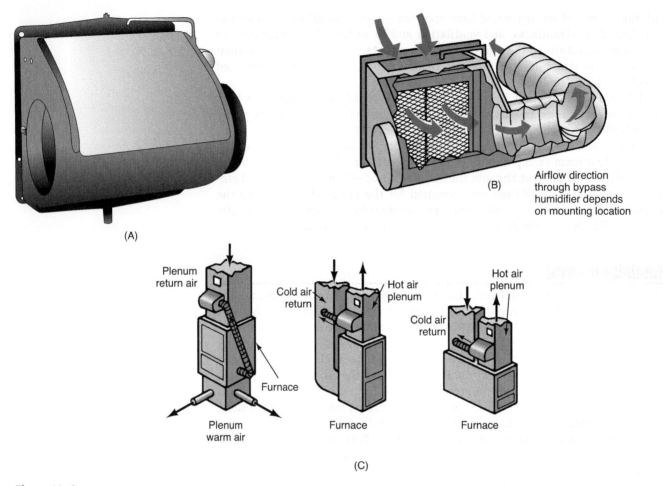

Figure 11–6
(A) The outside housing of a bypass humidifier; (B) a cutaway view of the humidifier, showing direction of airflow; (C) three typical installations. In each case, the hot air goes through the humidifier because hot air will absorb more moisture than cold air.

Safety Tip

Moist building materials are evidence of a moisture problem. Do not let such problems go unattended or uninvestigated. Moisture is one of the requirements for biological growth. If you are probing a moist area and find biological growth, stop your investigation and seek professional assistance to determine the extent of the problem. If there is no biological growth, attempt to identify and repair the source of the moisture.

problems for some people. Filtration systems alleviate many of the causes of these allergies. Additionally, cleaner air means less home and building maintenance. Good filtration systems can keep the interior of homes and buildings cleaner—they can even make the indoor air cleaner than the outdoor air! See Figure 11–7.

Tech Tip

Always check the filter for the directional arrow (see Figure 11–8) and install the filter with the arrow pointing in the same direction as the airflow—that is, pointing toward the return-air side of the blower.

Filter Rating

Various types of filters and filter media are depicted in Figures 11–9 through 11–14 (see Table 11–3 for summary specifications). With each filter type is associated a particular pressure drop. The system blower must generate enough static pressure to overcome the pressure drop for any filter in the system.

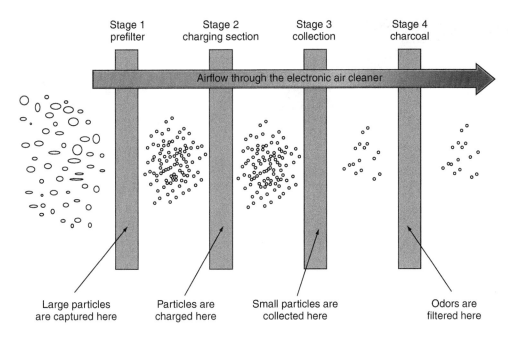

Figure 11–7
A four-stage filtration system consisting of individual filter sections.

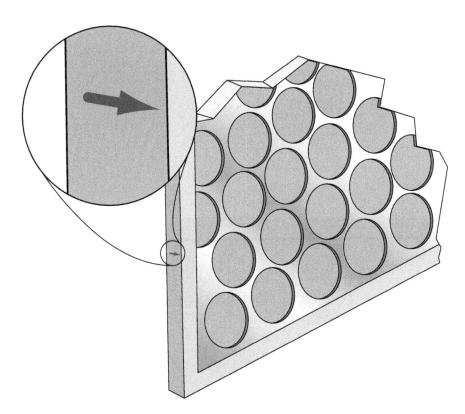

Figure 11–8
The printed arrow points in the direction of airflow.

Figure 11–9
Roll and pad filter media. (Courtesy of Aerostar Filtration Group)

Figure 11–10
Fiberglass filters. (Courtesy of Aerostar Filtration Group)

Figure 11–11
Pleated paper filters. (Courtesy of Aerostar Filtration Group)

Figure 11–12
A bag or pocket filter. (Courtesy of Aerostar Filtration Group)

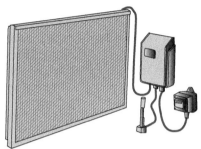

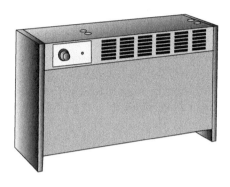

Figure 11–14
An activated carbon filter.

Figure 11–13
Electrostatic and electronic air filters.

Table 11–3 Filter Types by MERV Rating

MERV[a] Rating	Effectiveness	Arrestance[b]	Types of Filters
1–4	> 20%	60–80%	Disposable types • Panel filters • Metal filters • Fiberglass media • Foam media • Material roll
5	> 20%	80–90%	Pleated filters, panel filters, synthetic media
6	20–30%	90–95%	Cube filters, self-supported filters
6–7	25–30%	90–95%	Pleated panel filters
8	40–50%	95–98%	Pleated panel filters, panel filters, expanded surface filters, bag or pocket filters
9–10	50–60%	98%	Expanded surface filters, bag or pocket filters
10–11	60–70%	99%	Pleated panel filters, expanded surface filters, bag or pocket filters, rigid cell filters
12–14	80–90%	99%	Pleated filters, expanded surface filters, bag or pocket filters
14–15	90–95%	99%	HEPA pleated filters, expanded surface filters, bag or pocket filters, rigid cell filters
16	95%	n/a	Rigid cell filters
17–20	>95%	n/a	HEPA pleated filters, hybrid filters (capture particles of size $\geq$ 0.30 micron)

[a] Minimum Efficiency Reporting Value, a rating developed by ASHRAE, that evaluates the efficiency of the filter medium. Higher MERV numbers indicate better filtration (smaller particles will be trapped). Trapped particles are removed from the air stream, and the indoor becomes cleaner. Ironically, for a time, a filter will get better with use. As trapped particles fill voids in the filter, smaller and smaller particles will become trapped. However, increasing particles (dirt) will continue to plug the filter medium, causing a static pressure drop. The static pressure drop of the filter acts as a restriction to air flow, affecting system efficiencies and performance. The MERV rating is for a clean, newly installed filter. Each manufacturer has recommended change-out schedules that should be followed. Alternatively, sensors can be installed to monitor static pressure drop and provide a signal to change the filter.

[b] A measure of the ability of an air filter to remove a synthetic dust from the air. ASHRAE established the arrestance measure and developed a special procedure that tests a filter for arrestance.

Therefore, adding filtration to a system may reduce the air volume and the overall system performance. Always determine whether there is adequate blower static before retrofitting a system with high efficiency filters of the same face (height and width) dimensions.

Most air handling equipment used in residential and light commercial applications has a recommended maximum external static pressure of 0.5" w.c. If a high-efficiency filter has a pressure drop of 0.4" w.c. across it alone, that leaves very little pressure to overcome the resistance of the rest of the system, and a decrease in airflow will result.

Decreasing airflow has a tremendous negative effect on system performance. Adding a 2" pleated filter to an existing furnace may very well cut the airflow across the heat exchanger to harmful levels. Increasing the area of the filter will cut down the pressure drop and aid in maintaining good airflow. That may mean that two or more filters will need to be installed side by side to increase face area—with the duct system modified to accommodate these filters—to maintain adequate airflow. The homeowner who inadvertently buys a high-efficiency filter to be installed in his or her existing furnace system may very well be causing more harm than good.

UV Treatment

Ultraviolet light is used for treating the air in an air distribution system to reduce bacteria, mold, and other organisms that might be infectious. The effectiveness of UV treatment at killing bacteria, mold, and other organisms in a moving airstream in questionable. UV treatment is most effective when applied to the evaporator or the evaporator drain pan.

The wavelength of UV light energy is between 100 and 400 nanometers. There are three bandwidths, which are classified as UVA, UVB, and UVC. Most systems use UVC lamps, whose wavelengths range from 200 nm to 280 nm. The UV radiation disrupts the DNA of microorganisms, killing them or rendering them inactive.

Maintenance Procedures

Maintaining IAQ for most standard heating and cooling systems requires periodic system maintenance. The quality of the air delivered can make a big difference in interior living conditions. Table 11–4 lists some IAQ items to check for during maintenance of most standard heating and cooling systems.

As a service technician, your job is to work on the HVAC and related ventilation systems. However, as a best practice and as a service to your customer, you should always keep the customer aware of anything that could affect indoor environmental conditions. All information should be provided to the customer in a report after the service or maintenance has been performed. Here are some additional things to look for:

- Check for water intrusion or damage in the duct system and building.
- Look for sources of moisture (pool or spa areas, unvented bathrooms or ranges, etc.).
- Look for uncapped chemicals, pesticides, cleaning agents, etc.
- Look for rodent, bird, animal or pet waste products or activity.
- Look for insect activity (termites, ants, etc.).
- Smell for sewer gas, heating fuel, and other strange odors.
- Check attic and crawl space for the same conditions as the living space.
- Look for new carpets, construction material, or furniture that could create IAQ problems.
- Look for asbestos or peeling paint that may contain lead.

Table 11–4 IAQ Maintenance

Component	Maintenance Action
Humidifier	• Replace the humidifier pad (Figure 11–15) • Clean the pan or sump • Wash the interior of the humidifier housing • Make sure that the water make-up valve is clean and functional • Check and clean the overflow (if installed)
Heating system	• Check for high levels of CO • Check for proper combustion • Check vent connections and vent condition • Check blower wheels for dirt buildup; clean the wheel and scroll
Air conditioning coil	• Check the coil for buildup of dirt or mold; clean if necessary • Check and clean the condensate pan • Check the drain tube for blockage • Check the drain tube trap and where the tube empties
Air filter	• Replace the pre-filter (even if it doesn't look that dirty) as a best practice • Check and clean electronic filters • Replace final high-efficiency filters
Duct system	• Check diffusers for streaking or staining that might indicate dirty air • Check the duct system (supply and return) for buildup • Recommend more frequent filter changes as necessary • Recommend cleaning as necessary • Check for fresh-air or ventilation connections to the ducting system; adjust the damper as necessary • Check duct system for proper sealing

Figure 11–15
A bypass humidifier. (Courtesy of General Filters, Inc.)

AIR DELIVERY SYSTEMS

In addition to all of the interconnecting ductwork, air delivery systems include the furnace and air conditioning components, accessories such as humidifiers, supply diffusers and return grills, and any exhaust and ventilation equipment (e.g., bathroom vents and kitchen range hoods). All of these components affect air quality of the living space and must work together to create a healthy environment.

Cleaning and Sealing

Both duct cleaning and sealing may improve IAQ conditions. Duct cleaning rids the system of dust and dirt buildup that may harbor allergens affecting human comfort. However, the EPA states:

> Duct cleaning has never been shown to actually prevent health problems. Neither do studies conclusively demonstrate that particulate (e.g., dust) levels in homes increase because of dirty air ducts. This is because much of the dirt in air ducts adheres to duct surfaces and does not necessarily enter the living space. It is important to keep in mind that dirty air ducts are only one of many possible sources of particulates that are present in homes. Pollutants that enter the home both from outdoors and indoor activities such as cooking, cleaning, smoking, or just moving around can cause greater exposure to contaminants than dirty air ducts. Moreover, there is no evidence that a light amount of household dust or other particulate matter in *air ducts* poses any risk to your health.

According to the EPA, duct systems should be cleaned if:

1. There is concern that the amount of dirt and debris is blocking or restricting airflow
2. Ducts have visible evidence of rodent, animal, or insect activity or residue
3. There is evidence of mold growth

A service technician working for a large HVAC company is not usually expected to clean a ducting system, but he is expected to inspect and recommend cleaning, if necessary. The technician should also be able to explain the need for clean ducts to the customer and describe how the procedure is conducted (see Figure 11–16). Duct cleaning crews are typically specialized and perform only this service, so the technician need only be familiar with the process. Duct cleaning involves the containment of dirt within a section of the duct. The procedure,

Figure 11–16
The equipment used by duct cleaners for cleaning ducts.

Table 11–5 Duct Cleaning Procedure

Process	Equipment
Vacuum removes the dust and debris from the ducting system; the entire system is cleaned, from the return grill to the supply diffuser.	Truck-mounted or portable vacuum cleaner with extension hoses and nozzles
Dirt is knocked, brushed, or vibrated from each component and the duct system	Motorized dislodging equipment is manually or mechanically moved through the ductwork
Sanitizing kills bacteria and mold	Spray equipment is used to apply treatment to kill and prevent bacteria and fungal growth (*Note:* Application of these products is not recommended by the EPA.)

Figure 11–17
Mastic joint sealer is applied to a sheet-metal duct joint.

which is summarized in Table 11–5, minimizes the amount of dust that escapes into the living space during the cleaning.

Duct sealing is important to IAQ and system efficiency. Air leakage into the return duct can pull air from attics, crawl spaces, basements, and/or attached garages. Air from these spaces could be contaminated with mold, mildew, biological waste, or carbon monoxide. Duct sealing materials require meticulous application when a system is installed (see Figure 11–17). Existing duct systems can be sealed, but sometimes access to these ducts is difficult. Duct sealants can be used as described in Table 11–6.

Tech Tip

Attached garage walls and ceilings may have ducting installed in the ceiling or walls. Any penetration where ductwork is touching exterior walls of the dwelling the ducting must be completely sealed.

Table 11–6 Duct Sealant

Types of Duct Sealers	Supply or Return	Use
Duct mastic	Supply & Return	An adhesive paste that can be applied to fill joints, cracks, and gaps of up to 1/8". Fiberglass tape (labeled UL 181A or UL 181B) can be used to bridge larger gaps. Flexible ducts can be sealed by applying sealant to the duct connection, sliding the flexible duct over the metal connection, and then applying a tightening band.
Metal tape	Supply & Return	Adhesive-backed metal tape can be used on straight runs. For good adhesion, duct joints should be clean.
High-temperature RTV[a]	Supply & Return	Used on supply plenums and supply runs when they are being built or as a gasket where the supply plenum attaches to the furnace.
Caulk	Return	Urethane or silicone caulk is used when the system is built and to fill gaps in the materials used to build return air systems.
Nonexpanding foam	Return	Used to fill large gaps in building materials and in the structure components that are used for return air.

[a] RTV = room temperature vulcanizing.

Table 11–7 Return System Checklist

Return Component or Area	Work Needed
Ceiling joists	Block and seal joist where it commutes to cantilevered areas
Floor joists	Block and seal joist where it connects to the foundation
Ceiling/floor plate	Caulk and seal the plate to the drywall and completely finish the drywall seam
Wiring/pipe penetrations	Caulk and seal all penetrations in framing where joists are used for return air
Panning to duct	Seal all connections between panning connections and duct connections

In panned-in area return systems (which are not recommended), return air leaks can be prevalent. Table 11–7 lists those parts of the return system that often require special attention.

Cleaning and Maintenance of the Coil and the Drain Pan

Sometimes cleaning the air conditioning components of a system is not as easy as it sounds. The technician should expect that some system components will need

Field Problem

Low air volume. A commercial account requested a check of their heating system, complaining that the heating bills were too high. The account had not been active for two years; the customer had used a different service company for the past year. During the system check, air volumes were low and there was a large amount of staining on diffusers. This lead the technician to check the filtration system. She discovered that the filters were plugged with dirt and debris. In fact, one filter had collapsed altogether and was allowing dirty air directly into the heat exchanger. The technician spoke with the owner, described what she had found, and suggested that the system be cleaned and filters changed. A duct cleaning crew was dispatched to bring the system back to near-new cleanliness. Filters were changed, and airflow was verified to specification.

Tech Tip

There are several good sources of information about duct cleaning: the Environmental Protection Agency (EPA); Air Conditioning Contractors of America (ACCA), "Restoring the Cleanliness of HVAC Systems" (ANSI/ACCA 6 QR—2007); American Society for Heating Refrigeration and Air Conditioning Engineers (ASHRAE); National Air Duct Cleaners Association (NADCA), for metal duct systems; and North American Insulation Manufacturers Association (NAIMA). You can visit their respective websites for more information.

to be moved or removed in order to enable a good job of cleaning. For instance, the coil—which fits into or is part of the drain pan—is not easily accessible. Trying to clean the coil, in place, is difficult because it usually sits on top or in front of the heating system exchanger in a heating and cooling system (see Figure 11–18). The bottom or back (inlet) of the coil is blocked from view. Because the air moves through the heating exchanger and then through the air conditioning coil, the

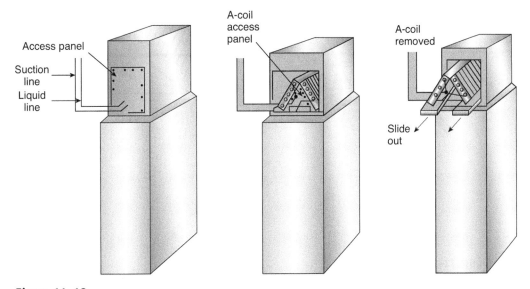

Figure 11–18
The A-coil sets atop the furnace. To access the A-coil inlet (bottom of coil), the coil may have to be pulled out until enough clearance is obtained to use a brush and vacuum hose.

bottom of the air conditioning coil is where dust, dirt, and other material will first accumulate. Technicians frequently are faced with such circumstances in the field. It is the job of smart technicians to learn innovative ways to clean systems.

When it's time to clean a coil, remember that a dirty coil may look clean when viewed from the plenum side (outlet) of the evaporator. But since dirt and other debris will accumulate where air *enters* the coil, you'll need access to the bottom side (inlet) of the evaporator. Most systems include an access panel (held in place with a few screws) that is large enough for the entire evaporator to be pulled from the plenum. Once this panel is removed, there may be another panel to remove for access to the inlet side of the evaporator. Otherwise, you may need to slide the coil partially out of the plenum to get at the bottom of the coil. With the coil partially moved out of the plenum, the technician can now inspect the inlet air side of the coil (see Figure 11–19); Table 11–8 provides the details.

During maintenance, pay close attention to evaporator drains and drain pans. Accumulations in pans and pan traps can cause blockage that leads to water accumulation in the pans, giving molds and bacteria a place to grow. If the pans overflow then insulation may become wet, providing additional environments in which these contaminating substances can grow. Many newer systems are designed with evaporator pans that are easily removed for cleaning. Ensure that pans are properly pitched for complete drainage during installation.

Measuring Pressure Drop

Pressure drop readings can be used to diagnose a system component. Pressure drop readings should be taken and compared to manufacturer specifications. If a pressure drop measures 0.2 inches of w.c., the technician should suspect there is

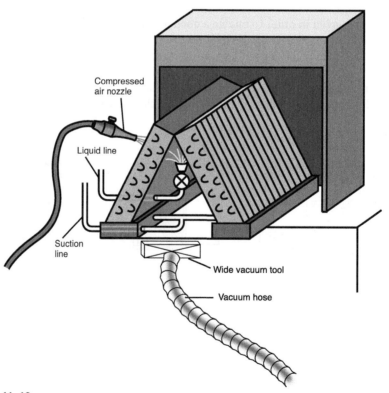

Figure 11–19
Cleaning with compressed air and a vacuum. The opening to the furnace heat exchanger can be shielded by sliding some cardboard between the drain pan and the coil mounting bracket in the plenum.

Table 11–8 Cleaning the Coil

Method	Instructions
Compressed air and vacuum	A vacuum can be used with a soft brush to clean heavy debris from the coil. The vacuum is held on the inlet side of the coil as compressed air is used to push debris toward a wide vacuum attachment, which catches the dust.
Coil cleaner	Always follow the instructions provided by the manufacturer of the cleaner. The cleaner is applied evenly to the surface of the evaporator and allowed to drain out of the evaporator condensation pan, where usually it is captured (unless the product is biodegradable). The cleaner is rinsed (although some chemicals do not require this) with plain water from a spray bottle. Both sides of the evaporator are cleaned using the same method. *Note:* Be sure that the evaporator pan does not overflow into the system.
Combination cleaning	A combination cleaning (brush, vacuum, and coil cleaner) removes both the large debris and the residue that adheres to the surface of the fins.

Tech Tip

Cleaning the A-coil. Compressed air or a compressed gas, such as nitrogen, can be used to dislodge dust from the coil. When doing this, the gas pressure should be at a safe psi level and the jet of gas should be perpendicular to the coil fins; taking these precautions will reduce the chance of bending the fins. A vacuum with a wide attachment should be used to capture the dust. A garbage bag can be taped to the bottom of the open portion of the A-coil and the vacuum hose then inserted through the bag. To keep dust from falling back into the heat exchanger, a piece of cardboard can be inserted between the A-coil drain pan and where the coil sits in the plenum. Be sure to vacuum the cardboard before removing it from the plenum. Also be sure to use proper PPE when cleaning the coil.

Field Problem

Air conditioning odors. The customer complaint was about the smell that came from the air conditioning system whenever it started. Once the system was running the smell seemed to dissipate, but it was overpowering at start-up. The customer demonstrated the problem: starting the system resulted in a heavy, musty smell, which gave the technician an idea of where to start. Gaining access to the drain pan, the technician found that the drain was plugged and that the condensate had been leaking down the inside of the furnace cabinet. The fiberglass insulation in the furnace cabinet had become saturated and was showing signs of biological growth. The furnace needed to be disassembled, the insulation replaced, and the cabinet scrubbed. With the customer's approval, the technician unplugged the condensate drain and finished the clean-up work while wearing a respirator. When he then put the system back into operation, the customer confirmed that the smell was gone.

a problem. Low air flow is symptomatic of a dirty coil whose pressure drop may be 0.2" w.c. or greater. Any heat exchanger, air filtration device, or other system component through which air moves can be measured for pressure drop. Once measurements are completed, the technician can compare the readings with the manufacturer specifications for the system or component. If the pressure drop is greater than called for by the specifications, then a cleaning may be necessary.

A technician who wishes to document the effectiveness of a cleaning process should take a pressure drop reading before and after cleaning. The pressure reading is taken across the coil with an inclined manometer (see Figure 11–20) or other pressure differential measuring device. The difference between the two pressure measurements (before cleaning and after cleaning) will indicate the effectiveness of the cleaning process.

After the coil is clean, another pressure drop reading will demonstrate how effective the cleaning process has been. The pressure drop should be what is recommended by the coil manufacturer and should always be less than the pressure drop before cleaning. Pressure drop readings may be required to calculate efficiency after cleaning or to document the cleaning process.

Tech Tip

Working pressure drops. A "working" pressure drop will be different than a pressure drop taken with only the blower operating. When it's working, the evaporator coil will condense moisture, which resists the flow of air. This is the difference between a dry coil and a wet coil. The working pressure drop will always be greater when the air conditioning system is working.

Chemical Treatment

There are some chemicals that can be applied to the surface of the coil to prevent debris buildup and mold growth. The chemical treatment allows the coil to oper-

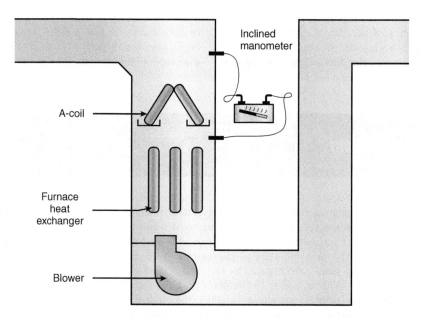

Figure 11–20
An inclined manometer is used to obtain direct readings of the pressure drop across the A-coil.

Table 11–9 Typical System Measurements

Measurement	Evaluate
Total static	Is there too much static pressure for the system blower to overcome?
CFM per supply diffuser	Is there enough air being delivered to each room?
CFM per return grille	Is the amount of air being returned the same as the amount of air being supplied? What is the leak rate?
Air temperature drop/rise	Is the air being cooled/heated according to the manufacturer's specifications?
Pressure drop across the heat exchanger	Is the "working" pressure drop what is recommended by the manufacturer?
Blower amperage	Is the blower amperage below the FLA (full load amperage) specifications on the motor nameplate?

ate longer without cleaning and also serves to make the cleaning process easier. In addition, the chemicals can protect the fins from deterioration caused by salts, acids, and other airborne corrosives.

After the coil is properly cleaned (according to the manufacturer's instructions), the protectant can be applied. It is usually applied with a spray bottle and then allowed to dry on the coil surface.

A different chemical can be applied to the condensate pan in order to prevent buildup of bacterial slime and molds. This chemical usually comes as a packet or pad that can be placed in the drain pan. Each pad will last 6–9 months before it needs to be replaced.

Air Testing and Evaluation

Air delivery systems are generally tested before and after work is done. By conducting a test of the system before work is done, the technician can document the system conditions as they are encountered. Table 11–9 describes the system measurements that are most commonly taken. This best practice also allows the technician to evaluate the entire system before making a final diagnosis. After all work is done, repeat testing and evaluation serves to document the new system conditions and thus any changes that resulted from the work being done. This documentation facilitates efficiency calculations, from which energy savings can be determined.

TECHNICIAN MEASUREMENTS

In addition to standard system operating measurements, the technician may be asked to check and document other conditions of the system. Examples include the amount of gases produced by the heating system and the amount of leakage into the occupied space. Another set of measurements determines the amount of air leakage or infiltration that a structure exhibits.

Evaluation of CO and CO$_2$

Carbon monoxide and carbon dioxide are produced by fuel-burning appliances. In a home or business there may be fuel-burning appliances besides the heating

Safety Tip

Before using any chemical, read and understand the manufacturer's instructions. Obtain and read the MSDS (material safety data sheet) for information on PPE (personal protective equipment) required.

Figure 11–21
A handheld indoor air quality monitor that records carbon dioxide, temperature, humidity, and gas levels.

Table 11–10 Limits for Exposure to CO and CO_2

Authority with Jurisdiction	CO	CO_2
ASHRAE	9 PPM/8 hr	1,000 PPM/continuous
EPA	9 PPM/8 hr; 35 PPM/hr	1,000 PPM/continuous
UL	30 PPM /30 day alarm limit (UL2034)	300 PPM/continuous (3%)
ACGIH	25 PPM/8 hr	5,000 PPM/TWA
OSHA	50 PPM/8 hr	10,000 PPM/TWA

Key: ACGIH, American Conference of Governmental Industrial Hygienists; ASHAE, American Society of Heating and Air-Conditioning Engineers; EPA, Environmental Protection Agency; OSHA, Occupational Safety and Health Administration; PPM, parts per million; TWA, time-weighted average; UL, Underwriters Laboratory.

system. These other appliances contribute to the overall condition within the building. Ovens, ranges, unvented heaters, and unvented fireplaces can all add carbon monoxide and carbon dioxide to the living space. Table 11–10 reports maximum allowed exposure as estimated by various authorities. Portable air quality monitors (see Figure 11–21) are available for evaluating these and other air conditions.

Some buildings require additional ventilation because of the number of people who exhale carbon dioxide within them. Insufficient ventilation or dilution of air will contribute to "sick building syndrome," adversely affecting the occupants. Some ventilation systems can be activated based on carbon dioxide monitoring.

Blower Door Testing

A *blower door test* consists of placing a blower in an outside door of the structure and using the blower to reduce static pressure within the building (see Figure 11–22). The amount of air passing continuously after the building reaches the prescribed static level is the amount of air that is leaking into the structure.

The blower door is used to create a pressure difference between the inside and outside of the structure to approximately 50 pascals (or 0.2" w.c.). Blower door technicians refer to this pressure as CFM50 (cubic feet per minute at 50 pascals pressure difference), which is considered the industry standard. Air moving through the blower at that pressure is measured (in cfm), and leakage rates are calculated based on the building size. For example, LEED (Leadership in Energy and Environmental Design) standards for multifamily dwelling units prescribe maximum infiltration rates of 1.25 square inches of leakage area per 100 square feet of enclosure area (all surfaces: floor, walls, and ceiling). The equivalent of 1.25 square inches of leakage is 0.0742 cfm/ft^2.

For example, in a 1,200-ft^2 home with an aspect ratio (length to width) of 2:1, the enclosure area is 3,576 square feet (approximately 49 ft long by 24.5 ft wide). Hence the home could have as much as 44.7 square inches of leakage area (3,576/100 × 1.25 = 44.7). If instead we calculate in units of cfm, then the limit would be 265.3 cfm of air leakage at CFM50 (3,576 × 0.0742 = 265.3).

If we normalize the readings based on the building's volume, then a new, air-tight home might have a leakage rate of 1–3 ACH50 (air changes per hour at 50 pascals); older homes may have as much as 10–15 ACH50. The objectives of the blower door test are first to determine how much air change is occurring and second to reduce the amount of air infiltration. Air leakage through cracks and holes in the building material or foundation may be eliminated with caulk and foam fillers. Air leakage around doors and windows may require upgrades.

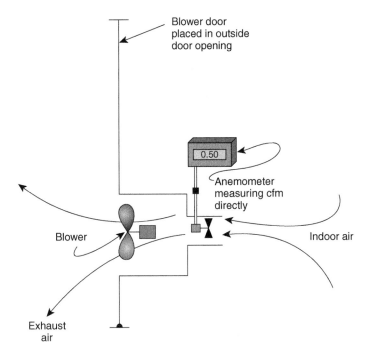

Figure 11–22
A simplified schematic of the blower door test, showing indoor air exhausted and the volume measured in cfm.

IAQ CHECKS AND MONITORING

Some HVAC systems have built-in monitoring controls to determine the level of CO_2 and moisture. These monitoring controls operate outdoor air dampers and air conditioning systems to reduce undesirable concentrations of both levels. Another monitoring device, the *static pressure control,* monitors the pressure difference between the inside and outside of the building. This monitoring control opens or closes outside air dampers or changes the blower motor speed to maintain positive pressure inside the building relative to the pressure outside. Each of these monitoring devices ensures that moisture, dilution air, and outside contaminants are controlled.

When called in to resolve an IAQ issue or simply to work on an HVAC system, the technician must conduct system checks to determine if any component or feature of the system, building, or occupancy condition could contribute to an IAQ problem. An IAQ management checklist can be found on the EPA website. Table 11–11 summarizes the elements that can contribute to poor IAQ.

OUTSIDE AGENCIES

Within the HVACR field, ACCA (Air Conditioning Contractors of America) and ASHRAE (American Society for Heating, Refrigeration, and Air Conditioning Engineers) set many of the standards that are adopted by outside agencies. Outside the field of HVACR, several other agencies issue guidelines or requirements that are related to IAQ. Some of these organizations are listed below.

- ACCA—the Air Conditioning Contractors of America sets the standards for Quality Restoration (QR).
- AHAM—the Association of Home Appliance Manufacturers set standards for appliances used in homes and businesses.

Table 11-11 Poor IAQ Checklist

Problem	Possible Cause
• HVAC system cleanliness	• Condensate drain pans need cleaning • Humidifier pads and sumps are dirty • Cooling tower or evaporative condenser systems show debris or biological growth • Blowers and system components are dirty
• Moisture/water problems	• Roof leaks or has leaked • Window/door leaks at the sill/threshold show moisture • Condensate leaks from the system • Condensation forming on pipes or ductwork • Ductwork is damp or shows signs of being wet • Biological growth • Drain(s) overflowing • Perimeter structural leaks • Recent flooding • Negative building pressure
• Air issues	• Problems with filters or filtration system • Blockage in heat exchanger • Leaking return duct(s) • Not enough dilution air • Carbon monoxide levels too high • Carbon dioxide levels too high • Smoking by occupants • No ventilation system or windows in bath or kitchen • Attached garage leakage to living space
• Building environment	• Open cleaning agents • Open chemical agents • Pests • High levels of pet hair and residue • Lack of building cleaning • Lack of trash disposal • Poor seals around doors and windows • Problems with building construction

- CDC—the Centers for Disease Control and Prevention set the standards for human habitation and toxic substances.
- DOE—the Department of Energy is the parent agency of the EPA.
- DOH or DEQ—the local departments of health or departments of environmental quality are charged with regulating the environmental quality of buildings and environments; in many cases, they are the agency of local jurisdiction.
- EPA—the Environmental Protection Agency is involved with energy, environments, fuels, and refrigerants.
- HUD—the U.S. Department of Housing and Urban Development sets the standards for housing projects.
- LEED—the Leadership in Energy and Environmental Design program sets standards for assessing the environmental impact of a new or existing building.

- NIOSH and OSHA—the National Institute for Occupational Safety and Health and the Occupational Safety and Health Administration set the standards for worker safety.

Each of these agencies can be accessed via the Internet. Using the browser of your choice, search for the acronym or the complete name.

SUMMARY

In this chapter we discussed many of the issues regarding IAQ. The purpose of conditioned air is to maintain the comfort and health of occupants; however, the "sick building syndrome" describes what happens when people become ill as a result of a poor indoor environment. The chapter also discussed the standards promulgated by various organizations and their impact on IAQ. Standards and measures are required to assess and compare different buildings and environments. By making comparisons, we can determine what is needed to make the indoor environment better.

Proper ventilation is an important function of a good HVAC system. Ventilation helps maintain breathable air, dilution reduces concentrations of CO_2 and other gases, and air movement through filtration systems helps to control particulate matter. Cooling the air reduces the amount of moisture; with too much moisture, buildings can grow mold.

Filtration reduces particulate matter, and filter types and their application were discussed and illustrated. Particle size is what determines how well filters work. We know that a filter performs better as it becomes dirty—but dirty filters can create other problems, including a greater pressure drop. Some bacteria and mold can be killed by UV filtration. We have outlined various maintenance procedures that have come to be considered best practices.

An air delivery system integrates all of the components of a HVAC system. Working as a system, it can deliver clean, conditioned air to every part of a building. This requires that the system be clean, to keep resistance low, and also sealed to prevent air from entering or escaping before it is delivered to the proper space. Coil and drain pan cleaning is an important maintenance procedure, and the change in pressure drop after a good cleaning can be measured. Taking measurements before and after maintenance work provides data for determining system efficiencies and energy savings.

Some information about other agencies involved with IAQ issues was provided. All of these agencies can provide additional information about IAQ problems and solutions.

REVIEW QUESTIONS

1. Describe "sick building syndrome" as if you were explaining it to a customer.
2. Provide one example of how the DOE affects IAQ issues.
3. Identify a contaminant and describe how an HVAC system could or should be operated to reduce or eliminate it.
4. Identify one VOC (volatile organic compound) that might be found in a residential building, and describe how to reduce or eliminate VOC concentration.
5. Identify an indoor pollutant and describe how an HVAC system would reduce or eliminate it.
6. Identify an ASHRAE standard and describe its application to IAQ.
7. What agency within the DOE is responsible for information about IAQ? List one of the services this agency provides.
8. Describe a customer complaint related to IAQ and explain what the service technician could do for the customer.
9. What is dilution air? How does it relate to ventilation requirements?
10. How can a technician reduce carbon dioxide levels in a building?

11. What are particulates? How can the level of particulates be reduced?
12. Why is humidity important in a living space? How can it be controlled?
13. Identify an HVAC system component and describe how its maintenance contributes to good IAQ.
14. Explain one method of sealing a duct system. Why should this be done?
15. Why is duct cleaning not recommended by the EPA?
16. Briefly explain how the bottom side (air inlet) of an A-coil is cleaned.
17. Why is it a best practice to take pressure drop readings before and after cleaning the air conditioning coil?
18. Identify an air measurement test that should be performed on an air delivery system, and briefly describe what is being evaluated.
19. Identify a gas that results from the combustion process, and list the amount of that gas permitted in a dwelling over an 8-hour period of time (state the agency that might have jurisdiction).
20. What is a blower door test, and how is it performed?
21. Name an outside agency concerned with IAQ, and describe the type of building environment that they are charged with protecting.

Answer Key

Chapter 1

1. Describe three ways in which gas and oil furnace checks are similar. *Blower checks, airflow checks, and cleaning*

2. Describe what is necessary to make the conversion from natural gas to LP. *A conversion kit from the manufacturer*

3. Why is combustion testing so important? *To reduce CO emissions, to optimize efficiency, and to maximize equipment life*

4. What needs to be done to conduct a complete combustion analysis? *Set up for the test; operate to temperature; conduct an initial test; readjust for highest efficiency*

5. Why is it necessary to bring in fresh air? *Fresh air is needed for people to breathe; it augments the oxygen depleted by fuel-burning equipment that uses interior air.*

6. What are the differences in checks made for air-source and ground-source cooling systems? *None, except for specific parts used by each system*

7. What are alternative energy systems? *Systems that use energy that is not derived from fossil fuel*

8. What are the similarities between solar hot-water systems and hydronic systems? *None*

Chapter 2

1. Describe the role of a system's terminal device. *It is the last point in a system where control can be applied to energy entering or leaving the system.*

2. List two meters or test instrumentation used for TAB. *Examples include inclined manometer (used to measure static pressure) and anemometer (used to measure velocity of air).*

3. What is a manometer, and how is it used? *A manometer is a device that measures in inches of water column; it is used to measure pressure.*

4. Describe the different types of manometers. *There are two types, fluid filled and mechanical.*

5. How are velocity and volume related? *As pressure increases, velocity and volume both increase.*

6. What purposes does a flow control serve in a hydronic system? *Flow controls are used as mixing valves, in setting circuit flow, and for regulating three-way flow.*

7. What is traversing? *Traversing of a duct means taking many measurements in a pattern to determine average flow.*

8. What is the first TAB step for an "air" system? *Operate all exhaust fans.*

9. What are TAB methods? *TAB methods are steps or processes in which measurements are made and flow control devices are positioned to balance the system.*

10. How is TAB data recorded? *TAB data can be recorded by hand or by computer. The data is typically stored in a computer database program.*

Chapter 3

1. What does it mean if one system has an SEER of 15 and another has an SEER of 13? *The higher SEER means that the system is more efficient.*

2. Can SEER and HSPF be compared? Why or why not? *No, because SEER applies to cooling equipment and HSPF to heating; however, both ratings may be applied to a heat pump.*

3. What is the purpose of energy efficiency ratings? *To compare similar systems from different manufacturers and to help calculate projected operating costs*

4. What makes the heat pump an interesting system for energy efficiency? *Its ability to move and concentrate existing heat that is available in air and the earth*

5. Why are gas forced air systems chosen for energy-efficient delivery systems? *Gas forced air systems heat and cool the air we live in directly. Return air temperature is low enough to extract heat from the flue gas until it condenses water.*

6. Why are hydronic systems often found in energy-efficient homes? *Their ability to heat floors at low temperatures means that most of the heat energy is extracted from flue gases, making it condense.*

7. Why do equipment manufacturers supply specifications? *Equipment specifications are used to select equipment that will match the needs of the building and the delivery system.*

8. Why are system delivery specifications developed? *Delivery specifications help to select components and equipment that will have an efficiency match or will work efficiently together.*

9. Describe one upgrade that qualifies as an energy-efficient retrofit. *For example, installing a vent damper to reduce off-cycle heat losses from the heat exchanger*

10. What is meant by "green building"? *Green building means building that is energy efficient; taking time to build in a way that leads to energy efficiency.*

Chapter 4

1. Describe the function of a heat transfer fluid. *It moves heat from one location to another.*

2. Identify the heat transfer fluids that are used to heat a room with a boiler. *Water in the boiler and air in the room*

3. Describe how air pressure is measured. *Air pressure is measured using a manometer.*

4. What are the units of measurement for air pressure? *Inches of water column (w.c.)*

5. What is velocity pressure, and how is it measured? *Velocity pressure is the speed of air and is measured in inches of water column (a formula is used to convert this value to speed).*

6. Describe how to convert velocity pressure to feet per minute. *Velocity pressure squared, multiplied by the constant 4,004, is equivalent to feet per minute.*

7. Describe how to convert gauge pressure to feet of head. *Multiply the gauge pressure by 2.31 to obtain feet of head.*

8. Describe the tool that is connected to an inclined manometer and used to measure pressure in a duct. *The tool, called a pitot tube, is designed to measure total and static pressures.*

9. Describe what happens to a fluid when the resistance to flow is increased (air or water can be used, and a curve may be used to diagram). *The flow of the fluid will drop when the resistance to flow is increased.*

10. Describe what will happen in a hydronic system if additional pipe and fittings are installed in a loop and the circulator is unchanged. *Fluid flow will decrease because the same circulator will need to push against additional resistance (feet of head).*

Chapter 5

1. Explain why PPE is used. *To protect against incidents that might occur on the job*

2. How is the correct PPE determined? *The hazards of any job must be assessed before the job is conducted, and the PPE selection is made at that time.*

3. Thoroughly describe one safe testing technique. *For example, deenergizing by turning off all live power and confirming that it is off (test with a voltmeter)*

4. What does CAT III mean for an HVAC technician? *It means that the test instrument is rated for HVAC type work.*

5. What types of electrical information are available on the equipment's nameplate? *Volts, amps, hertz, minimum ampacity, and maximum fuse (circuit breaker) sizing*

6. How much of an electrical print needs to be read before testing? *Only that part of the print that applies to the test*

7. What is logic wiring, and how is it read? *Logic wiring shows how one device interacts with another; it is read from one potential source, through devices, and to the other potential.*

8. What is AND logic? *It can be described as closing one switch AND closing another before a motor will run.*

9. What is "ruling out," and how does it work as a technique? *Ruling out is a process of eliminating as trouble sources those components that are working; this helps to pinpoint what is not working.*

10. What is a truth table? *It is a way of systematic testing and data collection to understand how a device or circuit works.*

11. How can testing for an open circuit be done with a voltmeter? *Testing across an open circuit with the power applied will generate a voltage reading on the meter.*

12. Describe how to find a good ground to which an electrical ground can be hooked. *You are looking for a ground source; testing from a known ground to a potential grounding point with a multimeter should show 2 ohms (or less) for a good ground.*

13. How does a megohmmeter test show problems with the compressor? *A megohmmeter reading may indicate winding insulation leakage, which could lead to a burnout.*

14. Why should electrical tests of equipment be documented? *So you can see trends in the data and repair potential problems.*

15. How many classifications of relay switching are there? Name one. *Six in this text (e.g., SPST)*

16. What is the output of a transformer (either step-up or step-down) called? *The secondary of the transformer*

17. To what are solid-state controls more susceptible than electromechanical ones? *Stray voltage and shorts*

18. Why do some three-phase motors require starting controls? *Large three-phase motors must be started in stages to reduce in-rush current.*

19. What is the difference between *current* and *potential* magnetic relays for starting motors? *Current relays monitor amperage, and their contacts are normally open (NO); potential relays monitor voltage, and their contacts are normally closed (NC).*

20. Why is phase protection necessary? *If one leg of a three-phase motor is lost, then the motor is left with only the single-phase power of the other two legs and so cannot operate. Phase protection disconnects all three lines of power when one leg is lost.*

Chapter 6

1. Describe why checks are done before a system is started for the first time. *To eliminate potential problems that may cost time and money*

2. Describe the process of ruling out parts of the system. *To eliminate those parts of the system that work*

3. How can cause and effect be used to identify problems in a refrigeration system? *Troubleshooting flow-charts are cause-and-effect diagrams used to identify problems.*

4. What does "scheduled maintenance" mean to a customer? *Scheduled maintenance usually means work performed under the terms of a service contract.*

5. Why is cleanliness so important to refrigeration service? *It minimizes contamination of the system.*

6. Describe the purpose of and procedures for purging the system with inert gas. *Purging removes oxygen from the system; it is used to prepare the system for brazing and to eliminate oxidation and flaking.*

7. What process must occur before recharging can be done? Why? *Evacuation; it is necessary to remove air and moisture*

8. How does pumping down a system isolate its low side? *Pumping down moves the refrigerant to the condensing unit; the condensing unit will then be isolated from the rest of the system, because liquid line or receiver service valves are frontseated prior to pumping down. Upon completion and after the compressor is turned off, the compressor discharge valve is shut.*

9. What are two of the largest causes of compressor failures? *Mechanical and electrical; or liquid-line slugging and high-temperature discharge*

10. What is one reason for compressor warranty claims that the technician can eliminate? *By verifying the mechanical or electrical fault, the technician can eliminate the unnecessary return of functional compressors.*

Chapter 7

1. What can be read at the sole of a psychrometric chart? *Dry-bulb temperatures and BTU/lb (enthalpy)*

2. What can be read at the heel of the psychrometric chart? *Pounds or grains of moisture; dew point*

3. What can be read at the toe of the psychrometric chart? *Nothing; this region of the chart is cut off because it is outside the range of normal comfort cooling.*

4. What can be read at the instep of the psychrometric chart? *Wet bulb and dew point*

5. Describe or demonstrate how to plot the following points on a psychrometric chart: (a) 68°F DB and 80% RH; (b) 73°F DB and 80% RH; (c) 80°F DB & 50% RH; (d) 80°F DB and 20% RH; (e) 68°F DB and 20% RH. *Plotting these points and drawing lines to connect the points will show the human comfort area.*

6. Describe or demonstrate how to plot the following points on a psychrometric chart: (a) 68°F DB and 80% RH; (b) 62°F WB and 100% RH. Describe or show the amount of moisture change between the two points. *There is little to no difference in moisture (depending on charts and plotting techniques).*

7. Describe or demonstrate how to plot the following points on a psychrometric chart: (a) 68°F DB and 80% RH; (b) 73°F DB and 80% RH; (c) 80°F DB and 50% RH; (d) 80°F DB and 20% RH; (e) 68°F DB and 20% RH. *Plotting these points and drawing lines to connect the points will show the human comfort area.*

8. Describe or plot 56.5°F WB and 80°F DB on a psychrometric chart, and then read the following conditions of air at that point: volume, total heat content, grains of moisture, and relative humidity. *The approximate answers are (depending on charts and plotting techniques) volume, 13.7 ft³/lb; total heat content, 23.9 BTU/lb; grains of moisture, 0.0043 pounds/lb or 30.7 grains/lb; relative humidity, 20%; dew point, 35.5°F WB.*

9. Describe what is happening to air that enters the return air grille of a comfort conditioning system at 70°F DB and 58.5°F WB but leaves the diffuser at 55°F DB and 51.5°F WB. *Comfort cooling and dehumidification; both the temperature and moisture content were reduced*

10. Describe what is happening to air that enters the return air grille of a comfort conditioning system at 70°F DB and 50°F WB but leaves the diffuser at 95°F DB and 68°F WB. *Comfort heating and humidification; both the temperature and moisture content were increased*

Chapter 8

1. Describe the function of the wetted surface of the cooling tower. *The surfaces of a cooling tower increase the surface area of the water, which allows it to evaporate more rapidly.*

2. Describe the function of the spray nozzles in a cooling tower. *Spray nozzles direct the water droplets to fill the air space evenly with water and to reduce the size of droplets, increasing the amount of water surface area and in some cases atomizing the water.*

3. Describe the function of a cooling tower fan. *The fan is designed to move a prescribed amount of air into or through the cooling tower to match the amount of cooling water generated.*

4. Describe the function of the water-level control. *The water-level control may be as simple as a float valve. Its job is to maintain the level of water in the base of the cooling tower by allowing new water to be introduced as water is evaporated.*

5. Describe the function of the bleed water control or valve. *Bleed water is water that is higher in mineral concentration and must be removed from the basin before it deposits on the surfaces of the cooling tower. The bleed water valve opens and allows the water with this concentrated mineral content to go to the drain.*

6. Describe how water is monitored for biological growth. *The surfaces of the cooling tower are visually*

inspected for biological growth, which is usually slimy and/or green in color.

7. Describe the function of a screen and what should be inspected. *The screen captures debris before it leaves the tower. At this point, biological growth as well as rust and tower parts can be caught.*

8. Describe the function of a cooling tower. *The explanation should incorporate the following elements: water evaporation, approach temperature, wet bulb, and air movement.*

9. Describe the structural difference between a cooling tower and an evaporative condenser. *The explanation should include location of the refrigerant tubing bundle, how water touches the bundle, and location of the evaporative condenser.*

10. Describe the benefits of good cooling tower maintenance. *The explanation should include examples of daily, weekly, monthly, and yearly maintenance; it should also indicate that maintenance increases the longevity of the cooling tower.*

11. A cooling tower is operating, and the outside air temperature is 90°F. How could water leaving the cooling tower measure 85°F? *The water in the cooling tower is being cooled to the outside air dew point or wet-bulb (WB) temperature. WB temperatures are typically below the outdoor dry-bulb (DB) condition. The only time WB and DB temperatures are the same is when the air is completely saturated with moisture—100% relative humidity (RH). If the WB temperature is 85°F or lower, the water leaving the coil would be 85°F.*

Chapter 9

1. Explain why centrifugal chillers are different from other chillers. *They use centrifugal compressors; they are usually large systems; they use low-pressure refrigerants.*

2. Explain what distinguishes a screw chiller from other types. *They use screw compressors; they are usually medium-large systems.*

3. Explain the benefits of a central plant system. *Central plants are cheaper to operate per ton; maintenance costs are reduced.*

4. Describe one troubleshooting check for metering devices. *For example, check superheat readings.*

5. Describe one troubleshooting check for compressors. *For example, check voltage and amperage.*

6. Describe one troubleshooting check for heat exchangers. *For example, check the pressure drop.*

7. Describe one troubleshooting check for solenoid valves. *For example, check for magnetic flux.*

8. Describe one troubleshooting check for reversing/diverting valves. *For example, check the operator (solenoid or motor).*

9. Describe one troubleshooting check for capacity controls. *For example, check the motor amperage for cylinder unloaders.*

10. Describe one troubleshooting check for sensors. *For example, measure the control input for electronic sensors.*

11. Describe one troubleshooting check for filter-driers. *For example, measure the temperature difference from inlet to outlet of a liquid-line filter-drier.*

12. Describe one troubleshooting check for liquid-level controls. *For example, check the level using a sight glass.*

13. Explain why maintenance records are important for service and maintenance of equipment. *They document changes in operation over long periods of time.*

Chapter 10

1. Describe the physical state of water that creates an acidic condition. *Water whose pH level is lower than neutral (7)*

2. Under what condition is calcium high? How is this detected? *A high calcium content creates hard water; the presence of this condition can be simply established by observing the amount of soap necessary to create soap bubbles.*

3. What is Legionella, and why is it a problem? *Legionella is a bacterium that causes flu-like symptoms and may lead to death; it's a problem because it grows in the warm, stagnant water found in cooling towers.*

4. What is biofilm? How can it be detected? *Biofilm is a colony of bacteria; it can be detected by touch (slimy feel) and by sight (film on water or large structures in advanced stages).*

5. Describe the treatment of water for high pH. *Water with high pH is alkaline and needs the addition of an acid to bring it back to neutral.*

6. A cooling water system needs to be blown down. Describe two conditions that might have made this procedure necessary. *(1) High turbidity; water has too many suspended solids. (2) Water chemicals are not effective in dropping out particles; chemical amounts may need to be adjusted.*

7. What is the difference between manual and automatic water treatment? *Manual treatment requires that a person add chemicals, whereas automatic treatment is done with timers or sensors.*

8. According to the maintenance schedule, how often should the system be drained and refilled? *Every three months*

9. Why is safety important for the environment? *The environment could be degraded by accidental spills or incorrect dosage rates.*

10. When troubleshooting a cooling water system, what is the first check? Why? *The first check is for proper pH level. If the water is not close to neutral, then the other chemistry may not work effectively.*

Chapter 11

1. Describe "sick building syndrome" as if you were explaining it to a customer. *In this condition, people's health is negatively affected by exposure to the building environment.*

2. Provide one example of how the DOE affects IAQ issues. *The EPA, which is part of the DOE, supplies information about IAQ.*

3. Identify a contaminant and describe how an HVAC system could or should be operated to reduce or eliminate it. *For example, carbon monoxide. Conduct combustion tests and adjust to reduce CO; make sure venting is occurring in accordance with manufacturer's specifications.*

4. Identify one VOC (volatile organic compound) that might be found in a residential building, and describe how to reduce or eliminate VOC concentration. *For example, out-gassing from a new carpet; increase the ventilation*

5. Identify an indoor pollutant and describe how an HVAC system would reduce or eliminate it. *For example, inoperable exhaust fans; repair and place back into service*

6. Identify an ASHRAE standard and describe its application to IAQ. *For example, Standard 62 for IAQ with regard to air changes per hour*

7. What agency within the DOE is responsible for information about IAQ? List one of the services this agency provides. *The EPA; for example, an electronic handbook on IAQ*

8. Describe a customer complaint related to IAQ and explain what the service technician could do for the customer. *The technician should document any complaint and attempt to determine the cause of the problem.*

9. What is dilution air? How does it relate to ventilation requirements? *Dilution air is provided by the ventilation system to reduce concentrations of gases in the living area.*

10. How can a technician reduce carbon dioxide levels in a building? *Increase ventilation rates or room air changes per hour.*

11. What are particulates? How can the level of particulates be reduced? *Particulates are particles or solids that are suspended or carried in an air stream. Use and maintain the right filtration system for the size of particle whose quantities are being reduced.*

12. Why is humidity important in a living space? How can it be controlled? *Moisture levels affect the comfort level; air conditioning is used to reduce the level of moisture in the air, and humidifiers are used to increase moisture levels.*

13. Identify an HVAC system component and describe how its maintenance contributes to good IAQ. *For example, the humidifier. Maintenance involves replacing the humidifier pad, cleaning the pan or sump, washing the interior of the humidifier housing, ensuring that the water make-up valve is clean and functional, and checking and cleaning the overflow (if installed).*

14. Explain one method of sealing a duct system. Why should this be done? *For example, nonexpanding foam to fill cracks in the building for return systems; return systems pull air through the crack or from the outside if not sealed*

15. Why is duct cleaning not recommended by the EPA? *Because no studies have demonstrated a positive effect from cleaning*

16. Briefly explain how the bottom side (air inlet) of an A-coil is cleaned. *The A-coil may need to be slid out from the plenum to reach the inlet side.*

17. Why is it a best practice to take pressure drop readings before and after cleaning the air conditioning coil? *To document the effect of cleaning or to establish that it required cleaning*

18. Identify an air measurement test that should be performed on an air delivery system, and briefly describe what is being evaluated. *For example, total static; this determines if the system blower has too much static pressure to overcome.*

19. Identify a gas that results from the combustion process, and list the amount of that gas permitted in a dwelling over an 8-hour period of time (state the agency that might have jurisdiction). *For example, CO, whose limits are 9 PPM/8 hr as established by ASHRAE*

20. What is a blower door test, and how is it performed? *A blower door is a device that exhausts indoor air to the outside; it is used to test for leaks in a building.*

21. Name an outside agency concerned with IAQ, and describe the type of building environment that they are charged with protecting. *For example, OSHA, which protects individuals in work environments*

Glossary

Absolute pressure. Gauge pressure plus the pressure of the atmosphere, normally 14.696 psi at sea level at 70°F.

Presión absoluta. La presión del calibrador más la presión de la atmósfera, que generalmente es 14.696 psi al nivel del mar a 70°F (21.11°C).

Absolute zero temperature. The lowest obtainable temperature where molecular motion stops, –460°F and –273°C.

Temperatura del cero absoluto. La temperatura más baja obtenible donde se detiene el movimiento molecular, –460°F y –273°C.

Absorption. The process by which one substance is absorbed by another.

Absorción. Proceso mediante el cual una sustancia es absorbida por otra.

Accumulator. A storage tank located in the suction line to a compressor. It allows small amounts of liquid refrigerant to boil away before entering the compressor. Sometimes used to store excess refrigerant in heat pump systems during the winter cycle.

Acumulador. Tanque de almacenaje ubicado en el conducto de aspiración a un compresor. Permite que pequeñas cantidades de refrigerante líquido se evaporen antes de entrar al compresor. Algunas veces se utiliza para almacenar exceso de refrigerante en sistemas de bombas de calor durante el ciclo de invierno.

Acetylene. A gas often used with air or oxygen for welding, brazing, or soldering applications.

Acetileno. Gas usado con frecuencia, junto con aire u oxígeno, en trabajos de soldadura y soldaduras de cobre.

Acid-contaminated system. A refrigeration system that contains acid due to contamination.

Sistema contaminado de ácido. Sistema de refrigeración que, debido a la contaminación, contiene ácido.

"A" coil. An evaporator coil that can be used for upflow, downflow, and horizontal-flow applications. It actually consists of two coils shaped like a letter "A."

Serpentín en forma de "A". Serpentín de evaporación que puede utilizarse para aplicaciones de flujo ascendente, descendente y horizontal. En realidad, consiste en dos serpentines en forma de "A".

ACR tubing. Air conditioning and refrigeration tubing that is very clean, dry, and normally charged with dry nitrogen. The tubing is sealed at the ends to contain the nitrogen.

Tubería ACR. Tubería para el acondicionamiento de aire y la refrigeración que es muy limpia y seca, y que por lo general está cargada de nitrógeno seco. La tubería se sella en ambos extremos para contener el nitrógeno.

Activated alumina. A chemical desiccant used in refrigerant driers.

Alúmina activada. Disecante químico utilizado en secadores de refrigerantes.

Activated charcoal. A substance manufactured from coal or coconut shells into pellets. It is often used to adsorb solvents, other organic materials, and odors.

Carbón activado. Sustancia fabricada utilizando carbón o cáscaras de coco, en forma de gránulos. Se emplea para adsorber disolventes, así como otros materiales orgánicos y olores.

Active recovery. Recovering refrigerant with the use of a recovery machine that has its own built-in compressor.

Recuperación activa. El hecho de recuperar refrigerante utilizando un recuperador que dispone de su propio compresor.

Adsorption. The process by which a thin film of a liquid or gas adheres to the surface of a solid substance.

Adsorción. Proceso mediante el cual una fina capa de líquido o gas se adhiere a la superficie de una sustancia sólida.

Air-acetylene. A mixture of air and acetylene gas that when ignited is used for soldering, brazing, and other applications.

Aire-acetilénico. Mezcla de aire y de gas acetileno que se utiliza en la soldadura, la broncesoldadura y otras aplicaciones al ser encendida.

Air conditioner. Equipment that conditions air by cleaning, cooling, heating, humidifying, or dehumidifying it. A term often applied to comfort cooling equipment.

Acondicionador de aire. Equipo que acondiciona el aire limpiándolo, enfriándolo, calentándolo, humidificándolo o deshumidificándolo. Término comúnmente aplicado al equípo de enfriamiento para comodidad.

Air conditioning. A process that maintains comfort conditions in a defined area.

Acondicionamiento de aire. Proceso que mantiene condiciones agradables en un área definida.

Air-cooled condenser. One of the four main components of an air-cooled refrigeration system. It receives hot gas from the compressor and rejects heat to a place where it makes no difference.

Condensador enfriado por aire. Uno de los cuatro componentes principales de un sistema de refrigeración enfriado por aire. Recibe el gas caliente del compresor y dirige el calor a un lugar donde no afecte la temperatura.

Air gap. The clearance between the rotating rotor and the stationary winding on an open motor. Known as a vapor gap in a hermetically sealed compressor motor.

Espacio de aire. Espacio libre entre el rotor giratorio y el devandado fijo en un motor abierto. Conocido como espacio de vapor en un motor de compresor sellado herméticamente.

Air handler. The device that moves the air across the heat exchanger in a forced air system—normally considered to be the fan and its housing.

Tratante de aire. Dispositivo que dirige el aire a través del intercambiador de calor en un sistema de aire forzado—considerado generalmente como el ventilador y su alojamiento.

Air heat exchanger. A device used to exchange heat between air and another medium at different temperature levels, such as air-to-air, air-to-water, or air-to-refrigerant.

Intercambiador de aire y calor. Dispositivo utilizado para intercambiar el calor entre el aire y otro medio, como por ejemplo aire y aire, aire y agua or aire y refrigerante, a diferentes niveles de temperatura.

Air loop. The heat pump's heating and cooling ducted air system, which exchanges heat with the refrigerant loop.

Circuito de aire. Sistema de tubería del aire de calentamiento y de refrigeración de la bomba de calor, que sirve para inter cambiar el calor con el circuito de refrigeración.

Air, standard. Dry air at 70°F and 14.696 psi, at which it has a mass density of 0.075 lb/ft^3 and a specific volume of 13.33 ft^3/lb, ASHRAE.

Aire, estándar. Aire seco a 70°F (21.11°C) y 14.696 psi (libra por pulgada cuadrada); a dicha temperatura tiene una densidad de masa de 0.075 libra/pies^3 y un volumen específico de 13.33 pies^3/libra, ASHRAE.

Air vent. A fitting used to vent air manually or automatically from a system.

Válvula de aire. Accesorio utilizado para darle al aire salida manual o automática de un sistema.

Ak factor. Area constant factor, a manufacturer's calculated free air opening of a duct or vent.

Factor Ak. Factor de área constante, la abertura neta calculada por el fabricante para un conducto o respiradero.

Algae. A form of green or black, slimy plant life that grows in water systems.

Alga. Tipo de planta legamosa de color verde o negro que crece en sistemas acuáticos.

Allen head. A recessed hex head in a fastener.

Cabeza allen. Cabeza de concavidad hexagonal en un asegurador.

All-weather system. System providing year-round conditioning of the air.

Sistema para todo el año. Sistema que proporciona una aclimatación ambiental todo el año.

Alternating current. An electric current that reverses its direction at regular intervals.

Corriente alterna. Corriente eléctrica que invierte su dirección a intervalos regulares.

Alternative refrigerant. One of the newer refrigerants that are replacing the traditional CFC or HCFC refrigerants that have been used for many years. Many of these refrigerants have very low ozone depletion and global warming indices. Some are completely chlorine-free.

Refrigerante alternativo. Cualquiera de los nuevos productos que sirve para sustituir a los refrigerantes basados en CFC o HCFC que han sido utilizados durante muchos años. Muchos de estos nuevos refrigerantes tienen un índice muy bajo de desgaste de la capa de ozono y de calentamiento de la superficie terrestre. Algunos no utilizan cloro.

Ambient temperature. The surrounding air temperature.

Temperatura ambiente. Temperatura del aire circundante.

American standard pipe thread. Standard thread used on pipe to prevent leaks.

Rosca estándar estadounidense para tubos. Rosca estándar utilizada en tubos para evitar fugas.

Ammeter. A meter used to measure current flow in an electrical circuit.

Amperímetro. Instrumento utilizado para medir el flujo de corriente en un circuito eléctrico.

Amperage. Amount (quantity) of electron or current flow (the number of electrons passing a point in a given time) in an electrical circuit.

Amperaje. Cantidad de flujo de electrones o de corriente (el número de electrones que sobrepasa un punto específico en un tiempo fijo) en un circuito eléctrico.

Ampere. Unit of current flow.

Amperio. Unidad de flujo de corriente.

Analog electronic devices. Devices that generate continuous or modulating signals within a certain control range.

Aparatos electrónicos analógicos. Aparatos que generan señales continuas o modulares adentro de cierto registro de control.

Analog VOM. A volt-ohm-milliampere meter constructed so that the meter indicator is a needle over a printed surface.

VOM analógico. Medidor de voltios-ohmios-miliamperios, construido de forma que el indicador consiste en una aguja que se mueve encima de una superficie impresa.

Angle valve. Valve with one opening at a 90° angle from the other opening.

Válvula en ángulo. Válvula con una abertura a un ángulo de 90° con respecto a la otra abertura.

Annual Fuel Utilization Efficiency (AFUE). The U.S. Federal Trade Commission requires furnace manufacturers to provide this rating so consumers may compare furnace performances before purchasing.

Eficacia de uso de combustible anual (AFUE en inglés). La Comisión Federal de Comercio (Federal Trade Commission) de los EE.UU. exige que los fabricantes de hornos indiquen este valor con el fin de que los consumidores puedan comparar los rendimientos de los hornos antes de adquirirlos.

Anode. A positively charged plate or terminal. A point from which electrons leave or move away.

Ánodo. Placa o terminal con carga positiva. Un punto desde donde parten o se alejan los electrones.

ANSI. Abbreviation for the American National Standards Institute.

ANSI. Acrónimo de American National Standards Institute (Instituto Nacional Americano de Normas).

Approach. The difference between the water temperature in the sump of a cooling tower and the wet-bulb temperature of air *entering*.

Aproximación. Diferencia entre la temperatura del agua en el colector de la torre de refrigeración y la temperatura del aire que ingresa, tomada con termómetro húmedo.

Arc flash. A short circuit through the air that flashes over from one exposed live conductor to another conductor or to ground.

Fogonazo de arco eléctrico. Circuito corto en el aire que provoca un fogonazo desde un conductor expuesto con corriente hasta otro conductor o hasta tierra.

Aspirating psychrometer. A tool that uses a fan to pull air over a wet- and dry-bulb sensor.

Sicrómetro aspirante. Herramienta que utiliza un ventilador para pasar aire por un sensor de bola seca.

Atmospheric pressure. The weight of the atmosphere's gases pressing down on the earth. Equal to 14.696 psi at sea level and 70°F.

Presión atmosférica. El peso de la presión ejercida por los gases de la atmósfera sobre la tierra, equivalente a 14.696 psi al nivel del mar a 70°F.

Atom. The smallest particle of an element.

Átomo. Partícula más pequeña de un elemento.

Atomize. Using pressure to change liquid to small particles of vapor.

Atomizar. Utilizar la presión para cambiar un líquido a partículas pequeñas de vapor.

Automatic changeover thermostat. A thermostat that changes from cool to heat automatically by room temperature.

Termostato de cambio automático. Un termostato que cambia de enfriar a calentar automáticamente con la temperatura del cuarto.

Automatic combination gas valve. A gas valve for gas furnaces that incorporates a manual control, gas supply for the pilot, adjustment and safety features for the pilot, pressure regulator, and the controls for and the main gas valve.

Válvula de gas de combinación automática. Válvula de gas para hornos de gas que incorpora un regulador manual, suministro de gas para la llama piloto, ajuste y dispositivos de seguridad, regulador de presión, la válvula de gas principal y los reguladores de la válvula.

Automatic control. Controls that react to a change in conditions to cause the condition to stabilize.

Regulador automático. Reguladores que reaccionan a un cambio en las condiciones para provocar la estabilidad de dicha condición.

Automatic expansion valve. A refrigerant control valve that maintains a constant pressure in an evaporator.

Válvula de expansión automática. Válvula de regulación del refrigerante que mantiene una presión constante en un evaporador.

Automatic pumpdown system. A control scheme in refrigeration consisting of a thermostat and liquid-line solenoid valve that clears refrigerant from the compressor's crankcase, evaporator, and suction line just before the compressor off cycle.

Sistema de bombeo hacia abajo automático. Un sistema de control en refrigeración que consiste de un termostato y una válvula de línea de líquido solenoide que vacía el refrigerante del cárter del cigüeñal del compresor, evaporador y línea de succión justo antes del ciclo de apagar del compresor.

Auxiliary drain pan. A separate drain pan that is placed under an air conditioner evaporator to catch condensate in the event that the primary drain pan runs over.

Plato de drenaje auxiliar. Un plato separado de drenaje debajo del evaporador de un aire acondicionador para recoger la condensación en caso de que el plato principal se derrame.

Back electromotive force (BEMF). This is the voltage-generating effect of an electric motor's rotor turning within the motor.

Fuerza contraelectromotriz (BEMF en inglés). El efecto de generar tensión provocado por el rotor de un motor eléctrico al girar dentro del motor.

Back pressure. The pressure on the low-pressure side of a refrigeration system (also known as suction pressure).

Contrapresión. La presión en el lado de baja presión de un sistema de refrigeración (conocido también como presión de aspiración).

Backseat. The position of a refrigeration service valve when the stem is turned away from the valve body and seated, shutting off the service port.

Asiento trasero. Posición de una válvula de servicio de refrigeración cuando el vástago está orientado fuera del cuerpo de la válvula y aplicado sobre su asiento, cerrando así la apertura de servicio.

Baffle. A plate used to keep fluids from moving back and forth at will in a container.

Deflector. Placa utilizada para evitar el libre movimiento de líquidos en un recipiente.

Balanced-port TXV. A valve that will meter refrigerant at the same rate when the condenser head pressure is low.

Válvula electrónica de expansión con conducto equilibrado. Válvula que medirá el refrigerante a la misma proporción cuando la presión en la cabeza del condensador sea baja.

Ball check valve. A valve with a ball-shaped internal assembly that only allows fluid flow in one direction.

Válvula de retención de bolas. Válvula con un conjunto interior en forma de bola que permite el flujo de fluido en una sola dirección.

Ball valve. A valve with an internal part that is shaped like a sphere with a hole through the center. When turned 90°, the hole is crossways of the flow and stops the flow.

Válvula de bola. Una válvula con una parte interna en forma de esfera con un hueco a través del centro. Cuando vira 90°, el hueco es transversal al flujo y el flujo para.

Baseboard heating. Convection heaters providing whole-house, spot, or individual room heating. The heat is normally provided by electrical resistance or hot water.

Calefacción de zócalo. Calentadores por convección que proporcionan calefacción a toda la casa, en un punto específico o en una sola habitación. Normalmente, el calor se obtiene mediante una resistencia eléctrica o agua caliente.

Battery. A device that produces electricity from the interaction of metals and acid.

Pila. Dispositivo que genera electricidad de la interacción entre metales y el ácido.

Bearing. A device that surrounds a rotating shaft and provides a low-friction contact surface to reduce wear from the rotating shaft.

Cojinete. Dispositivo que rodea un árbol giratorio y provee una superficie de contacto de baja fricción para disminuir el desgaste de dicho árbol.

Bearing washout. A cleaning of the compressor's bearing surfaces, which causes lack of lubrication. It is usually caused by liquid refrigerant mixing with the compressor's crankcase oil due to liquid floodback or migration.

Derrubio del cojinete. Una limpieza de las superficies del cojinete del compresor, lo que lleva a falta de lubricación. Generalmente causado por refrigerante líquido mezclado con aceite del cárter del cigüeñal del compresor debido a inundación o migración del líquido.

Bellows. An accordion-like device that expands and contracts when internal pressure changes.

Fuelles. Dispositivo en forma de acordeón con pliegues que se expanden y contraen cuando la presión interna sufre cambios.

Bellows seal. A method of sealing a rotating shaft or valve stem that allows rotary movement of the shaft or stem without leaking.

Cierre hermético de fuelles. Método de sellar un árbol giratorio o el vástago de una válvula que permite el movimiento giratorio del árbol o del vástago sin producir fugas.

Belly-band mount motor. An electric motor mounted with a strap around the motor secured with brackets on the strap.

Motor con barriguera. Motor eléctrico que se monta colocando una cincha a su alrededor y fijándola con abrazaderas.

Bending spring. A coil spring that can be fitted inside or outside a piece of tubing to prevent its walls from collapsing when being formed.

Muelle de flexión. Muelle helicoidal que puede acomodarse dentro o fuera de una pieza de tubería para evitar que sus paredes se doblen al ser formadas.

Bimetal. Two dissimilar metals fastened together to create a distortion of the assembly with temperature changes.

Bimetal. Dos metales distintos fijados entre sí para producir una distorción del conjunto al ocurrir cambios de temperatura.

Bimetal strip. Two dissimilar metal strips fastened back-to-back.

Banda bimetálica. Dos bandas de metales distintos fijadas entre sí en su parte posterior.

Binary. Consisting of 1s and 0s. The 1s and 0s represent numbers, words, or signals that can be stored in the computer's memory for future use. Calculators and computers are digital systems.

Binario. Consiste de 1s y 0s. Los 1s y 0s representan números, palabras o señales que pueden almacenarse en la memoria de la computadora para uso futuro. Las calculadoras y computadoras son sistemas digitales.

Biofilm. A collection of microorganisms growing on wet surfaces and forming a slimy mass.

Biofilm. Ecosistema de microorganismos que crecen sobre superficies húmedas y forman una masa viscosa.

Biofuel. Fuel derived from living or recently living matter. This matter or fuel can be in the form of wood, vegetable oil, or animal waste. It may also be processed from this matter and take other forms (e.g., gases like methane).

Biocombustible. Combustible derivado de materia viva o recientemente muerta. Esta materia o combustible puede presentarse en forma de madera, aceite vegetal o desechos animales. También puede procesarse usando esta materia y tomar otras formas (por ejemplo, gases tales como el metano).

Biological growth. A growth of algae or bacteria that is generally visible as surface scum, clouding, physical growth, or mass in or around a water source, sump, or holding tank.

Crecimiento biológico. Crecimiento de algas o bacterias, generalmente visible en forma de nata espumosa, enturbiamiento, crecimiento físico o masa dentro o alrededor de una fuente de agua, un colector o un tanque de almacenamiento.

Bleeding. Allowing pressure to move from one pressure level to another very slowly.

Sangradura. Proceso a través del cual se permite el movimiento de presión de un nivel a otro de manera muy lenta.

Bleed valve. A valve with a small port usually used to bleed pressure from a vessel to the atmosphere.

Válvula de descarga. Válvula con un conducto pequeño utilizado normalmente para purgar la presión de un depósito a la atmósfera.

Blocked suction. A method of cylinder unloading. The suction line passage to a cylinder in a reciprocating compressor is blocked, thus causing that cylinder to stop pumping.

Aspiración obturada. Método de descarga de un cilindro. El paso del conducto de aspiración a un cilindro en un compresor alternativo se obtura, provocando así que el cilindro deje de bombear.

Blowdown. A system in a cooling tower whereby some of the circulating water is bled off and replaced with fresh water to dilute the sediment in the sump.

Vaciado. Sistema en una torre de refrigeración por medio del cual se purga parte del agua circulante y se reemplaza con agua fresca para diluir el sedimento en el sumidero.

Boiler. A container in which a liquid may be heated using any heat source. When the liquid is heated to the point that vapor forms and is used as the circulating medium, it is called a steam boiler.

Cardera. Recipiente en el que se puede calentar un líquido utilizando cualquier fuente de calor. Cuando se calienta el líquido al punto en que se produce vapor y se utiliza éste como el medio para la circulación, se llama caldera de vapor.

Boiling point. The temperature level of a liquid at which it begins to change to a vapor. The boiling temperature is controlled by the vapor pressure above the liquid.

Punto de ebullición. El nivel de temperatura de un líquido al que el líquido empieza a convertirse en vapor. La temperatura de ebullición se regula por medio de la presión del vapor sobre el líquido.

Boiling temperature. The boiling temperature of the liquid can be controlled by controlling the pressure. The standard boiling pressure for water is an atmospheric pressure of 29.92 in. Hg (mercury) where water boils at 212°F.

Temperatura de ebullición. La temperatura de ebullición de un líquido puede controlarse controlando la presión. La presión de ebullición estándar para agua es una presión atmosférica de 29.92 pulgadas de mercurio donde el agua hierve a 212°F.

Booster pump.   An additional pump that is used to build the pressure above what the primary pump can accomplish.

Bomba promotora.   Una bomba adicional que se usa para aumentar la presión por encima de lo que la bomba primaria puede.

Boot.   The connection between the branch line duct and the floor register. It transitions the branch duct to the register size. It may be from rectangular to another size of rectangle or from round to rectangular.

Manguito.   La conexión entre el conducto de la línea ramal y el contador en el piso. Él hace la transición de la línea ramal al tamaño del contador. Puede conectar un rectángulo a otro rectángulo de tamaño diferente o círculo a un rectángulo.

Bore.   The inside diameter of a cylinder.

Calibre.   Diámetro interior de un cilindro.

Bourdon tube.   C-shaped tube manufactured of thin metal and closed on one end. When pressure is increased inside, it tends to straighten. It is used in a gauge to indicate pressure.

Tubo Bourdon.   Tubo en forma de C fabricado de metal delgado y cerrado en uno de los extremos. Al aumentarse la presión en su interior, el tubo tiende a enderezarse. Se utiliza dentro de un calibrador para indicar la presión.

Brazing.   High-temperature (above 800°F) melting of a filler metal for the joining of two metals.

Broncesoldadura.   El derretir de un metal de relleno para fusionar dos metales a temperaturas altas (sobre los 800°F o 430°C).

Breaker.   A heat-activated electrical device used to open an electrical circuit to protect it from excessive current flow.

Interruptor.   Dispositivo eléctrico activado por el calor que se utiliza para abrir en circuito eléctrico a fin de protegerlo de un flujo excesivo de corriente.

British thermal unit.   The amount (quantity) of heat required to raise the temperature of 1 lb of water 1°F.

Unidad térmica británica.   Cantidad de calor necesario para elevar en 1°F (−17.56°C) la temperatura de una libra inglesa de agua.

BTU.   Abbreviation for British thermal unit.

BTU.   Abreviatura de unidad térmica británica.

Bubble point.   The refrigerant temperature at which bubbles begin to appear in a saturated liquid.

Punto de burbujear.   La temperatura refrigerada en que burbujas empiezan a aparecer en un liquído saturado.

Bulb sensor.   The part of a sealed automatic control used to sense temperature.

Bombilla sensora.   Pieza de un regulador automático sellado que se utiliza para advertir la temperatura.

Burner.   A device used to prepare and burn fuel.

Quemador.   Dispositivo utilizado para la preparación y la quema de combustible.

Burr.   Excess material squeezed into the end of tubing or pipe after a cut has been made. This burr must be removed.

Rebaba.   Exceso de material introducido por fuerza en el extremo de una tubería después de hacerse un corte. Esta rebaba debe removerse.

Bypass.   A circuit in which fluid or electricity (depending on the device) is allowed to go around a component or several components, bypassing them without effect.

Tubo de derivación.   Circuito donde se permite que el fluido o la electricidad (dependiendo del dispositivo) circunvenga un componente o varios componentes, evitando pasar por ellos, sin provocar efecto alguno.

Calibrate.   To adjust instruments or gauges to the correct setting for known conditions.

Calibrar.   Ajustar instrumentos o calibradores en posición correcta para su operación en condiciones conocidas.

Calibrated orifice plate.   A plate that is drilled with a specific size hole that will allow only a certain rate of flow under a particular pressure or range of pressure.

Placa de orificio calibrada.   Placa en la que se perfora un agujero de un tamaño específico, que permite sólo una cierta velocidad de fluido de acuerdo a una presión o rango de presión específicos.

Calorimeter.   An instrument of laboratory-grade quality used to measure heat absorbed into a substance.

Calorímetro.   Instrumento de calidad laboratorio que sirve para medir el calor absorbido por una sustancia.

Capacitance.   The term used to describe the electrical storage ability of a capacitor.

Capacitancia.   Término utilizado para describir la capacidad de almacenamiento eléctrico de un capacitador.

Capacitive circuit.   When the current in a circuit leads the voltage by 90°.

Circuito capacitivo.   Un circuito en que el corriente mueve el tensíon por un 90°.

Capacitor.   An electrical storage device used to start motors (start capacitor) and to improve the efficiency of motors (run capacitor).

Capacitador.   Dispositivo de almacenamiento eléctrico utilizado para arrancar motores (capacitador de arranque) y para mejorar el rendimiento de motores (capacitador de funcionamiento).

Capacitor-start–capacitor-run motor.   A single-phase motor that has a start capacitor in series with the start winding that is disconnected after start-up and a run capacitor that is also in parallel with the start windings that stays in the circuit while running. This capacitor is built for full-time duty and uses the potential voltage generated by the start winding to give the run winding more efficiency.

Motor de arranque capacitivo-marcha capacitiva.   Motor de fase sencilla que tiene un condensador de arranque en serie con la bobina de arranque que se desconecta después del encendido, y un condensador de marcha que está en paralelo con las bobinas de arranque y que permanece en el circuito mientras está en marcha. Este condensador está hecho para servicio a tiempo completo y usa el voltaje que genera la bobina de arranque para darle mayor eficiencia a la bobina de marcha.

Capacitor-start motor.   A single-phase motor with a start and run winding that has a capacitor in series with the start winding, which remains in the circuit until the motor gets up to about 75% the run speed.

Motor de arranque capacitivo.   Motor de fase sencilla con una bobina de arranque y marcha que tiene un condensador en serie con la bobina de arranque, el cual permanece en el circuito hasta que el motor alcanza alrededor de 75% de la velocidad de marcha.

Capacity.   The rating system of equipment used to heat or cool substances.

Capacidad.   Sistema de clasificación de equipo utilizado para calentar o enfriar sustancias.

Capillary attraction.   The attraction of a liquid material between two pieces of material such as two pieces of copper or copper and brass. For instance, in a joint made up of copper tubing and a brass fitting, the solder filler material has a greater attraction to the copper and brass than to itself and is drawn into the space between them.

Atracción capilar.   Atracción de un material líquido entre dos piezas de material, como por ejemplo dos piezas de cobre o cobre y latón. Por ejemplo, en una junta fabricada de tubería de cobre y un accesorio de latón, el material de relleno de la soldadura tiene mayor atracción al cobre y al latón que a sí mismo y es arrastrado hacia el espacio entre éstos.

Capillary tube.   A fixed-bore metering device. This is a small-diameter tube that can vary in length from a few inches to several feet. The amount of refrigerant flow needed is predetermined and the length and diameter of the capillary tube is sized accordingly.

Tubo capilar.   Dispositivo de medición de calibre fijo. Éste es un tubo de diámetro pequeño cuyo largo puede oscilar entre unas cuantas pulgadas y varios pies. La cantidad de flujo de refrigerante requerida es predeterminada y, de acuerdo a esto, se fijan el largo y el diámetro del tubo capilar.

Carbon dioxide.   A by-product of natural gas combustion that is not harmful.

Dióxido de carbono.   Subproducto de la combustión del gas natural que no es nocivo.

Carbon monoxide.   A poisonous, colorless, odorless, tasteless gas generated by incomplete combustion.

Monóxido de carbono.   Gas mortífero, inodoro, incoloro e insípido que se desprende en la combustión incompleta del carbono.

Cathode.   A negatively charged plate or terminal. A point that collects electrons.

Cátodo.   Placa o terminal con carga negativa. Punto que recolecta electrones.

Cavitation.   A vapor formed due to a drop in pressure in a pumping system. Vapor at a pump inlet may be caused at a cooling tower if the pressure is low and water is turned to vapor.

Cavitación.   Vapor producido como consecuencia de una caída de presión en un sistema de bombeo. El vapor a la entrada de una bomba puede ser producido en una torre de refrigeración si la presión es baja y el agua se convierte en vapor.

Cellulose.   A substance formed in wood plants from glucose or sugar.

Celulosa.   Sustancia presente en plantas de madera y que se forma a partir de glucosa y azúcar.

Celsius scale.   A temperature scale with 100-degree graduations between water freezing (0°C) and water boiling (100°C).

Escala Celsio.   Escala dividida en cien grados, con el cero marcado a la temperatura de fusión del hielo (0°C) y el cien a la de ebullición del agua (100°C).

Centigrade scale.   See Celsius scale.

Centígrado.   Véase Escala Celsio.

Centrifugal compressor.   A compressor used for large refrigeration systems that uses centrifugal force to accomplish compression. It is not positive displacement, but it is similar to a blower.

Compresor centrífugo.   Compresor utilizado en sistemas grandes de refrigeración que usa fuerza centrífuga para lograr compresión. No es desplazamiento positivo, pero es similar a un soplador.

Centrifugal pump.   A pump that uses centrifugal force to move a fluid. An impeller is rotated rapidly within the pump, causing the fluid to fly away from the center, which forces the fluid through a piping system.

Bomba centrífuga.   Bomba que utiliza la fuerza centrífuga para desplazar un fluido. Un propulsor dentro de la bomba gira a alta velocidad alejando el líquido del centro e impulsándolo a través de una tubería.

Centrifugal switch.   A switch that uses a centrifugal action to disconnect the start windings from the circuit.

Conmutador centrífugo.   Conmutador que utiliza una acción centrífuga para desconectar los devanados de arranque del circuito.

cfm.   Cubic feet per minute, a measurement of the volume of a gas or liquid passing a particular point.

pcm (cfm, por su sigla en inglés).   Pies cúbicos por minuto. Sistema de medición del volumen de un gas o líquido que pasa por un punto específico.

Change of state.   The condition that occurs when a substance changes from one physical state to another, such as ice to water and water to steam.

Cambio de estado.   Condición que ocurre cuando una sustancia cambia de un estado físico a otro, como por ejemplo el hielo a agua y el agua a vapor.

Charge of refrigerant.   The quantity of refrigerant in a system.

Carga de refrigerante.   Cantidad de refrigerante en un sistema.

Charging curve.   A graphical method of assisting a service technician with charging an air conditioning or heat pump system.

Curva de recarga.   Un método gráfico para asistir al técnico de servicio con el recargar de un sistema de aire acondicionado o de bomba de calor.

Charging cylinder.   A device that allows the technician to accurately charge a refrigeration system with refrigerant.

Cilindro cargador.   Dispositivo que le permite al mecánico cargar correctamente un sistema de refrigeración con refrigerante.

Charging scale.   A scale used to weigh refrigerant when charging a refrigeration or air conditioning system.

Báscula de plancha.   Báscula que se utiliza para pesar el refrigerante durante la carga de un sistema de refrigeración o de aire acondicionado.

Check valve. A device that permits fluid flow in one direction only.

Válvula de retención. Dispositivo que permite el flujo de fluido en una sola dirección.

Chilled water system. An air conditioning system that circulates refrigerated water to the area to be cooled. The refrigerated water picks up heat from the area, thus cooling the area.

Sistema de agua enfriada. Sistema de acondicionamiento de aire que hace circular agua refrigerada al área que será enfriada. El agua refrigerada atrapa el calor del área y la enfria.

Chiller. A machine that removes heat from a liquid by vapor compression or absorption.

Refrigerador. Artefacto que elimina el calor de un líquido usando compresión de vapor o absorción.

Chiller purge unit. A system that removes air or noncondensables from a low-pressure chiller.

Unidad enfriadora de purga. Sistema que remueve el aire o sustancias no condensables de un enfriador de baja presión.

Chill factor. A factor or number that is a combination of temperature, humidity, and wind velocity that is used to compare a relative condition to a known condition.

Factor de frío. Factor o número que es una combinación de la temperatura, la humedad y la velocidad del viento utilizado para comparar una condición relativa a una condición conocida.

Chimney. A vertical shaft used to convey flue gases above the rooftop.

Chimenea. Cañón vertical utilizado para conducir los gases de combustión por encima del techo.

Chimney effect. A term used to describe air or gas when it expands and rises when heated.

Efecto de chimenea. Término utilizado para describir el aire oel gas cuando se expande y sube al calentarse.

Chlorofluorocarbons (CFCs). Those refrigerants thought to contribute to the depletion of the ozone layer.

Clorofluorocarburos (CFCs en inglés). Líquidos refrigerantes que, según algunos, han contribuido a la reducción de la capa de ozono.

Circuit. An electron or fluid-flow path that makes a complete loop.

Circuito. Electrón o trayectoria del flujo de fluido que hace un ciclo completo.

Circuit breaker. A device that opens an electric circuit when an overload occurs.

Interruptor para circuitos. Dispositivo que abre un circuito eléctrico cuando ocurre una sobrecarga.

Clamp-on ammeter. An instrument that can be clamped around one conductor in an electrical circuit and measure the current.

Amperímetro fijado con abrazadera. Instrumento que puede fijarse con una abrazadera a un conductor en un circuito eléctrico y medir la corriente.

Clearance volume. The volume at the top of the stroke in a reciprocating compressor cylinder between the top of the piston and the valve plate.

Volumen de holgura. Volumen en la parte superior de una carrera en el cilindro de un compresor recíproco entre la parte superior del pistón y la placa de una válvula.

Closed circuit. A complete path for electrons to flow on.

Circuito cerrado. Circuito de trayectoria ininterrumpida que permite un flujo continuo de electrones.

Closed-circuit cooling tower. May have a wet/dry mode, an adiabatic mode, and a dry mode.

Torre de enfriamiento de circuito cerrado. Puede tener un modo seco/mojado, un modo adiabático y un modo seco.

Closed loop. Piping circuit that is complete and not open to the atmosphere.

Ciclo cerrado. Circuito de tubería completo y no abierto a la atmósfera.

Closed-loop heat pump. Heat pump system that reuses the same heat transfer fluid, which is buried in plastic pipes within the earth or within a lake or pond for the heat source.

Bomba de calor de circuito cerrado. Sistema de bomba de calor que reutiliza el mismo líquido de transferencia térmica que se encuentra en tubos de plástico enterrados o sumergidos en un lago o estanque de los que obtiene el calor.

CO_2 indicator. An instrument used to detect the quantity of carbon dioxide in flue gas for efficiency purposes.

Indicador del CO_2. Instrumento utilizado para detectar la cantidad de dióxido de carbono en el gas de combustión a fin de lograr un mejor rendimiento.

Coaxial heat exchanger. A tube-within-a-tube liquid heat exchanger. Typically it is used for water-source heat pumps and small water-cooled air conditioners.

Intercambiador de calor coaxial. Intercambiador de calor líquido con un tubo dentro de un tubo. Típicamente se usa para bombas de calor en fuentes de agua y acondicionadores de aire pequeños enfriados por agua.

Code. The local, state, or national rules that govern safe installation and service of systems and equipment for the purpose of safety of the public and trade personnel.

Código. Reglamentos locales, estatales o federales que rigen la instalación segura y el servicio de sistemas y equipo con el propósito de garantizar la seguridad del personal público y profesional.

Coefficient of performance (COP). The ratio of usable output energy divided by input energy.

Coeficiente de rendimiento (COP en inglés). Relación de la de energía de salida utilizable dividida por la energía de entrada.

Cold. The word used to describe heat at lower levels of intensity.

Frío. Término utilizado para describir el calor a niveles de intensidad más bajos.

Cold anticipator. A fixed resistor in a thermostat that is wired in parallel with the cooling contacts. This starts the cooling system before the thermostat calls for cooling, which allows the system to get up to capacity before the cooling is actually needed.

Anticipador de frío. Resistor fijo en un termostato, conectado en paralelo con los contactos de enfriamiento.

Dicho resistor pone en marcha el sistema de enfriamiento antes de que lo haga el termostato, permitiendo así que el sistema alcance su plena capacidad antes de que realmente se necesite el enfriamiento.

Cold wall.   The term used in comfort heating to describe a cold outside wall and its effect on human comfort.

Pared fría.   Término utilizado en la calefacción para comodidad que describe una pared exterior fría y sus efectos en la comodidad de una persona.

Combustion.   A reaction called rapid oxidation or burning produced with the right combination of a fuel, oxygen, and heat.

Combustión.   Reacción conocida como oxidación rápida o quema producida con la combinación correcta de combustible, oxígeno y calor.

Combustion analyzer.   An instrument used to measure oxygen concentrations within flue gases. This analyzer can test smoke and test for carbon monoxide and other gases.

Analizador de combustión.   Instrumento que se utiliza para medir la concentración del oxígeno en los gases de escape. Este tipo de analizador permite medir el contenido de monóxido de carbono y otros gases en los humos.

Comfort.   People are said to be comfortable when they are not aware of the ambient air surrounding them. They do not feel cool or warm or sweaty.

Comodidad.   Se dice que las personas están cómodas cuandos no están conscientes del aire ambiental que les rodea. No sienten frío ni calor, ni están sudadas.

Comfort chart.   A chart used to compare the relative comfort of one temperature and humidity condition to another condition.

Esquema de comodidad.   Esquema utilizado para comparar la comodidad relativa de una condición de temperatura y humedad a otra condición.

Compound gauge.   A gauge used to measure the pressure above and below the atmosphere's standard pressure. It is a Bourdon tube sensing device and can be found on all gauge manifolds used for air conditioning and refrigeration service work.

Calibrador compuesto.   Calibrador utilizado para medir la presión mayor y menor que la presión estándar de la atmósfera. Es un dispositivo sensor de tubo Bourdon que puede encontrarse en todos los distribuidores de calibrador utilizados para el servicio de sistemas de acondicionamiento de aire y de refrigeración.

Compression.   A term used to describe a vapor when pressure is applied and the molecules are compacted closer together.

Compresión.   Término utilizado para describir un vapor cuando se aplica presión y se compactan las moléculas.

Compression ratio.   A term used with compressors to describe the actual difference in the low- and high-pressure sides of the compression cycle. It is absolute discharge pressure divided by absolute suction pressure.

Relación de compresión.   Término utilizado con compresores para describir la diferencia real en los lados de baja y alta presión del ciclo de compresión. Es la presión absoluta de descarga dividida por la presión absoluta de aspiración.

Compressor.   A vapor pump that pumps vapor (refrigerant or air) from one pressure level to a higher pressure level.

Compresor.   Bomba de vapor que bombea el vapor (refrigerante o aire) de un nivel de presión a un nivel de presión más alto.

Compressor crankcase.   The internal part of the compressor that houses the crankshaft and lubricating oil.

Cárter del compresor.   Parte interna del compresor donde se aloja el cigüeñal y el aceite de lubricación.

Compressor displacement.   The internal volume of a compressor's cylinders, used to calculate the pumping capacity of the compressor.

Desplazamiento del compresor.   Volumen interno de los cilindros de un compresor, utilizado para calcular la capacidad de bombeo del mismo.

Compressor head.   The component that sits on top of the compressor cylinder and holds the components together.

Cabeza del compresor.   El componente que está en la parte superior del cilindro y que mantiene a los componentes unidos.

Compressor oil cooler.   One or more piping systems used for cooling the crankcase oil.

Enfriador del aceite del compresor.   Uno o varios sistemas de tubería que sirven para enfriar el aceite del cárter.

Compressor shaft seal.   The seal that prevents refrigerant inside the compressor from leaking around the rotating shaft.

Junta de estanqueidad del árbol del compresor.   La junta de estanqueidad que evita la fuga, alrededor del árbol giratorio, del refrigerante en el interior del compresor.

Condensate.   The moisture collected on an evaporator coil.

Condensado.   Humedad acumulada en la bobina de un evaporador.

Condensate pump.   A small pump used to pump condensate to a higher level.

Bomba para condensado.   Bomba pequeña utilizada para bombear el condensado a un nivel más alto.

Condensation.   Liquid formed when a vapor condenses.

Condensación.   El líquido formado cuando se condensa un vapor.

Condense.   Changing a vapor to a liquid.

Condensar.   Convertir un vapor en líquido.

Condenser.   The component in a refrigeration system that transfers heat from the system by condensing refrigerant.

Condensador.   Componente en un sistema de refrigeración que transmite el calor del sistema al condensar el refrigerante.

Condenser flooding.   An automatic method of maintaining the correct head pressure in mild weather by using refrigerant from an auxiliary receiver.

Inundación del condensador.   Método automático de mantener una presión correcta en la cabeza en un tiempo suave utilizando refrigerante de un receptor auxiliar.

Condensing pressure.   The pressure that corresponds to the condensing temperature in a refrigeration system.

Presión para condensación.   La presión que corresponde a la temperatura de condensación en un sistema de refrigeración.

Condensing temperature.   The temperature at which a vapor changes to a liquid.

Temperatura de condensación.   Temperatura a la que un vapor se convierte en líquido.

Condensing unit.   A complete unit that includes the compressor and the condensing coil.

Conjunto del condensador. Unidad completa que incluye el compresor y la bobina condensadora.

Conduction. Heat transfer from one molecule to another within a substance or from one substance to another.

Conducción. Transmisión de calor de una molécula a otra dentro de una sustancia o de una sustancia a otra.

Conductivity. The ability of a substance to conduct electricity or heat.

Conductividad. Capacidad de una sustancia de conducir electricidad o calor.

Conductivity meter. A meter that measures the ability of a fluid to conduct electricity. The ability to conduct electricity is inversely related to the resistance of the fluid; lower resistance means higher conductivity, and higher resistance means lower conductivity.

Medidor de conductividad. Medidor que mide la capacidad de un fluido de conducir electricidad. La capacidad de conducir electricidad está inversamente relacionada con la resistencia del fluido; las resistencias menores indican mayor conductividad y las resistencias mayores, menor conductividad.

Conductor. A path for electrical energy to flow on.

Conductor. Trayectoria que permite un flujo continuo de energía eléctrica.

Connecting rod. A rod that connects the piston to the crankshaft.

Barra conectiva. Barra que conecta al pistón con el cigüeñal.

Contactor. A larger version of the relay. It can be repaired or rebuilt and has movable and stationary contacts.

Contactador. Versión más grande del relé. Puede ser reparado o reconstruido. Tiene contactos móviles y fijos.

Contaminant. Any substance in a refrigeration system that is foreign to the system, particularly if it causes damage.

Contaminante. Cualquier sustancia en un sistema de refrigeración extraña a éste, principalmente si causa averías.

Control. A device to stop, start, or modulate flow of electricity or fluid to maintain a preset condition.

Regulador. Dispositivo para detener, poner en marcha o modular el flujo de electricidad o de fluido a fin de mantener una condición establecida con anticipación.

Controlled device. May be any control that stops, starts, or modulates fuel, fluid flow, or air to provide expected conditions in the conditioned space.

Aparato controlado. Un aparato que puede ser cualquier control que detiene, enciende o modula el combustible, flujo de fluido o aire para proveer las condiciones esperadas en un espacio acondicionado.

Controller. A device that provides the output to the controlled device.

Controlador. Aparato que provee la salida para el aparato controlado.

Control system. A network of controls to maintain desired conditions in a system or space.

Sistema de regulación. Red de reguladores que mantienen las condiciones deseadas en un sistema o un espacio.

Convection. Heat transfer from one place to another using a fluid.

Convección. Transmisión de calor de un lugar a otro por medio de un fluido.

Conversion factor. A number used to convert from one equivalent value to another.

Factor de conversión. Número utilizado en la conversión de un valor equivalente a otro.

Cooler. A walk-in or reach-in refrigerated box.

Nevera. Caja refrigerada donde se puede entrar o introducir la mano.

Cooling tower. The final device in many water-cooled systems, which rejects heat from the system into the atmosphere by evaporation of water.

Torre de refrigeración. Dispositivo final en muchos sistemas enfriados por agua, que dirige el calor del sistema a la atmósfera por medio de la evaporación de agua.

Copper plating. Small amounts of copper are removed by electrolysis and deposited on the ferrous metal parts in a compressor.

Encobrado. Remoción de pequeñas cantidades de cobre por medio de electrólisis que luego se colocan en las piezas de metal férreo en un compresor.

Corrosion. A chemical action that eats into or wears away material from a substance.

Corrosión. Acción química que carcoma o desgasta el material de una sustancia.

Cotter pin. Used to secure a pin. The cotter pin is inserted through a hole in the pin, and the ends spread to retain it.

Pasador de chaveta. Se usa para asegurar una clavija. El pasador se inserta a través de un roto en la clavija y sus extremos se abren para asegurarla.

Counter EMF. Voltage generated or induced above the applied voltage in a single-phase motor.

Contra EMF. Tensión generada o inducida sobre la tensión aplicada en un motor unifásico.

Counterflow. Two fluids flowing in opposite directions.

Contraflujo. Dos fluidos que fluyen en direcciones opuestas.

Coupling. A device for joining two fluid-flow lines. Also the device connecting a motor driveshaft to the driven shaft in a direct-drive system.

Acoplamiento. Dispositivo utilizado para la conexión de dos conductos de flujo de fluido. Es también el dispositivo que conecta un árbol de mando del motor al árbol accionado en un sistema de mando directo.

CPVC (chlorinated polyvinyl chloride). Plastic pipe similar to PVC except that it can be used with temperatures up to 180°F at 100 psig.

CPVC (cloruro de polivinilo clorado). Tubo plástico similar al PVC, pero que puede utilizarse a temperaturas de hasta 180°F (82°C) a 100 psig [indicador de libras por pulgada cuadrada].

Crackage. Small spaces in a structure that allow air to infiltrate the structure.

Formación de grietas. Espacios pequeños en una estructura que permiten la infiltración del aire dentro de la misma.

Cradle-mount motor. A motor with a mounting cradle that fits the motor end housing on each end and is held down with a bracket.

Motor montado con cuña.   Motor equipado de una cuña adaptada a la caja en los dos extremos y sujetado por fijaciones.

Crankcase heat.   Heat provided to the compressor crankcase.

Calor para el cárter del cigüeñal.   Calor suministrado al cárter del cigüeñal del compresor.

Crankcase pressure regulator (CPR).   A valve installed in the suction line, usually close to the compressor. It is used to keep a low-temperature compressor from overloading on a hot pull down by limiting the pressure to the compressor.

Regulador de la presión del cárter del cigüeñal (CPR en inglés).   Válvula instalada en el conducto de aspiración, normalmente cerca del compresor. Se utiliza para evitar la sobrecarga en un compresor de temperatura baja durante un arrastre caliente hacia abajo limitando la presión al compresor.

Crankshaft.   In a reciprocating compressor, the crankshaft changes the round-and-round motion into the reciprocating back-and-forth motion of the pistons using off-center devices called throws.

Cigüeñal.   En un compresor alternativo, el cigüeñal cambia el movimiento circular en un movimiento alternativo hacia delante y hacia atrás de los pistones usando aparatos descentrados llamados cigüeñas.

Crankshaft seal.   Same as the compressor shaft seal.

Junta de estanqueidad del árbol del cigüeñal.   Exactamente igual que la junta de estanqueidad del árbol del compresor.

Crankshaft throw.   The off-center portion of a crankshaft that changes rotating motion to reciprocating motion.

Excentricidad del cigüeñal.   Porción descentrada de un cigüeñal que cambia el movimiento giratorio a un movimiento alternativo.

Cross charge.   A control with a sealed bulb that contains two different fluids that work together for a common specific condition.

Carga transversal.   Regulador con una bombilla sellada compuesta de dos fluidos diferentes que pueden funcionar juntos para una condición común específica.

Cross liquid charge bulb.   A type of charge in the sensing bulb of the TXV that has different characteristics from the system refrigerant. This is designed to help prevent liquid refrigerant from flooding to the compressor at start-up.

Bombilla de carga del líquido transversal.   Tipo de carga en la bombilla sensora de la válvula electrónica de expansión que tiene características diferentes a las del refrigerante del sistema. La carga está diseñada para ayudar a evitar que el refrigerante líquido se derrame dentro del compresor durante la puesta en marcha.

Cross vapor charge bulb.   Similar to the vapor charge bulb but contains a fluid different from the system refrigerant. This is a special-type charge and produces a different pressure-temperature relationship under different conditions.

Bombilla de carga del vapor transversal.   Similar a la bombilla de carga del vapor pero contiene un fluido diferente al del refrigerante del sistema. Ésta es una carga de tipo especial y produce una relación diferente entre la presión y la temperatura bajo condiciones diferentes.

Crystallization.   When a salt solution becomes too concentrated and part of the solution turns to salt.

Cristalización.   Condición que ocurre cuando una solución salina se concentra demasiado y una parte de la solución se convierte en sal.

Current, electrical.   Electrons flowing along a conductor.

Corriente eléctrica.   Electrones que fluyen a través de un conductor.

Current relay.   An electrical device activated by a change in current flow.

Relé para corriente.   Dispositivo eléctrico accionado por un cambio en el flujo de corriente.

Current sensing relay.   An inductive relay coil usually located around a wire used to sense current flowing through the wire. Its action usually opens or closes a set of contacts in the starting of single phase induction motors.

Relé detector de corriente.   Una bobina de relé inductiva que generalmente está ubicada cerca de un cable y se usa para detectar el flujo de corriente a través del cable. Su acción generalmente abre o cierra una serie de contactos al encender los motores de fase sencilla.

Cut-in and cut-out.   The two points at which a control opens or closes its contacts based on the condition it is supposed to maintain.

Puntos de conexión y desconexión.   Los dos puntos en los que un regulador abre o cierra sus contactos según las condiciones que debe mantener.

Cycle.   A complete sequence of events (from start to finish) in a system.

Ciclo.   Secuencia completa de eventos, de comienzo a fin, que ocurre en un sistema.

Cycles of concentration.   A term that is used to describe the relationship of water conditions after use to the conditions of new water being added to a system. After water is used in a cooling tower, the mineral content of the water has increased per volume of water. The amount of mineral content is compared to an equal volume of new water. One cycle means that the concentration has doubled.

Ciclos de concentración.   Término que se utiliza para describir la relación del estado del agua después del uso con el estado del agua nueva que se agrega a aun sistema. Después de que el agua circula por una torre de refrigeración, aumenta el contenido mineral del agua por volumen. La cantidad de contenido mineral se compara con un volumen equivalente de agua nueva. Un ciclo significa que se ha duplicado la concentración.

Cylinder.   A circular container with straight sides used to contain fluids or to contain the compression process (the piston movement) in a compressor.

Cilindro.   Recipiente circular con lados rectos, utilizado para contener fluidos o el proceso de compresión (movimiento del pistón) en un compresor.

Cylinder, compressor.   The part of the compressor that contains the piston and its travel.

Cilindro del compresor.   Pieza del compresor que contiene el pistón y su movimiento.

Cylinder head, compressor.   The top to the cylinder on the high-pressure side of the compressor.

Culata del cilindro del compresor.   Tapa del cilindro en el lado de alta presión del compresor.

Cylinder, refrigerant. The container that holds refrigerant.
Cilindro del refrigerante. El recipiente que contiene el refrigerante.

Cylinder unloading. A method of providing capacity control by causing a cylinder in a reciprocating compressor to stop pumping.
Descarga del cilindro. Método de suministrar regulación de capacidad provocando que el cilindro en un compresor alternativo deje de bombear.

Damper. A component in an air distribution system that restricts airflow for the purpose of air balance.
Desviador. Componente en un sistema de distribución de aire que limita el flujo de aire para mantener un equilibrio de aire.

DB. See Dry-bulb temperature.
BS. Véase Temperatura de bombilla seca.

DC converter. A type of rectifier that changes alternating current (AC) to direct current (DC).
Convertidor CD. Tipo de rectificador que cambia la corriente alterna (CA) a corriente directa (CD).

DC motor. A motor that operates on direct current (DC).
Motor CD. Un motor que opera con corriente directa (CD).

Deep vacuum. An attained vacuum that is below 250 microns.
Vacío profundo. Un vacío que se obtiene lo cual es menor de 250 micrones.

Defrost. Melting of ice.
Descongelar. Convertir hielo en líquido.

Defrost condensate. The condensate or water from a defrost application of a refrigeration system.
Condensado de descongelación. Condensado o agua causada por el dispositivo de descongelación en un sistema de refrigeración.

Dehumidify. To remove moisture from air.
Deshumidificar. Remover la humedad del aire.

Dehydrate. To remove moisture from a sealed system or a product.
Deshidratar. Remover la humedad de un sistema sellado o un producto.

Delta-T. The temperature difference at two different points, such as the inlet and outlet temperature difference across a water chiller.
Delta-T. La diferencia en temperatura en dos puntos diferentes, tales como la diferencia de temperatura entre la entrada y la salida de un enfriador de agua.

Demand metering. In this system, the power company charges the customer based on the highest usage for a prescribed period of time during the billing period. The prescribed time for demand metering may be any 15- or 30-minute period within the billing period.
Medición por demanda. Un sistema utilizado por la compañía de electricidad en el cual cobran al consumidor por el período de facturación basado en el uso más alto durante un período de tiempo prescrito durante el período de facturación. El tiempo prescrito para la medición por demanda puede ser cualquier período de 15 o 30 minutos durante el período de facturación.

Density. The weight per unit of volume of a substance.

Densidad. Relación entre el peso de una sustancia y su volumen.

Department of Transportation (DOT). The governing body of the U.S. government that makes the rules for transporting items, such as volatile liquids.
Departamento de Transportación (DOT en inglés). El cuerpo regente del gobierno de los Estados Unidos que crea las reglas para transportar artículos tales como líquidos volátiles.

Desiccant. Substance in a refrigeration system drier that collects and holds moisture.
Disecante. Sustancia en el secador de un sistema de refrigeración que acumula y guarda la humedad.

Desiccant drier. A device that dehumidifies compressed air for use in controls or processing.
Secador desecante. Un aparato que se usa para deshumedecer el aire comprimido que se usa en los controles o procesos.

Desiccant wheel. A rotating wheel of dry desiccant (water absorbent material), which passes through incoming air where the moisture is adsorbed and then through outgoing air where the moisture is expelled.
Rueda desecante. Rueda giratoria con desecante seco (material que absorbe agua), que pasa por el aire que ingresa, donde se absorbe la humedad, y luego por el aire que egresa, donde se expulsa la humedad.

Design pressure. The pressure at which the system is designed to operate under normal conditions.
Presión de diseño. Presión a la que el sistema ha sido diseñado para funcionar bajo condiciones normales.

De-superheating. Removing heat from the superheated hot refrigerant gas down to the condensing temperature.
Des sobrecalentamiento. Reducir el calor del gas caliente del refrigerante sobrecalentado hasta alcanzar la temperatura de condensación.

Detector. A device used to search and find.
Detector. Dispositivo de búsqueda y detección.

Detent or snap action. The quick opening and closing of an electrical switch.
Acción de detén o de encaje. El abrir y cerrar rápido de un interruptor eléctrico.

Dew. Moisture droplets that form on a cool surface.
Rocío. Gotitas de humedad que se forman en una superficie fría.

Dew point. The exact temperature at which moisture begins to form.
Punto de rocío. Temperatura exacta a la que la humedad comienza a formarse.

Diaphragm. A thin flexible material (metal, rubber, or plastic) that separates two pressure differences.
Diafragma. Material delgado y flexible, como por ejemplo el metal, el caucho o el plástico, que separa dos presiones diferentes.

Die. A tool used to make an external thread such as on the end of a piece of pipe.
Troquel. Herramienta utilizada para formar una rosca externa, como por ejemplo en el extremo de un tubo.

Dielectric connection. A connection that does not electrically conduct from one side of the connection to the other. Typically, there is an insulating separation between two metallic connectors, made of plastic or plastic-like material.

Conexión dieléctrica.   Conexión que no conduce electricidad desde un lado de la conexión al otro. Por lo general, hay un aislante, de plástico o de un material similar al plástico, que separa dos conectores.

Differential.   The difference in the cut-in and cut-out points of a control, pressure, time, temperature, or level.

Diferencial.   Diferencia entre los puntos de conexión y desconexión de un regulador, una presión, un intervalo de tiempo, una temperatura o un nivel.

Diffuser.   The terminal or end device in an air distribution system that directs air in a specific direction using louvers.

Placa difusora.   Punto o dispositivo terminal en un sistema de distribución de aire que dirige el aire a una dirección específica, utilizando aberturas tipo celosía.

Diffuse radiation.   Radiation from the sun that reaches the earth after it is reflected from other substances, such as moisture or other particles in space.

Radiación difusa.   Radiación solar que alcanza la tierra después de haber sido reflejada por otras sustancias como gotas de agua u otras partículas presentes en el aire.

Digital electronic devices.   Devices that generate strings of data or groups of logic consisting of 1s and 0s.

Aparatos electrónicos digitales.   Aparatos que generan cadenas de data o grupos de lógica que consisten de unos y ceros.

Digital electronic signal.   An electrical signal, usually 0 to 10 V DC or 0 to 20 milliamps DC, that is used to control system conditions.

Señal electrónica digital.   Una señal eléctrica, generalmente de 0 a 10 voltios CD o de 0 a 20 miliamperes CD, que se usa para controlar las condiciones del sistema.

Digital VOM.   A volt-ohm-milliampere meter that displays the reading in digits or numbers.

VOM digital.   Medidor de voltios-ohmios-miliamperios que indica la lectura en dígitos o números.

Diode.   A solid-state device composed of both P-type and N-type material. When connected in a circuit one way, current will flow. When the diode is reversed, current will not flow.

Diodo.   Dispositivo de estado sólido compuesto de material P y de material N. Cuando se conecta a un circuito de una manera, la corriente fluye. Cuando la dirección del diodo cambia, la corriente deja de fluir.

DIP (dual inline pair) switch.   A very small low-amperage, single-pole, double-throw switch used in electronic circuits to set up the program in the circuit.

Interruptor de doble paquete en línea (DIP en inglés).   Un interruptor muy pequeño, de bajo amperaje, unipolar de doble tiro que se usa en los circuitos electrónicos para preparar el programa en un circuito.

Direct current.   Electricity in which all electron flow is continuously in one direction.

Corriente continua.   Electricidad en la que todos los electrones fluyen continuamente en una sola dirección.

Direct digital control (DDC).   Very low-voltage control signal, usually 0 to 10 V DC or 0 to 20 milliamps DC.

Control digital directo (DDC en inglés).   Señal control de voltaje bien bajo generalmente 0 a 20 voltios CD o de 0 a 20 miliamperes CD.

Direct-drive compressor.   A compressor that is connected directly to the end of the motor shaft. No pulleys are involved.

Compresor de conducción directa.   Un compresor que está conectado directamente a un extremo del eje de un motor, sin usar poleas.

Direct-drive motor.   A motor that is connected directly to the load, such as an oil burner motor or a furnace fan motor.

Motor de conducción directa.   Un motor que está conectado directamente a la carga, tal como un motor de un quemador de aceite o el motor de un ventilador en un calefactor.

Direct expansion.   The term used to describe an evaporator with an expansion device other than a low-side float type.

Expansión directa.   Término utilizado para describir un evaporador con un dispositivo de expansión diferente al tipo de dispositivo flotador de lado bajo.

Direct radiation.   The energy from the sun that reaches the earth directly.

Radiación directa.   Energía solar que alcanza la tierra directamente.

Discharge pressure.   The pressure on the high-pressure side of a compressor.

Presión de descarga.   La presión en el extremo de alta presión de un compresor.

Discharge valve.   The valve at the top of a compressor cylinder that shuts on the downstroke to prevent high-pressure gas from reentering the refrigerant cylinder, allowing low-pressure gas to enter.

Válvula de descarga.   La válvula que está en la parte superior del cilindro de un compresor que se cierra en el recorrido hacia abajo del pistón para evitar que el gas a alta presión regrese al cilindro del refrigerante y permitir que el gas a baja presión entre.

Distributor.   A component installed at the outlet of the expansion valve that distributes the refrigerant to each evaporator circuit.

Distribuidor.   Componente instalado a la salida de la válvula de expansión que distribuye el refrigerante a cada circuito del evaporador.

Double flare.   A connection used on copper, aluminum, or steel tubing that folds tubing wall to a double thickness.

Abocinado doble.   Conexión utilizada en tuberías de cobre, aluminio o acera que pliega la pared de la tubería y crea un espesor doble.

Dowel pin.   A pin, which may or may not be tapered, used to align and fasten two parts.

Pasador de espiga.   Pasador, que puede o no ser cónico, utilizado para alinear y fijar dos piezas.

Downflow furnace.   This furnace sometimes is called a counterflow furnace. The air intake is at the top, and the discharge air is at the bottom.

Horno de corriente descendente.   También conocido como horno de contracorriente. La entrada del aire está en la parte superior y la salida en la parte inferior.

Drier.   A device used in a refrigerant line to remove moisture.

Secador.   Dispositivo utilizado en un conducto de refrigerante para remover la humedad.

Drip pan.   A pan shaped to collect moisture condensing on an evaporator coil in an air conditioning or refrigeration system.

Colector de goteo.   Un colector formado para acumular la humedad que se condensa en la bobina de un evaporador en un sistema de acondicionamiento de aire o de refrigeración.

Dry-bulb temperature.   The temperature measured using a plain thermometer.

Temperatura de bombilla seca.   Temperatura que se mide con un termómetro sencillo.

Dry well.   A well used for the discharged water in an open-loop geothermal heat pump.

Pozo seco.   Pozo que se utiliza para depositar agua de descarga en una bomba de calor geotérmica de circuito abierto.

Duct.   A sealed channel used to convey air from the system to and from the point of utilization.

Conducto.   Canal sellado que se emplea para dirigir el aire del sistema hacia y desde el punto de utilización.

Dust mites.   Microscopic spiderlike insects. Dust mites and their remains are thought to be a primary irritant to some people.

Ácaros del polvo.   Insectos microscópicos parecidos a arañas. Los ácaros del polvo y sus restos se consideran irritantes principales para algunas personas.

Eccentric.   An off-center device that rotates in a circle around a shaft.

Excéntrico.   Dispositivo descentrado que gira en un círculo alrededor de un árbol.

ECM.   An electronically commutated motor. This DC motor uses electronics to commutate the rotor instead of brushes. It is typically built for under 1 hp.

CEM (ECM en inglés).   Un motor conmutado electrónicamente. Este motor CD usa electrónica para conmutar el rotor en lugar de escobillas. Típicamente están hechos para menos de 1 caballo de fuerza.

Economizer.   A device or system that uses outside environmental conditions to cool an interior space without the use of an HVAC system. It takes advantage of the outdoor temperatures to offset or economize the cost of cooling.

Economizador.   Dispositivo o sistema que utiliza las condiciones ambientales exteriores para refrigerar un espacio interior sin utilizar un sistema de calefacción, ventilación y aire acondicionado (HVAC, por su sigla en inglés). Aprovecha las temperaturas exteriores para compensar o economizar el costo de refrigerar.

Eddy current test.   A test with an instrument to find potential failures in evaporator or condenser tubes.

Prueba para la corriente de Foucault.   Prueba que se realiza con un instrumento para detectar posibles fallas en los tubos del evaporador o del condensador.

Effective temperature.   Different combinations of temperature and humidity that provide the same comfort level.

Temperatura efectiva.   Diferentes combinaciones de temperatura y humedad que proveen el mismo nivel de comodidad.

Electrical power.   Electrical power is measured in watts. One watt is equal to one ampere flowing with a potential of one volt. Watts = Volts × Amperes (P = E × I).

Potencia eléctrica.   La potencia eléctrica se mide en watios. Un watio equivale a un amperio que fluye con una potencia de un voltio. Watios = Voltios × Amperios (P = E × I).

Electrical shock.   When an electrical current travels through a human body.

Sacudida eléctrica.   Paso brusco de una corriente eléctrica a través del cuerpo humano.

Electric forced air furnace.   An electrical resistance type of heating furnace used with a duct system to provide heat to more than one room.

Horno eléctrico de aire soplado.   Horno con resistencia eléctrica que se utiliza con un sistema de conductos para proporcionar calefacción a varias habitaciones.

Electric heat.   The process of converting electrical energy, using resistance, into heat.

Calor eléctrico.   Proceso de convertir energía eléctrica en calor a través de la resistencia.

Electric hydronic boiler.   A boiler using electrical resistance heat, which often has a closed-loop piping system to distribute heated water for space heating.

Caldera hidrónica eléctrica.   Caldera que utiliza calor proporcionado por una resistencia eléctrica. A menudo cuenta con un sistema cerrado para distribuir agua caliente para usos de calefacción.

Electrodes.   Electrodes carry high voltage to the tips, where an arc is created for the purpose of ignition for oil or gas furnaces.

Electrodos.   Los electrodos llevan alto voltaje hasta las puntas, donde se crea un arco con el propósito de encender los calefactores de aceite o de gas.

Electromagnet.   A coil of wire wrapped around a soft iron core that creates a magnet.

Electroimán.   Bobina de alambre devanado alrededor de un núcleo de hierro blando que crea un imán.

Electromechanical controls.   Electromechanical controls convert some form of mechanical energy to operate an electrical function, such as a pressure-operated switch.

Controles electromecánicos.   Los controles electromecánicos convierten algún tipo de energía mecánica para operar una función eléctrica, tal como un interruptor operado por presión.

Electromotive force.   A term often used for voltage indicating the difference of potential in two charges.

Fuerza electromotriz.   Término empleado a menudo para el voltaje, indicando la diferencia de potencia entre dos cargas.

Electron.   The smallest portion of an atom that carries a negative charge and orbits around the nucleus of an atom.

Electrón.   La parte más pequeña de un átomo, con carga negativa y que sigue una órbita alrededor del núcleo de un átomo.

Electronic air filter.   A filter that charges dust particles using a high-voltage direct current and then collects these particles on a plate of an opposite charge.

Filtro de aire electrónico.   Filtro que carga partículas de polvo utilizando una corriente continua de alta tensión y luego las acumula en una placa de carga opuesta.

Electronic charging scale.   An electronically operated scale used to accurately charge refrigeration systems by weight.

Escala electrónica para carga.   Escala accionada electrónicamente que se utiliza para cargar correctamente sistemas de refrigeración por peso.

Electronic circuit board.   A phenolic type of plastic board that electronic components are mounted on. Typically, the circuits are routed on the back side of the board and the components are mounted on the front with prongs of wire that are soldered to the circuits on the back. These can be mass-produced and coated with a material that keeps the circuits separated if moisture and dust accumulate.

Tarjeta de circuitos electrónicos.   Una tarjeta plástica tipo fenólico en la cual se montan los componentes electrónicos. La tarjeta generalmente tiene los circuitos trazados en la parte de atrás de la tarjeta y los componentes se colocan en el frente con cables que se sueldan al circuito por detrás. Éstas pueden producirse en masa y pueden recubrirse con un material que mantiene separado a los circuitos si se acumula humedad y polvo.

Electronic controls.   Controls that use solid-state semiconductors for electrical and electronic functions.

Controles electrónicos.   Controles que usan semiconductores de estado sólido para las funciones eléctricas y electrónicas.

Electronic expansion valve (EXV).   A metering valve that uses a thermistor as a temperature-sensing element that varies the voltage to a heat motor-operated valve.

Válvula electrónica de expansión (EXV en inglés).   Válvula de medición que utiliza un termistor como elemento sensor de temperatura para variar la tensión a una válvula de calor accionada por motor.

Electronic leak detector.   An instrument used to detect gases in very small portions by using electronic sensors and circuits.

Detector electrónico de fugas.   Instrumento que se emplea para detectar cantidades de gases sumamente pequeñas utilizando sensores y circuitos electrónicos.

Electronic or programmable thermostat.   A space thermostat that is electronic in nature with semiconductors that provide different timing programs for cycling the equipment.

Termostato electrónico o programable.   Un termostato de espacio que es de naturaleza electrónica con semiconductores que proveen diferentes programas de cronometraje para ciclar el equipo.

Electronic relay.   A solid-state relay with semiconductors used to stop, start, or modulate power in a circuit.

Relé electrónico.   Un relé de estado sólido con semiconductores para detener, iniciar o modular la electricidad en un circuito.

Electronics.   The use of electron flow in conductors, semiconductors, and other devices.

Electrónica.   La utilización del flujo de electrones en conductores, semiconductores y otros dispositivos.

Electrostatic precipitator.   Another term for an electronic air cleaner.

Precipitador electrostático.   Otro término para un limpiador eléctrico del aire.

Emitter.   A terminal on a semiconductor.

Emisor.   Punto terminal en un semiconductor.

End bell.   The end structure of an electric motor that normally contains the bearings and lubrication system.

Extremo acampanado.   Estructura terminal de un motor eléctrico que generalmente contiene los cojinetes y el sistema de lubricación.

End-mount motor.   An electric motor mounted with tabs or studs fastened to the motor housing end.

Motor con montaje en los extremos.   Motor eléctrico con lengüetas o espigas de montaje en el extremo de su caja.

End play.   The amount of lateral travel in a motor or pump shaft.

Holgadura.   Amplitud de movimiento lateral en un motor o en el árbol de una bomba.

Energy.   The capacity for doing work.

Energía.   Capacidad para realizar un trabajo.

Energy efficiency ratio (EER).   An equipment efficiency rating that is determined by dividing the output in BTU/h by the input in watts. This does not take into account the start-up and shutdown for each cycle.

Relación del rendimiento de energía (EER en inglés).   Clasificación del rendimiento de un equipo que se determina al dividir la salida en BTU/h por la entrada en watios. Esto no toma en cuenta la puesta en marcha y la parada de cada ciclo.

Energy management.   The use of computerized or other methods to manage the power to a facility. This may include cycling off nonessential equipment, such as water fountain pumps or lighting, when it may not be needed. The air conditioning and heating system is also operated at optimum times when needed instead of around the clock.

Manejo de energía.   El uso de métodos computarizados o de otro tipo para manejar el consumo de energía en la facilidad. Esto puede incluir apagar los equipos no esenciales en ciclos, tales como las bombas de las fuentes de agua o las luces cuando no son necesarias. Los sistemas de aire acondicionado y de calefacción también se operan en tiempos óptimos cuando son necesarios en vez de todo el tiempo.

Enthalpy.   The amount of heat a substance contains from a predetermined base or point.

Entalpía.   Cantidad de calor que contiene una sustancia, establecida desde una base o un punto predeterminado.

Environment.   Our surroundings, including the atmosphere.

Medio ambiente.   Nuestros alrededores, incluyendo la atmósfera.

Environmental Protection Agency (EPA).   A branch of the federal government dealing with the control of ozone-depleting refrigerants and other chemicals and the overall welfare of the environment.

Agencia de Protección Ambiental (EPA en inglés).   Una rama del gobierno federal que trata con el control de los refrigerantes y otros químicos que repletan el ozono, y el bienestar completo del ambiente.

EPA.   Abbreviation for the Environmental Protection Agency.

EPA.   Acrónimo en inglés de Environmental Protection Agency (Agencia de protección ambiental).

ERV. Energy recovery ventilator, a device that reclaims energy from exhaust airflows.

VRE (ERV, por su sigla en inglés). Ventilador de recuperación de energía, un dispositivo que recupera energía de las corrientes de aire de los tubos de escape.

Ester.   A popular synthetic lubricant that performs best with HFCs and HFC-based blends.

Ester.   Lubricante sintético de uso común que da óptimos resultados con HFC y mezclas basadas en HFC.

Ethane gas.   The fossil fuel, natural gas, used for heat.

Gas etano. Combustible fósil, gas natural, utilizado para generar calor.

Evacuation. The removal of any gases not characteristic to a system or vessel.

Evacuación. Remoción de los gases no característicos de un sistema o depósito.

Evaporation. The condition that occurs when heat is absorbed by liquid and it changes to vapor.

Evaporación. Condición que ocurre cuando un líquido absorbe calor y se convierte en vapor.

Evaporative condenser (cooling tower). A combination water cooling tower and condenser. The refrigerant from the compressor is routed to the cooling tower where the tower evaporates water to cool the refrigerant. In the evaporative condenser the refrigerant is routed to the tower and the water circulates only in the tower. In a cooling tower, the water is routed to the condenser at the compressor location.

Condensador evaporatorio (torre de enfriamiento). Una combinación de una torre de enfriamiento de agua y un condensador. El refrigerante del compresor se desvía a la torre de enfriamiento donde la torre evapora agua para enfriar al refrigerante. En el condensador evaporatorio el refrigerante se lleva a la torre y el agua circula sólo en la torre. En una torre de enfriamiento, el agua se pasa por el condensador donde está ubicado el compresor.

Evaporative cooling. Devices that provide this type of cooling use fiber mounted in a frame with water slowly running down the fiber. Fresh air is drawn in and through the water-soaked fiber and cooled by evaporation to a point close to the wet-bulb temperature of the air.

Enfriamiento por formación de vapor. Los dispositivos que proporcionan este tipo de enfriamiento utilizan agua que corre sobre fibra colocada en un marco. Se pasa aire nuevo a través de la fibra húmeda, el aire se enfría por evaporación a una temperatura próxima a la de una bombilla húmeda.

Evaporator. The component in a refrigeration system that absorbs heat into the system and evaporates the liquid refrigerant.

Evaporador. El componente en un sistema de refrigeración que absorbe el calor hacia el sistema y evapora el refrigerante líquido.

Evaporator fan. A forced convector used to improve the efficiency of an evaporator by air movement over the coil.

Abanico del evaporador. Convector forzado que se utiliza para mejorar el rendimiento de un evaporador por medio del movimiento de aire a través de la bobina.

Evaporator pressure regulator (EPR). A mechanical control installed in the suction line at the evaporator outlet that keeps the evaporator pressure from dropping below a certain point.

Regulador de presión del evaporador (EPR en inglés). Regulador mecánico instalado en el conducto de aspiración de la salida del evaporador; evita que la presión del evaporador caiga hasta alcanzar un nivel por debajo del nivel específico.

Evaporator types. Flooded—an evaporator where the liquid refrigerant level is maintained to the top of the heat exchange coil. Dry type—an evaporator coil that achieves the heat exchange process with a minimum of refrigerant charge.

Clases de evaporadores. Inundado—un evaporador en el que se mantiene el nivel del refrigerante líquido en la parte superior de la bobina de intercambio de calor. Seco—una bobina de evaporador que logra el proceso de intercambio de calor con una mínima cantidad de carga de refrigerante.

Even parallel system. Parallel compressors of equal sizes mounted on a steel rack and controlled by a microprocessor.

Sistema paralelo homogéneo. Compresores de capacidades iguales, montados en paralelo en un bastidor de acero y controlados mediante un microprocesador.

Exhaust valve. The movable component in a refrigeration compressor that allows hot gas to flow to the condenser and prevents it from refilling the cylinder on the downstroke.

Válvula de escape. Componente móvil en un compresor de refrigeración que permite el flujo de gas caliente al condensador y evita que este gas rellene el cilindro durante la carrera descendente.

Expansion joint. A flexible portion of a piping system or building structure that allows for expansion of the materials due to temperature changes.

Junta de expansión. Parte flexible de un sistema de tubería o de la estructura de un edificio que permite la expansión de los materiales debido a cambios de temperatura.

Expansion (metering) device. The component between the high-pressure liquid line and the evaporator that feeds the liquid refrigerant into the evaporator.

Dispositivo de (medición) de expansión. Componente entre el conducto de líquido de alta presión y el evaporador que alimenta el refrigerante líquido hacia el evaporador.

Explosion-proof motor. A totally sealed motor and its connections that can be operated in an explosive atmosphere, such as in a natural gas plant.

Motor a prueba de explosión. Un motor totalmente sellado y con conexiones que puede operarse en una atmósfera explosiva, tal como dentro de una planta de gas natural.

External drive. An external type of compressor motor drive, as opposed to a hermetic compressor.

Motor externo. Motor tipo externo de un compresor, en comparación con un compresor hermético.

External equalizer. The connection from the evaporator outlet to the bottom of the diaphragm on a thermostatic expansion valve.

Equilibrador externo. Conexión de la salida del evaporador a la parte inferior del diafragma en una válvula de expansión termostática.

External heat defrost. A defrost system for a refrigeration system where the heat comes from some external source. It might be an electric strip heater in an air coil, or water in the case of an ice maker. External means other than hot gas defrost.

Descongelador de calor externo. Un sistema de descongelación para un sistema de refrigeración en el cual el calor viene de una fuente externa. La misma puede ser un calentador de tira eléctrico en un serpentín o agua en caso de una hielera. Externo se refiere a otro tipo de descongelación aparte del de gas caliente.

External motor protection.   Motor overload protection that is mounted on the outside of the motor.

Protección externa para motor.   Protección para un motor contra la sobrecarga y que está montada en el exterior del motor.

Fahrenheit scale.   The temperature scale that places the boiling point of water at 212°F and the freezing point at 32°F.

Escala Fahrenheit.   Escala de temperatura en la que el punto de ebullición del agua se encuentra a 212°F y el punto de fusión del hielo a 32°F.

Fan.   A device that produces a pressure difference in air to move it.

Abanico.   Dispositivo que produce una diferencia de presión en el aire para moverlo.

Fan cycling.   The use of a pressure control to turn a condenser fan on and off to maintain a correct pressure within the system.

Funcionamiento cíclico.   La utilización de un regulador de presión para poner en marcha y detener el abanico de un condensador a fin de mantener una presión correcta dentro del sistema.

Fan laws.   Also known as affinity laws, this set of equations describes the proportional relationships between speed, flow, pressure, and power of centrifugal fans.

Leyes de los ventiladores.   También conocidas como leyes de afinidad, este conjunto de ecuaciones describe la relación proporcional entre la velocidad, el flujo, la presión y la potencia de los ventiladores centrífugos.

Fan relay coil.   A magnetic coil that controls the starting and stopping of a fan.

Bobina de relé del ventilador.   Bobina magnética que regula la puesta en marcha y la parada de un ventilador.

Farad.   The unit of capacity of a capacitor. Capacitors in our industry are rated in microfarads.

Faradio.   Unidad de capacidad de un capacitador. En nuestro medio, los capacitadores se clasifican en microfaradios.

Feedback loop.   The circular data route in a control loop that usually travels from the control medium's sensor to the controller, then to the controlled device, and back into the controlled process to the sensor again as a change in the control point.

Circuito de retroalimentación.   Ruta circular de la data en un circuito de control que generalmente va desde el sensor del medio de control al controlador y luego al aparato controlado, y de regreso al proceso controlado y al sensor nuevamente como un cambio en el punto de control.

Female thread.   The internal thread in a fitting.

Rosca hembra.   Rosca interna en un accesorio.

Fill or wetted-surface method.   Water in a cooling tower is spread out over a wetted surface while air is passed over it to enhance evaporation.

Método de relleno o de superficie mojada.   El agua en una torre de refrigeración se extiende sobre una superficie mojada mientras el aire se dirige por encima de la misma para facilitar la evaporación.

Film factor.   The relationship between the medium giving up heat and the heat exchange surface (evaporator). This relates to the velocity of the medium passing over the evaporator. When the velocity is too slow, the film between the air and the evaporator becomes greater and becomes an insulator, which slows the heat exchange.

Factor de capa.   Relación entre el medio que emite calor y la superficie del intercambiador de calor (evaporador). Esto se refiere a la velocidad del medio que pasa sobre el evaporador. Cuando la velocidad es demasiado lenta, la capa entre el aire y el evaporador se expande y se convierte en un aislador, disminuyendo así la velocidad del intercambio del calor.

Filter.   A fine mesh or porous material that removes particles from passing fluids.

Filtro.   Malla fina o material poroso que remueve partículas de los fluidos que pasan por él.

Filter-drier.   A type of refrigerant filter that includes a desiccant material that has an attraction for moisture. The filter drier will remove particles and moisture from refrigerant and oil.

Secador de filtro.   Un tipo de filtro de refrigerante que incluye un material desecante el cual tiene una atracción a la humedad. El secador de filtro removerá las partículas y la humedad del aceite y del refrigerante.

Fin comb.   A hand tool used to straighten the fins on an air-cooled condenser.

Herramienta para aletas.   Herramienta manual utilizada para enderezar las aletas en un condensador enfriado por aire.

Finned-tube evaporator.   A copper or aluminum tube that has fins, usually made of aluminum, pressed onto the copper lines to extend the surface area of the tubes.

Evaporador de tubo con aletas.   Un tubo de cobre o aluminio que tiene aletas, generalmente de aluminio, colocadas a presión contra las líneas de cobre para extender el área de superficie de los tubos.

Fixed-bore device.   An expansion device with a fixed diameter that does not adjust to varying load conditions.

Dispositivo de calibre fijo.   Dispositivo de expansión con un diámetro fijo que no se ajusta a las condiciones de carga variables.

Fixed resistor.   A nonadjustable resistor. The resistance cannot be changed.

Resistor fijo.   Resistor no ajustable. La resistencia no se puede cambiar.

Flapper valve.   See Reed valve.

Chapaleta.   Véase Válvula de lámina.

Flare.   The angle that may be fashioned at the end of a piece of tubing to match a fitting and create a leak-free connection.

Abocinado.   Ángulo que puede formarse en el extremo de una pieza de tubería para emparejar un accesorio y crear una conexión libre de fugas.

Flare nut.   A threaded connector used in a flare assembly for tubing.

Tuerca abocinada.   Conector de rosca utilizado en un conjunto abocinado para tuberías.

Flash gas.   A term used to describe the pressure drop in an expansion device when some of the liquid passing through the valve is changed quickly to a gas and cools the remaining liquid to the corresponding temperature.

Gas instantáneo. Término utilizado para describir la caída de la presión en un dispositivo de expansión cuando una parte del líquido que pasa a través de la válvula se convierte rápidamente en gas y enfría el líquido restante a la temperatura correspondiente.

Floating head pressure. Letting the head pressure (condensing pressure) fluctuate with the ambient temperature from season to season for lower compression ratios and better efficiencies.

Fluctuar la presión de la carga. Hecho de dejar que la presión de la carga (presión de condensación) fluctúe con la temperatura del ambiente en cada temporada del año, para obtener relaciones de compresión más bajas y mejores rendimientos.

Float valve or switch. An assembly used to maintain or monitor a liquid level.

Válvula o conmutador de flotador. Conjunto utilizado para mantener o controlar el nivel de un líquido.

Flooded evaporator. A refrigeration system operated with the liquid refrigerant level very close to the outlet of the evaporator coil for improved heat exchange.

Sistema inundado. Sistema de refrigeración que funciona con el nivel del refrigerante líquido bastante próximo a la salida de la bobina del evaporador para mejorar el intercambio de calor.

Flooding. The term applied to a refrigeration system when the liquid refrigerant reaches the compressor.

Inundación. Término aplicado a un sistema de refrigeración cuando el nivel del refrigerante líquido llega al compresor.

Flue. The duct that carries the products of combustion out of a structure for a fossil or a solid-fuel system.

Conducto de humo. Conducto que extrae los productos de combustión de una estructura en sistemas de combustible fósil o sólido.

Flue-gas analysis instruments. Instruments used to analyze the operation of fossil fuel-burning equipment such as oil and gas furnaces by analyzing the flue gases.

Instrumentos para el análisis del gas de combustión. Instrumentos utilizados para llevar a cabo un análisis del funcionamiento de los quemadores de combustible fósil, como por ejemplo hornos de aceite pesado o gas, a través del estudio de los gases de combustión.

Fluid. The state of matter of liquids and gases.

Fluido. Estado de la materia de líquidos y gases.

Fluid expansion device. Using a bulb or sensor, tube, and diaphragm filled with fluid, this device will produce movement at the diaphragm when the fluid is heated or cooled. A bellows may be added to produce more movement. These devices may contain vapor and liquid.

Dispositivo para la expansión del fluido. Utilizando una bombilla o sensor, un tubo y un diafragma lleno de fluido, este dispositivo generará movimiento en el diafragma cuando se caliente o enfríe el fluido. Se le puede agregar un fuelle para generar aún más movimiento. Dichos dispositivos pueden contener vapor y líquido.

Flush. The process of using a fluid to push contaminants from a system.

Descarga. Proceso de utilizar un fluido para remover los contaminantes de un sistema.

Flux. A substance applied to soldered and brazed connections to prevent oxidation during the heating process.

Fundente. Sustancia aplicada a conexiones soldadas y broncesoldadas para evitar la oxidación durante el proceso de calentamiento.

Foaming. A term used to describe oil when it has liquid refrigerant boiling out of it.

Espumación. Término utilizado para describir el aceite cuando el refrigerante líquido se derrama del mismo.

Foot-pound. The amount of work accomplished by lifting 1 lb of weight 1 ft; a unit of energy.

Libra-pie. Medida de la cantidad de energía o fuerza que se requiere para levantar una libra a una distancia de un pie; unidad de energía.

Force. Energy exerted.

Fuerza. Energía ejercida sobre un objeto.

Forced convection. The movement of fluid by mechanical means.

Convección forzada. Movimiento de fluido por medios mecánicos.

Forced-draft cooling tower. A water cooling tower that has a fan on the side of the tower that pushes air through the tower, as opposed to an induced-draft tower, which has the fan on the side and draws air through the tower.

Torre de enfriamiento de ventilación forzada. Una torre de enfriamiento que tiene un ventilador en el lado de la torre y empuja aire a través de la torre, contrario a una torre de ventilación inducida que tiene un ventilador en el lado de la torre y jala el aire a través de la torre.

Forced-draft evaporator. An evaporator over which air is forced to spread the cooling more efficiently. This term usually refers to a domestic refrigerator or freezer.

Evaporador de tiro forzado. Evaporador encima del cual se envía aire soplado para obtener una distribución más eficaz del frío. Este término suele aplicarse a refrigeradores o congeladores domésticos.

Fossil fuels. Natural gas, oil, and coal formed millions of years ago from dead plants and animals.

Combustibles fósiles. El gas natural, el petróleo y el carbón que se formaron hace millones de años de plantas y animales muertos.

Four-way valve. The valve in a heat pump system that changes the direction of the refrigerant flow between the heating and cooling cycles.

Válvula con cuatro vías. Válvula en un sistema de bomba de calor que cambia la dirección del flujo de refrigerante entre los ciclos de calentamiento y enfriamiento.

Fractionation. When a zeotropic refrigerant blend phase changes, the different components in the blend all have different vapor pressures. This causes different vaporization and condensation rates and temperatures as they phase-change.

Fraccionación. Cuando se produce un cambio en la fase de una mezcla zeotrópica de refrigerantes, los diferentes componentes que forman la mezcla tienen presiones de vapor diferentes. Esto provoca diferentes temperaturas y tasas de vaporización y de condensación según cambien de fase.

Freezer burn.   The term applied to frozen food when it becomes dry and hard from dehydration due to poor packaging.

Quemadura del congelador.   Término aplicado a la comida congelada cuando se seca y endurece debido a la deshidratación ocacionada por el empaque de calidad inferior.

Freeze-up.   Excess ice or frost accumulation on an evaporator to the point that airflow may be affected.

Congelación.   Acumulación excesiva de hielo o congelación en un evaporador a tal extremo que el flujo de aire puede ser afectado.

Freezing.   The change of state of water from a liquid to a solid.

Congelamiento.   Cambio de estado del agua de líquido a sólido.

Freon.   The previous trade name for refrigerants manufactured by E. I. du Pont de Nemours & Co., Inc.

Freón.   Marca registrada previa para refrigerantes fabricados por la compañía E. I. du Pont de Nemours, S.A.

Frequency.   The cycles per second (cps) of the electrical current supplied by the power company. This is normally 60 cps in the United States.

Frecuencia.   Ciclos por segundo (cps), generalmente 60 cps en los Estados Unidos, de la corriente eléctrica suministrada por la empresa de fuerza motriz.

Frequency drive.   An electronic control device that uses frequency (hertz) to change the speed of a motor. Most motors operate on 60 Hz (or cycles per second) power. The frequency drive increases or decreases cycles per second, which changes the speed of the motor to match the frequency.

Impulsor de frecuencia.   Dispositivo de control electrónico que utiliza la frecuencia (hertzio) para modificar la velocidad de un motor. La mayoría de los motores funciona a una potencia de 60 Hz (o ciclos por segundo). El impulsor de frecuencia aumenta o disminuye los ciclos por segundo, lo que modifica la velocidad del motor, igualándola a la frecuencia.

Friction loss.   The loss of pressure in a fluid flow system (air or water) due to the friction of the fluid rubbing on the sides. It is typically measured in feet of equivalent loss.

Pérdida por fricción.   La pérdida de presión en un sistema de flujo de fluido (aire o agua) debido a la fricción del fluido al frotar contra los lados. Típicamente se mide en pies de pérdida equivalente.

Frontseated.   A position on a service valve that will not allow refrigerant flow in one direction.

Sentado delante.   Posición en una válvula de servicio que no permite el flujo de refrigerante en una dirección.

Frost back.   A condition of frost on the suction line and even on the compressor body.

Obturación por congelación.   Condición de congelación que ocurre en el conducto de aspiración e inclusive en el cuerpo del compresor.

Frostbite.   When skin freezes.

Quemadura por frío.   Congelación de la piel.

Frozen.   The term used to describe water in the solid state; also used to describe a rotating shaft that will not turn.

Congelado.   Término utilizado para describir el agua en un estado sólido; utilizado también para describir un árbol giratorio que no gira.

Fuel oil.   The fossil fuel used for heating; a petroleum distillate.

Aceite pesado.   Combustible fósil utilizado para calentar; un destilado de petróleo.

Full-load amperage (FLA).   The current an electric motor draws while operating under a full-load condition. Also called the run-load amperage and rated-load amperage.

Amperaje de carga total (FLA en inglés).   Corriente que un motor eléctrico consume mientras funciona en una condición de carga completa. Conocido también como amperaje de carga de funcionamiento y amperaja de carga estándar.

Furnace.   Equipment used to convert heating energy, such as fuel oil, gas, or electricity, to usable heat. It usually contains a heat exchanger, a blower, and the controls to operate the system.

Horno.   Equipo utilizado para la conversión de energía calórica, como por ejemplo el aceite pesado, el gas o la electricidad, en calor utilizabe. Normalmente contiene un intercambiador de calor, un soplador y los reguladores para accionar el sistema.

Fuse.   A safety device used in electrical circuits for the protection of the circuit conductor and components.

Fusible.   Dispositivo de seguridad utilizado en circuitos eléctricos para la protección del conductor y los componentes del circuito.

Fusible link.   An electrical safety device normally located in a furnace that burns and opens the circuit during an overheat situation.

Cartucho de fusible.   Dispositivo eléctrico de seguridad ubicado por lo general en un horno, que quema y abre el circuito en caso de sobrecalentamiento.

Fusible plug.   A device (made of low-melting-temperature metal) used in pressure vessels that is sensitive to high temperatures and relieves the vessel contents in an overheating situation.

Tapón de fusible.   Dispositivo utilizado en depósitos en presión, hecho de un metal que tiene una temperatura de fusión baja. Este dispositivo es sensible a temperaturas altas y alivia el contenido del depósito en caso de sobrecalentamiento.

Gas.   The vapor state of matter.

Gas.   Estado de vapor de una materia.

Gas gun.   A furnace burner designed to burn LP or natural gas in place of oil in an existing oil heat system.

Quemador de gas.   Quemador de calderas diseñado para quemar gas de petróleo líquido o gas natural, en vez de aceite, en sistemas de calefacción a aceite existentes.

Gasket.   A thin piece of flexible material used between two metal plates to prevent leakage.

Guarnición.   Pieza delgada de material flexible utilizada entre dos piezas de metal para evitar fugas.

Gas-pressure switch.   Used to detect gas pressure before gas burners are allowed to ignite.

Conmutador de presión del gas. Utilizado para detectar la presión del gas antes de que los quemadores de gas puedan encenderse.

Gas valve. A valve used to stop, start, or modulate the flow of natural gas.

Válvula de gas. Válvula utilizada para detener, poner en marcha o modular el flujo de gas natural.

Gauge. An instrument used to indicate pressure.

Calibrador. Instrumento utilizado para indicar presión.

Gauge manifold. A tool that may have more than one gauge with a valve arrangement to control fluid flow.

Distribuidor de calibrador. Herramienta que puede tener más de un calibrador con las válvulas arregladas a fin de regular el flujo de fluido.

Gauge port. The service port used to attach a gauge for service procedures.

Orificio de calibrador. Orificio de servicio utilizado con el propósito de fijar un calibrador para procedimientos de servicio.

Geothermal. This term directly translates as "earth heat" and refers to the amount of heat energy available in the earth.

Geotérmico. Este térmico significa "calor de la tierra" y se refiere a la cantidad de energía calórica disponible en la tierra.

Geothermal well. A well dedicated to a geothermal heat pump that draws water from the top of the water column and returns the same water to the bottom of the water column.

Pozo geotermal. Pozo utilizado por una bomba de calor geotérmica que extrae agua de la parte superior de la columna de agua y la devuelve en la parte inferior de la misma columna.

Global warming. An earth-warming process caused by the atmosphere's absorption of the heat energy radiated from the earth's surface.

Calentamiento de la tierra. Proceso de calentamiento de la tierra provocado por la absorción de la energía radiada de la superficie de la tierra, por la atmósfera, en forma de calor.

Global warming potential (GWP). An index that measures the direct effect of chemicals emitted into the atmosphere.

Potencial de calentamiento de la tierra (GWP en inglés). Índice que mide el efecto directo de los productos químicos que se emiten a la atmósfera.

Globe valve. A type of valve with a movable disk and a stationary ring seat in a generally spherical body. It is used for regulating flow in a pipe.

Válvula esférica. Válvula con un disco móvil y un anillo estacionario dentro de un cuerpo, generalmente esférico. Se utiliza para regular el flujo en un conducto o en un tubo.

Glow coil. A device that automatically reignites a pilot light if it goes out.

Bobina encendedora. Dispositivo que automáticamente vuelve a encender la llama piloto si ésta se apaga.

Glycol. Antifreeze solution used in the water loop of geothermal heat pumps.

Glicol. Líquido anticongelante que se emplea en el circuito de agua de las bombas de calor geotérmicas.

gpm. Gallons per minute, a measurement of flow or quantity of a liquid within one minute of time.

gpm. Galones por minuto. Sistema para medir el flujo o la cantidad de líquido durante un un minuto.

Graduated cylinder. A cylinder with a visible column of liquid refrigerant used to measure the refrigerant charged into a system. Refrigerant temperatures can be dialed on the graduated cylinder.

Cilindro graduado. Cilindro con una columna visible de refrigerante líquido utilizado para medir el refrigerante inyectado al sistema. Las temperaturas del refrigerante pueden marcarse en el cilindro graduado.

Grain. Unit of measure. One pound = 7,000 grains.

Grano. Unidad de medida. Una libra equivale a 7000 granos.

Gram. Metric measurement term used to express weight.

Gramo. Término utilizado para referirse a la unidad básica de peso en el sistema métrico.

Grille. A louvered, often decorative, component in an air system at the inlet or the outlet of the airflow.

Rejilla. Componente con celosías, comúnmente decorativo, en un sistema de aire que se encuentra a la entrada o a la salida del flujo de aire.

Grommet. A rubber, plastic, or metal protector usually used where wire or pipe goes through a metal panel.

Guardaojal. Protector de caucho, plástico o metal normalmente utilizado donde un alambre o un tubo pasa a través de una base de metal.

Ground, electrical. A circuit or path for electron flow to earth ground.

Tierra eléctrica. Circuito o trayectoria para el flujo de electrones a la puesta a tierra.

Ground fault circuit interrupter (GFCI). A circuit breaker that can detect very small leaks to ground, which, under certain circumstances, could cause an electrical shock. This small leak, which may not be detected by a conventional circuit breaker, will cause the GFCI circuit breaker to open the circuit.

Disyuntor por pérdidas a tierra (GFCI en inglés). Disyuntor capaz de detectar fugas muy pequeñas hacia tierra, que en determinadas circunstancias pueden provocar sacudidas eléctricas. Existe la posibilidad de que un disyuntor convencional no sea capaz de detectar las fugas pequeñas, en cuyo caso el disyuntor GFCI abre el circuito.

Ground loop. These loops of plastic pipe are buried in the ground in a closed-loop geothermal heat pump system and contain a heat transfer fluid.

Circuito de tierra. Circuitos de tubos de plástico enterrados y que forman parte de un sistema de bomba geotérmica de circuito cerrado. Dichos tubos contienen un fluido que permite la transferencia de calor.

Ground wire. A wire from the frame of an electrical device to be wired to the earth ground.

Alambre a tierra. Alambre que va desde el armazón de un dispositivo eléctrico para ser conectado a la puesta a tierra.

Guide vanes. Vanes used to produce capacity control in a centrifugal compressor. Also called prerotation guide vanes.

Paletas directrices. Paletas utilizadas para producir la regulación de capacidad en un compresor centrífugo. Conocidas también como paletas directrices para prerotación.

Halide refrigerants. Refrigerants that contain halogen chemicals: R-12, R-22, R-500, and R-502 are among them.

Refrigerantes de hálido.   Refrigerantes que contienen productos químicos de halógeno: entre ellos se encuentran el R-12, R-22, R-500 y R-502.

Halide torch.   A torch-type leak detector used to detect the halogen refrigerants.

Soplete de hálido.   Detector de fugas de tipo soplete utilizado para detectar los refrigerantes de halógeno.

Halogens.   Chemical substances found in many refrigerants containing chlorine, bromine, iodine, and fluorine.

Halógenos.   Sustancias químicas presentes en muchos refrigerantes que contienen cloro, bromo, yodo y flúor.

Hand truck.   A two-wheeled piece of equipment that can be used for moving heavy objects.

Vagoneta para mano.   Equipo con dos ruedas que puede utilizarse para transportar objetos pesados.

Hanger.   A device used to support tubing, pipe, duct, or other components of a system.

Soporte.   Dispositivo utilizado para apoyar tuberías, tubos, conductos u otros componentes de un sistema.

Head.   Another term for pressure, usually referring to gas or liquid.

Carga.   Otro término para presión, refiriéndose normalmente a gas o líquido.

Header.   A pipe or containment to which other pipe lines are connected.

Conductor principal.   Tubo o conducto al que se conectan otras conexiones.

Head pressure control.   A control that regulates the head pressure in a refrigeration or air conditioning system.

Regulador de la presión de la carga.   Regulador que controla la presión de la carga en un sistema de refrigeración o de acondicionamiento de aire.

Heat.   Energy that causes molecules to be in motion and to raise the temperature of a substance.

Calor.   Energía que ocasiona el movimiento de las moléculas provocando un aumento de temperatura en una sustancia.

Heat anticipator.   A device that anticipates the need for cutting off the heating system prematurely so the system does not overshoot the set point temperature.

Anticipador de calor.   Dispositivo que anticipa la necesidad de detener la marcha del sistema de calentamiento para que el sistema no exceda la temperatura programada.

Heat coil.   A device made of tubing or pipe designed to transfer heat to a cooler substance by using fluids.

Bobina de calor.   Dispositivo hecho de tubos, diseñado para transmitir calor a una sustancia más fría por medio de fluidos.

Heaters.   Replaceable devices in a motor starter that trip a starter open if they sense high amp draw.

Calentadores.   Dispositivos reemplazables en el reóstato de arranque del motor que abren el reóstato de arranque si detectan una corriente de amperios alta.

Heat exchanger.   A device that transfers heat from one substance to another.

Intercambiador de calor.   Dispositivo que transmite calor de una sustancia a otra.

Heat fusion.   A process that will permanently join sections of plastic pipe together.

Fusión térmica.   Proceso mediante el cual se unen dos piezas de tubo de plástico permanentemente.

Heat of compression.   That part of the energy from the pressurization of a gas or a liquid converted to heat.

Calor de compresión.   La parte de la energía generada de la presurización de un gas o un líquido que se ha convertido en calor.

Heat of fusion.   The heat released when a substance is changing from a liquid to a solid.

Calor de fusión.   Calor liberado cuando una sustancia se convierte de líquido a sólido.

Heat of respiration.   When oxygen and carbon hydrates are taken in by a substance or when carbon dioxide and water are given off. Associated with fresh fruits and vegetables during their aging process while stored.

Calor de respiración.   Cuando se admiten oxígeno e hidratos de carbono en una sustancia o cuando se emiten dióxido de carbono y agua. Se asocia con el proceso de maduración de frutas y legumbres frescas durante su almacenamiento.

Heat pump.   A refrigeration system used to supply heat or cooling using valves to reverse the refrigerant gas flow.

Bomba de calor.   Sistema de refrigeración utilizado para suministrar calor o frío mediante válvulas que cambian la dirección del flujo de gas del refrigerante.

Heat reclaim.   Using heat from a condenser for purposes such as space and domestic water heating.

Reclamación de calor.   La utilización del calor de un condensador para propósitos tales como la calefacción de espacio y el calentamiento doméstico de agua.

Heat recovery ventilator.   Units that recover heat only, used primarily in winter.

Ventilador de recuperación de calor.   Unidades que recuperan calor solamente, usados principalmente en el invierno.

Heat sink.   A low-temperature surface to which heat can transfer.

Fuente fría.   Superficie de temperatura baja a la que puede transmitírsele calor.

Heat tape.   Electric resistance wires embedded into a flexible housing usually wrapped around a pipe to keep it from freezing.

Cinta calefactora.   Resistencia eléctrica incrustada en una cubierta flexible normalmente instalada alrededor de un tubo para impedir su congelación.

Heat transfer.   The transfer of heat from a warmer to a colder substance.

Transmisión de calor.   Cuando se transmite calor de una sustancia más caliente a una más fría.

Helix coil.   A bimetal formed into a helix-shaped coil that provides longer travel when heated.

Bobina en forma de hélice.   Bimetal encofrado en una bobina en forma de hélice que provee mayor movimiento al ser calentado.

HEPA filter.   An abbreviation for high-efficiency particulate arrestor. These filters are used when a high degree of filtration is desired or required.

Filtro HEPA.   Abreviatura para filtro de partículas de alto rendimiento. Este tipo de filtro se utiliza cuando se requiere un elevado grado de filtrado.

Hermetic compressor.   A motor and compressor that are totally sealed by being welded in a container.

Compresor hermético.   Un motor y un compresor que están totalmente sellados al ser soldados al contenedor.

Hermetic system.   An enclosed refrigeration system where the motor and compressor are sealed within the same system with the refrigerant.

Sistema hermético.   Sistema de refrigeracíon cerrado donde el motor y el compresor se obturan dentro del mismo sistema con el refrigerante.

Hertz.   Cycles per second.

Hertz.   Ciclos por segundo.

Hg.   Abbreviation for the element mercury.

Hg.   Abreviatura del elemento mercurio.

High-pressure control.   A control that stops a boiler heating device or a compressor when the pressure becomes too high.

Regulador de alta presión.   Regulador que detiene la marcha del dispositivo de calentamiento de una caldera o de un compresor cuando la presión alcanza un nivel demasiado alto.

High side.   A term used to indicate the high-pressure or condensing side of the refrigeration system.

Lado de alta presión.   Término utilizado para indicar el lado de alta presión o de condensación del sistema de refrigeración.

High-temperature refrigeration.   A refrigeration temperature range starting with evaporator temperatures no lower than 35°F, a range usually used in air conditioning (cooling).

Refrigeración a temperatura alta.   Margen de la temperatura de refrigeración que comienza con temperaturas de evaporadores no menores de 35°F (2°C). Este margen se utiliza normalmente en el acondicionamiento de aire (enfriamiento).

High-vacuum pump.   A pump that can produce a vacuum in the low micron range.

Bomba de vacío alto.   Bomba que puede generar un vacío dentro del margen de micrón bajo.

Horsepower.   A unit equal to 33,000 ft-lb of work per minute.

Potencia en caballos.   Unidad equivalente a 33,000 libras-pies de trabajo por minuto.

Hot gas.   The refrigerant vapor as it leaves the compressor. This is often used to defrost evaporators.

Gas caliente.   El vapor del refrigerante al salir del compresor. Esto se utiliza con frecuencia para descongelar evaporadores.

Hot gas bypass.   Piping that allows hot refrigerant gas into the cooler low-pressure side of a refrigeration system, usually for system capacity control.

Desviación de gas caliente.   Tubería que permite la entrada de gas caliente del refrigerante en el lado más frío de baja presión de un sistema de refrigeración, normalmente para la regulación de la capacidad del sistema.

Hot gas defrost.   A system where the hot refrigerant gases are passed through the evaporator to defrost it.

Descongelación con gas caliente.   Sistema en el que los gases calientes del refrigerante se pasan a través del evaporador para descongelarlo.

Hot gas line.   The tubing between the compressor and the condenser.

Conducto de gas caliente.   Tubería entre el compresor y el condensador.

Hot water heat.   A heating system using hot water to distribute the heat.

Calor de agua caliente.   Sistema de calefacción que utiliza agua caliente para la distribución del calor.

Hot wire.   The wire in an electrical circuit that has a voltage potential between it and another electrical source or between it and ground.

Conductor electrizado.   Conductor en un circuito eléctrico a través del cual fluye la tensión entre éste y otra fuente de electricidad o entre éste y la tierra.

HSPF.   Heating seasonal performance factor, a measure of the heating efficiency of a heat pump over a season of operation. It is calculated by dividing the estimated seasonal heating output in BTUs by the amount of consumed energy in watt-hours.

FDEC (HSPF, por su sigla en inglés).   Factor de desempeño estacional del calentador. Medición del desempeño de una bomba de calor durante una temporada de funcionamiento. Se calcula dividiendo la salida de calor estacional aproximada en UTB (unidades térmicas británicas) por la cantidad de energía consumida en horas vatio.

Humidifier.   A device used to add moisture to the air.

Humedecedor.   Dispositivo utilizado para agregarle humedad al aire.

Humidistat.   A control operated by a change in humidity.

Humidistato.   Regulador activado por un cambio en la humedad.

Humidity.   Moisture in the air.

Humedad.   Vapor de agua existente en el ambiente.

Hunting.   The open and close throttling of a valve that is searching for its set point.

Caza.   El abrir y cerrar de una válvula que está buscando su punto de ajuste.

Hydraulics.   Producing mechanical motion by using liquids under pressure.

Hidráulico.   Generación de movimiento mecánico por medio de líquidos bajo presión.

Hydrocarbons.   Organic compounds containing hydrogen and carbon found in many heating fuels.

Hidrocarburos.   Compuestos orgánicos que contienen el hidrógeno y el carbón presentes en muchos combustibles de calentamiento.

Hydrochlorofluorocarbons (HCFCs).   Refrigerants containing hydrogen, chlorine, fluorine, and carbon, thought to contribute to the depletion of the ozone layer, although not to the extent of chlorofluorocarbons.

Hidroclorofluorocarburos (HCFCs en inglés).   Líquidos refrigerantes que contienen hidrógeno, cloro, flúor y carbono, y que, según algunos, han contribuido a la reducción de la capa de ozono aunque no en tal grado como los cloroflurocarburos.

Hydrofluorocarbon (HFC).   A chlorine-free refrigerant containing hydrogen, fluorine, and carbon with zero ozone depletion potential.

Hidroflurocarbono (HFC en inglés).   Refrigerante libre de cloro, compuesto de hidrógeno, fluoro y carbono que no tiene efectos perjudiciales sobre la capa de ozono.

Hydrometer.   An instrument used to measure the specific gravity of a liquid.

Hidrómetro. Instrumento utilizado para medir la gravedad específica de un líquido.

Hydronic. Usually refers to a hot water heating system.

Hidrónico. Normalmente se refiere a un sistema de calefacción de agua caliente.

Hygrometer. An instrument used to measure the amount of moisture in the air.

Higrómetro. Instrumento utilizado para medir la cantidad de humedad en el aire.

I=B=R. Institute of Boiler and Radiator Manufacturers, an organization that provides ratings for boilers, baseboard radiation, finned tube radiation, and indirect-fired water heaters. The I=B=R is also a nationally recognized boiler and radiator certification organization.

I=B=R. Sigla inglesa del Instituto de Fabricantes de Calderas y Radiadores, una organización que proporciona normas para calderas, radiadores de zócalo, radiadores de tubos de aletas y calentadores de agua de llama indirecta. El I=B=R es, también, una organización de certificación de calderas y radiadores nacionalmente reconocida.

Impedance. A form of resistance in an alternating current circuit.

Impedancia. Forma de resistencia en un circuito de corriente alterna.

Impeller. The rotating part of a pump that causes the centrifugal force to develop fluid flow and pressure difference.

Impulsor. Pieza giratoria de una bomba que hace que la fuerza centrífuga desarrolle flujo de fluido y una diferencia en presión.

Inclined water manometer. Indicates air pressures in very low-pressure systems.

Manómetro de agua inclinada. Señala las presiones de aire en sistemas de muy baja presión.

Indoor air quality (IAQ). This term generally refers to the study or research of air quality within buildings and the procedures used to improve air quality.

Calidad del aire en el interior (IAQ en inglés). Generalmente, este término hace referencia al estudio o investigación de la calidad del aire en el interior de los edificios, así como a los procesos empleados para su mejora.

Induced magnetism. Magnetism produced, usually in a metal, from another magnetic field.

Magnetismo inducido. Magnetismo generado, normalmente en un metal, desde otro campo magnético.

Inductance. An induced voltage producing a resistance in an alternating current circuit.

Inductancia. Tensión inducida que genera una resistencia en un circuito de corriente alterna.

Induction motor. An alternating current motor where the rotor turns from induced magnetism from the field windings.

Motor inductor. Motor de corriente alterna donde el rotor gira debido al magnetismo inducido desde los devanados inductores.

Inductive circuit. When the current in a circuit lags the voltage by 90°.

Circuito inductivo. Cuando la corriente en un circuito está atrasada al voltaje por 90°.

Inductive reactance. A resistance to the flow of an alternating current produced by an electromagnetic induction.

Reactancia inductiva. Resistencia al flujo de una corriente alterna generada por una inducción electromagnética.

Inefficient equipment. Equipment that is not operating at its design level of capacity because of some fault in the equipment, such as a cylinder not pumping in a multicylinder compressor.

Equipo ineficiente. Equipo que no está operando al nivel de capacidad al que fue diseñado debido a alguna falla en el equipo, tal como que un cilindro no esté bombeando en un compresor de múltiples cilindros.

Inert gas. A gas that will not support most chemical reactions, particularly oxidation.

Gas inerte. Gas incapaz de resistir la mayoría de las reacciones químicas, especialmente la oxidación.

Infiltration. Air that leaks into a structure through cracks, windows, doors, or other openings due to less pressure inside the structure than outside the structure.

Infiltración. Penetración de aire en una estructura a través de grietas, ventanas, puertas u otras aberturas debido a que la presión en el interior de la estructura es menor que en el exterior.

Infrared humidifier. A humidifier that has infrared lamps with reflectors to reflect the infrared energy onto the water. The water evaporates rapidly into the duct airstream and is carried throughout the conditioned space.

Humidificador por infrarrojos. Humidificador equipado con lámparas de infrarrojos cuyos reflectores reflejan la energía infrarroja sobre el agua haciendo que ésta se evapore rápidamente hacia el conducto de aire para ser transportada en el espacio acondicionado.

Infrared rays. The rays that transfer heat by radiation.

Rayos infrarrojos. Rayos que transmiten calor por medio de la radiación.

Inherent motor protection. This is provided by internal protection such as a snap-disc or a thermistor.

Protección de motor inherente. Ésta es provista por una protección interna tal como un disco de encaje o un termisor.

In. Hg vacuum. The atmosphere will support a column of mercury 29.92 in. high. To pull a complete vacuum in a refrigeration system, the pressure inside the system must be reduced to 29.92 in. Hg vacuum.

Vacío en mm Hg. La atmósfera soporta una columna de mercurio de 760 mm. Para poder crear un vacío completo en un sistema de refrigeración, la presión interna debe descender a 760 mm Hg.

In-phase. When two or more alternating current circuits have the same polarity at all times.

En fase. Cuando dos o más circuitos de corriente alterna tienen siempre la misma polaridad.

Inrush current. The maximum input current drawn by an electrical device when first turned on. An electrical load that pulls an inrush current when energized is known as an inductive load.

Corriente de entrada. La corriente de entrada máxima generada por un dispositivo eléctrico cuando recién se enciende. Una carga eléctrica que genera una corriente de entrada cuando se energiza se conoce como una carga inductiva.

Insulation, electric.   A substance that is a poor conductor of electricity.

Aislamiento eléctrico.   Sustancia que es un conductor pobre de electricidad.

Insulation, thermal.   A substance that is a poor conductor of the flow of heat.

Aislamiento térmico.   Sustancia que es un conductor pobre de flujo de calor.

Insulator.   A material with several electrons in the outer orbit of the atom making them poor conductors of electricity or good insulators. Examples are glass, rubber, and plastic.

Aislante.   Material con varios electrones en la órbita exterior del átomo, que los convierte en malos conductores de electricidad o en buenos aislantes, por ejemplo vidrio, caucho y plástico.

Interlocking components.   Mechanical and electrical interlocks that are used to prevent a piece of equipment from starting before it is safe to start. For example, the chilled water pump and the condenser water pump must both be started before the compressor in a water-cooled chilled water system.

Componentes con enclavamiento.   Enclavamientos mecánicos y eléctricos se usan para evitar que una pieza de equipo se encienda antes de ser seguro que encienda. Por ejemplo, la bomba de agua enfriada y la bomba de agua del condensador deben encenderse antes que el compresor en un sistema de enfriamiento de agua enfriado por agua.

Internal motor overload.   An overload that is mounted inside the motor housing, such as a snap-disc or thermistor.

Sobrecarga interna del motor.   Una sobrecarga que se monta dentro del cárter del motor tal como disco de encaje o un termisor.

Inverter.   A device that alters the frequency of an electronically altered sine wave, which will affect the speed of an alternating current motor.

Inversor.   Dispositivo que hace alternar la frecuencia de una onda de signos alternados electrónicamente. Esta inversión tiene un efecto sobre la velocidad de un motor de corriente alterna.

Joule.   Metric measurement term used to express the quantity of heat.

Joule.   Término utilizado para referirse a la unidad básica de cantidad de calor en el sistema métrico.

Jumper.   A physical electrical connection that bypasses a control by electrically connecting points on either side of it with a temporary conductor.

Conector de empalme.   Una conexión eléctrica física que circunvala un control conectando eléctricamente puntos que están a cada lado del control con un conductor temporal.

Junction box.   A metal or plastic box within which electrical connections are made.

Caja de empalme.   Caja metálica o plástica dentro de la cual se nacen conexiones eléctricas.

Kelvin.   A temperature scale where absolute 0 equals 0 or where molecular motion stops at 0. It has the same graduations per degree of change as the Celsius scale.

Escala absoluta.   Escala de temperaturas donde el cero absoluto equivale a 0 o donde el movimiento molecular se detiene en 0. Tiene las mismas graduaciones por grado de cambio que la escala Celsio.

Kilopascal.   A metric unit of measurement for pressure used in the air conditioning, heating, and refrigeration field. There are 6.89 kilopascals in 1 psi.

Kilopascal.   Unided métrica de medida de presión utilizada en el ramo del acondicionamiento de aire, calefacción y refrigeración. 6.89 kilopascales equivalen a 1 psi.

Kilowatt.   A unit of electrical power equal to 1,000 watts.

Kilowatio.   Unidad eléctrica de potencia equivalente a 1,000 watios.

Kilowatt-hour.   1 kilowatt (1,000 watts) of energy used for 1 hour.

Kilowatio hora.   Unidad de energía equivalente a la que produce un kilowatio durante una hora.

King valve.   A service valve at the liquid receiver's outlet in a refrigeration system.

Válvula maestra.   Válvula de servicio ubicada en el receptor del líquido.

Ladder diagram.   A type of electrical print that depicts the electrical circuit in vertical planes similar to the rungs of a ladder (also referred to as an across-the-line diagram).

Diagrama en escalera.   Impreso eléctrico que indica el circuito eléctrico en planos verticales, similares a los escalones de una escalera (también conocido como diagrama de una línea).

Lag shield anchors.   Used with lag screws to secure screws in masonry materials.

Anclajes de tornillos barraqueros.   Se usan con tornillos barraqueros para asegurar tornillos en materiales de albañilería.

Latent heat.   Heat energy absorbed or rejected when a substance is changing state and there is no change in temperature.

Calor latente.   Energía calórica absorbida o rechazada cuando una sustancia cambia de estado y no se experimentan cambios de temperatura.

Latent heat of condensation.   The latent heat given off when refrigerant condenses.

Calor latente de la condensación.   Calor latente producido por la condensación del refrigerante.

Latent heat of vaporization.   The latent heat absorbed when refrigerant evaporates.

Calor latente de vaporización.   Calor latente absorbido por la evaporación del refrigerante.

Leads.   Extended surfaces inside a heat exchanger used to enhance the heat transfer qualities of the heat exchanger.

Extensiones.   Superficies extendidas dentro de un intercambiador de calor que se usan para mejorar las cualidades de transferencia de calor del intercambiador de calor.

Leak detector.   Any device used to detect leaks in a pressurized system.

Detector de fugas.   Cualquier dispositivo utilizado para detectar fugas en un sistema presurizado.

Limit control.   A control used to make a change in a system, usually to stop it when predetermined limits of pressure or temperature are reached.

Regulador de límite.   Regulador utilizado para realizar un cambio en un sistema, normalmente para detener su marcha cuando se alcanzan niveles predeterminados de presión o de temperatura.

Limit switch.   A switch that is designed to stop a piece of equipment before it does damage to itself or the surroundings, for example, a high limit on a furnace or an amperage limit on a motor.

Interruptor de límite.   Interruptor que está diseñado para detener una pieza de equipo antes que ésta se haga daño o dañe sus alrededores, por ejemplo, un límite alto en un calefactor o un límite de amperaje en un motor.

Line set.   A term used for tubing sets furnished by the manufacturer.

Juego de conductos.   Término utilizado para referise a los juegos de tubería suministrados por el fabricante.

Line tap valve.   A device that may be used for access to a refrigerant line.

Válvula de acceso en línea.   Dispositivo que puede utilizarse para acceder al tubo del refrigerante.

Line voltage thermostat.   A thermostat that switches line voltage. For example, it is used for electric baseboard heat.

Termostato de voltaje de línea.   Un termostato que interrumpe el voltaje de línea. Por ejemplo, para los calentadores eléctricos de rodapié.

Line wiring diagram.   Sometimes called a ladder diagram, this type of diagram shows the power-consuming devices between the lines. Usually, the right side of the diagram consists of a common line.

Diagrama del cableado de línea.   También conocido como diagrama de escalera, este tipo de diagrama muestra los dispositivos de consumo de corriente que hay entre las líneas. Generalmente, la línea común está en el lado derecho del diagrama.

Liquefied petroleum.   Liquefied propane, butane, or a combination of these gases. The gas is kept as a liquid under pressure until ready to use.

Petróleo licuado.   Propano o butano licuados, o una combinación de estos gases. El gas se mantiene en estado líquido bajo presión hasta que se encuentre listo para usar.

Liquid.   A substance where molecules push outward and downward and seek a uniform level.

Líquido.   Sustancia donde las moléculas empujan hacia afuera y hacia abajo y buscan un nivel uniforme.

Liquid bypass.   A method of capacity control in which liquid moves around the evaporator to decrease flow to the evaporator and maintain liquid flow to the expansion valve.

Derivación de líquido.   Método utilizado para controlar la capacidad en el que el líquido se deriva, sin pasar por el evaporador, para disminuir el flujo al evaporador y mantener el flujo de líquido a la válvula de expansión.

Liquid charge bulb.   A type of charge in the sensing bulb of the thermostatic expansion valve. This charge is characteristic of the refrigerant in the system and contains enough liquid so that it will not totally boil away.

Bombilla de carga líquida.   Tipo de carga en la bombilla sensora de la válvula de expansión termostática. Esta carga es característica del refrigerante en el sistema y

contiene suficiente líquido para que el mismo no se evapore completamente.

Liquid-filled remote bulb.   A remote bulb thermostat that is completely liquid filled, such as the mercury bulb on some gas furnace pilot safety devices.

Bombillo remoto lleno de líquido.   Un termostato de bombillo remoto que está completamente lleno con líquido, tal como el bulbo de mercurio en algunos dispositivos de seguridad de los calefactores de gas.

Liquid floodback.   Liquid refrigerant returning to the compressor's crankcase during the running cycle.

Regreso de líquido.   El regresar del refrigerante líquido al cárter del cigüeñal del compresor durante el ciclo de marcha.

Liquid hammer.   The momentum force of liquid causing a noise or a disturbance when hitting against an object.

Martillo líquido.   La fuerza mecánica de un líquido que causa un ruido o un disturbio cuando choca con un objeto.

Liquid line.   A term applied in the industry to refer to the tubing or piping from the condenser to the expansion device.

Conducto de líquido.   Término aplicado en nuestro medio para referirse a la tubería que va del condensador al dispositivo de expansión.

Liquid nitrogen.   Nitrogen in liquid form.

Nitrógeno líquido.   Nitrógeno en forma líquida.

Liquid receiver.   A container in the refrigeration system where liquid refrigerant is stored.

Receptor del líquido.   Recipiente en el sistema de refrigeración donde se almacena el refrigerante líquido.

Liquid refrigerant charging.   The process of allowing liquid refrigerant to enter the refrigeration system through the liquid line to the condenser and evaporator.

Carga para refrigerante líquido.   Proceso de permitir la entrada del refrigerante líquido al condensador y al evaporador en el sistema de refrigeración a través del conducto de líquido.

Liquid refrigerant distributor.   This device is used between the expansion valve and the evaporator on multiple circuit evaporators to evenly distribute the refrigerant to all circuits.

Distribuidor de refrigerante líquido.   Este aparato se usa entre la válvula de expansión y el evaporador en los evaporadores de múltiples circuitos para distribuir el refrigerante a todos los circuitos.

Liquid slugging.   A large amount of liquid refrigerant in the compressor cylinder, usually causing immediate damage.

Relleno de líquido.   Acumulación de una gran cantidad de refrigerante líquido en el cilindro del compresor, que normalmente provoca una avería inmediata.

Lithium-bromide.   A type of salt solution used in an absorption chiller.

Bromuro de litio.   Tipo de solución salina utilizada en un enfriador por absorción.

Load matching.   Trying to always match the capacity of the refrigeration or air conditioning system with that of the heat load put on the evaporators.

Adaptación de carga.   Hecho de intentar adaptar la capacidad del sistema de refrigeración o aire acondicionado a la carga térmica que deben soportar los evaporadores.

Load shed. Part of an energy management system where various systems in a structure may be cycled off to conserve energy.

Despojo de carga. Parte de un sistema de manejo de energía en el cual varios sistemas en una estructura pueden apagarse para conservar energía.

Locked-rotor amperage (LRA). The current an electric motor draws when it is first turned on. This is normally five times the full-load amperage.

Amperaje de rotor bloqueado (LRA en inglés). Corriente que un motor eléctrico consume al ser encendido, la cual generalmente es cinco veces mayor que el amperaje de carga completa.

Low ambient control. Various types of controls that are used to control head pressure in air-cooled air conditioning and refrigeration systems that must operate year-round or in cold weather.

Control de ambiente bajo. Varios tipos de controles que se usan para controlar la presión en los sistemas de aire acondicionado y refrigeración, enfriados por aire, y que tienen que operar todo el año o en climas fríos.

Low-boy furnace. This furnace is approximately 4 ft high, and the air intake and discharge are both at the top.

Horno bajo. Este tipo de horno tiene una altura de aproximadamente 1.2 metros, con la toma y evacuación del aire situadas en la parte de arriba.

Low-loss fitting. A fitting that is fastened to the end of a gauge manifold that allows the technician to connect and disconnect gauge lines with a minimum of refrigerant loss.

Acoplamiento de poca pérdida. Un tipo de acoplamiento que se conecta en un extremo de un colector de calibración y que le permite al técnico conectar y desconectar las líneas de calibración con una pérdida mínima de refrigerante.

Low-pressure control. A pressure switch that can provide low charge protection by shutting down the system on low pressure. It can also be used to control space temperature.

Regulador de baja presión. Conmutador de presión que puede proveer protección contra una carga baja al detener el sistema si éste alcanza una presión demasiado baja. Puede utilizarse también para regular la temperatura de un espacio.

Low side. A term used to refer to that part of the refrigeration system that operates at the lowest pressure, between the expansion device and the compressor.

Lado bajo. Término utilizado para referirse a la parte del sistema de refrigeración que funciona a niveles de presión más baja, entre el dispositivo de expansión y el compresor.

Low-temperature refrigeration. A refrigeration temperature range starting with evaporator temperatures no higher than 0°F for storing frozen food.

Refrigeración a temperatura baja. Margen de la temperatura de refrigeración que comienza con temperaturas de evaporadores no mayores de 0°F (–18°C) para almacenar comida congelada.

Low-voltage thermostat. The typical thermostat used for residential and commercial air conditioning and heating equipment to control space temperature. The supplied voltage is 24 V.

Termostato de bajo voltaje. El termostato típico que se usa en el equipo de aire acondicionado y de calefacción comercial y residencial para controlar la temperatura de un espacio. El voltaje suplido es de 24 voltios.

LP fuel. Liquefied petroleum, propane, or butane. A substance used as a gas for fuel. It is transported and stored in the liquid state.

Combustible PL. Petróleo licuado, propano o butano. Sustancia utilizada como gas para combustible. El petróleo licuado se transporta y almacena en estado líquido.

Magnetic field. A field or space where magnetic lines of force exist.

Campó magnético. Campo o espacio donde existen líneas de fuerza magnética.

Magnetic flux. A measure of quantity of magnetism describing the strength and extent of an object's interaction with a magnetic field.

Flujo magnético. Medición de la cantidad de magnetismo que describe la potencia y cantidad de interacción de un objeto con un campo magnético.

Magnetic overload protection. This protection reads the actual current draw of the motor and is able to shut it off based on actual current, versus the heat-operated thermal overloads, which are sensitive to the ambient heat of a hot cabinet.

Protección de sobrecarga magnética. Esta protección lee la toma de corriente actual del motor y es capaz de apagarlo basado en la corriente actual; esto es contrario a las sobrecargas termales operadas por calor, las cuales son sensibles al calor ambiental de un gabinete caliente.

Magnetism. A force causing a magnetic field to attract ferrous metals, or where like poles of a magnet repel and unlike poles attract each other.

Magnetismo. Fuerza que hace que un campo magnético atraiga metales férreos, o cuando los polos iguales de un imán se rechazan y los opuestos se atraen.

Make-up air. Air, usually from outdoors, provided to make up for the air used in combustion.

Aire de compensación. Aire, normalmente procedente del exterior, que se utiliza para compensar aquél utilizado en la combustión.

Make-up water. Water that is added back into any circulating water system due to loss of water. Make-up water in a cooling tower may be quite a large volume.

Agua de compensación. Agua que se añade a cualquier sistema de circulación de agua debido a la pérdida de agua. El agua de compensación en una torre de enfriamiento puede ser un volumen bastante grande.

Male thread. A thread on the outside of a pipe, fitting, or cylinder; an external thread.

Rosca macho. Rosca en la parte exterior de un tubo, accesorio o cilindro; rosca externa.

Manifold. A device where multiple outlets or inlets can be controlled with valves or other devices. Our industry typically uses a gas manifold with orifices for gas-burning appliances and gauge manifolds used by technicians.

Colector. Aparato desde el cual se pueden controlar varias entradas y salidas con válvulas u otros aparatos. Nuestra industria generalmente usa un colector de gas con orificios para los enseres que queman gas y colectores de calibración que los usan los técnicos.

Manometer.   An instrument used to check low vapor pressures. The pressures may be checked against a column of mercury or water.

Manómetro.   Instrumento utilizado para revisar las presiones bajas de vapor. Las presiones pueden revisarse comparándolas con una columna de mercurio o de agua.

Manual reset.   A safety control that must be reset by a person, as opposed to automatically reset, to call attention to the problem. An electrical breaker is a manual reset device.

Reinicio manual.   Un control de seguridad que una persona tiene que reiniciar, contrario a un reinicio automático, para llamar atención al problema. Un cortacircuito eléctrico es un aparato de reinicio manual.

Mapp gas.   A composite gas similar to propane that may be used with air.

Gas Mapp.   Gas compuesto similar al propano que puede utilizarse con aire.

Mass.   Matter held together to the extent that it is considered one body.

Masa.   Materia compacta que se considera un solo cuerpo.

Mass spectrum analysis.   An absorption machine factory leak test performed using helium.

Análisis del límite de masa.   Prueba para fugas y absorción llevada a cabo en la fábrica utilizando helio.

Matter.   A substance that takes up space and has weight.

Materia.   Sustancia que ocupa espacio y tiene peso.

MBH.   A unit of measure equal to 1,000 BTU/hr.

MBH.   Unidad de medida equivalente a 1000 Unidades Térmicas Británicas por hora.

Mechanical controls.   A control that has no connection to power, such as a water-regulating valve or a pressure relief valve.

Controles mecánicos.   Un control que no tiene conexión a corriente, tales como una válvula reguladora de agua o una válvula de alivio de presión.

Medium-temperature refrigeration.   Refrigeration where evaporator temperatures are 32°F or below, normally used for preserving fresh food.

Refrigeración a temperatura media.   Refrigeración, donde las temperaturas del evaporador son 32°F (0°C) o menos, utilizada generalmente para preservar comida fresca.

Megohmmeter.   An instrument that can detect very high resistances, in millions of ohms. A megohm is equal to 1,000,000 ohms.

Megaohmnímetro.   Un instrumento que puede detectar resistencias muy altas, de millones de ohmios. Un megaohmio es equivalente a 1,000,000 de ohmios.

Melting point.   The temperature at which a substance will change from a solid to a liquid.

Punto de fusión.   Temperatura a la que una sustancia se convierte de sólido a líquido.

Mercury bulb.   A glass bulb containing a small amount of mercury and electrical contacts used to make and break the electrical circuit in a low-voltage thermostat.

Bombilla de mercurio.   Bombilla de cristal que contiene una pequeña cantidad de mercurio y que funciona como contacto eléctrico, utilizada para conectar y desconectar el circuito eléctrico en un termostato de baja tensión.

Metering device.   A valve or small fixed-size tubing or orifice that meters liquid refrigerant into the evaporator.

Dispositivo de medida.   Válvula o tubería pequeña u orificio que mide la cantidad de refrigerante líquido que entra en el evaporador.

Metering orifice plate.   A plate with a hole in the middle mounted inside a pipe. The fluid is accelerated as it flows through the hole, and the flow rate is measured by taking pressure readings on either side of the orifice.

Placa con orificio dosificador.   Placa con un orificio en el medio que se monta en el interior de un tubo o conducto. El fluido se acelera a medida que circula por el orificio, y la velocidad de fluido se mide tomando la presión a cada lado del orificio.

Methane.   Natural gas composed of 90% to 95% methane, a combustible hydrocarbon.

Metano.   El gas natural se compone de un 90% a un 95% de metano, un hidrocarburo combustible.

Metric system.   System International (SI); system of measurement used by most countries in the world.

Sistema métrico.   Sistema internacional; el sistema de medida utilizado por la mayoría de los países del mundo.

Micro.   A prefix meaning $\frac{1}{1,000,000}$.

Micro.   Prefijo que significa una parte de un millón.

Microfarad.   Capacitor capacity equal to $\frac{1}{1,000,000}$ of a farad.

Microfaradio.   Capacidad de un capacitor equivalente a $\frac{1}{1,000,000}$ de un faradio.

Micrometer.   A precision measuring instrument.

Micrómetro.   Instrumento de precisión utilizado para medir.

Micron.   A unit of length equal to $\frac{1}{1,000}$ of a millimeter or $\frac{1}{1,000,000}$ of a meter.

Micrón.   Unidad de largo equivalente a $\frac{1}{1000}$ de un milímetro, o $\frac{1}{1,000,000}$ de un metro.

Micron gauge.   A gauge used when it is necessary to measure pressure close to a perfect vacuum.

Calibrador de micrón.   Calibrador utilizado cuando es necesario medir la presión de un vacío casi perfecto.

Microprocessor.   A small, preprogrammed, solid-state microcomputer that acts as a main controller.

Microprocesador.   Un microordenador de estado sólido preprogramado que actúa de controlador principal.

Midseated (cracked).   A position on a service valve that allows refrigerant flow in all directions.

Sentado en el medio (agrietado).   Posición en una válvula de servicio que permite el flujo de refrigerante en cualquier dirección.

Migration of oil or refrigerant.   When the refrigerant moves to some place in the system where it is not supposed to be, such as when oil migrates to an evaporator or when refrigerant migrates to a compressor crankcase.

Migración de aceite o refrigerante.   Cuando el refrigerante se mueve a cualquier lugar en el sistema donde no debe estar, como cuando aceite migra a un evaporador o cuando refrigerante se transplanta al cárter del cigüeñal del compresor.

Milli.   A prefix meaning $\frac{1}{1,000}$.

Mili.   Prefijo que significa una parte de mil.

Mineral oil.   A traditional refrigeration lubricant used in CFC and HCFC systems.

Aceite mineral.   Lubricante de refrigeración utilizado tradicionalmente en los sistemas de CFC y HCFC.

Minimum efficiency reporting value (MERV).   Air filter rating system ranging from 1 to 20, with upper levels providing the most filtering.

Valor mínimo de reporte de eficacia (MERV en inglés).   Sistema de clasificación de filtros de aire que va desde el 1 al 20 con los niveles mayores indicando más filtración.

Modulating flow.   Controlling the flow between maximum or no flow. For example, the accelerator on a car provides modulating flow.

Flujo modulante.   Controlado el flujo entre flujo que no es flujo máximo o ningún flujo. Por ejemplo, el acelerador de un carro provee un flujo modulante.

Modulator.   A device that adjusts by small increments or changes.

Modulador.   Dispositivo que se ajusta por medio de incrementos o cambios pequeños.

Moisture indicator.   A device for determining moisture.

Indicador de humedad.   Dispositivo utilizado para determinar la humedad.

Mold.   A fungus found where there is moisture that develops and releases spores. Can be harmful to humans.

Moho.   Un hongo encontrado donde hay humedad que se desarrolla y libera esporas. Puede ser nocivo a los seres humanos.

Molecular motion.   The movement of molecules within a substance.

Movimiento molecular.   Movimiento de moléculas dentro de una sustancia.

Molecule.   The smallest particle that a substance can be broken into and still retain its chemical identity.

Molécula.   La partícula más pequeña en la que una sustancia puede dividirse y aún conservar sus propias características.

Monochlorodifluoromethane.   The refrigerant R-22.

Monoclorodiflorometano.   El refrigerante R-22.

Montreal Protocol.   An agreement signed in 1987 by the United States and other countries to control the release of ozone-depleting gases.

Protocolo de Montreal.   Un acuerdo firmado en 1987 por los Estados Unidos y otros países para controlar la liberación de gases que destruyen el ozono.

Motor service factor.   A factor above an electric motor's normal operating design parameters, indicated on the nameplate, under which it can operate.

Factor de servicio del motor.   Factor superior a los parametros de diseño normales de funcionamiento de un motor eléctrico, indicados en el marbete; este factor indica su nivel de funcionamiento.

Motor starter.   Electromagnetic contactors that contain motor protection and are used for switching electric motors on and off.

Arrancador de motor.   Contactadores electromagnéticos que contienen protección para el motor y se utilizan para arrancar y detener motores eléctricos.

Motor temperature-sensing thermostat.   A thermostat that monitors the motor temperature and shuts it off for the motor's protection.

Termostato que detecta la temperatura del motor.   Un termostato que vigila la temperatura del motor y lo apaga para la protección del motor.

Muffler, compressor.   Sound absorber at the compressor.

Silenciador del compresor.   Absorbedor de sonido ubicado en el compresor.

Mullion.   Stationary frame between two doors.

Parteluz.   Armazón fijo entre dos puertas.

Mullion heater.   Heating element mounted in the mullion of a refrigerator to keep moisture from forming on it.

Calentador del parteluz.   Elemento de calentamiento montado en el parteluz de un refrigerador para evitar la formación de humedad en el mismo.

Multimeter.   An instrument that will measure voltage, resistance, and milliamperes.

Multímetro.   Instrumento que mide la tensión, la resistencia y los miliamperios.

Multiple circuit coil.   An evaporator or condenser coil that has more than one circuit because of the coil length. When the coil is too long, there will be an unacceptable pressure drop and loss of efficiency.

Serpentín de circuito múltiple.   Un serpentín de evaporador o condensador que tiene más de un circuito por causa de la longitud del serpentín. Cuando el serpentín es demasiado largo, habrá una pérdida de presión y eficacia inaceptable.

Multiple evacuation.   A procedure for evacuating a system. A vacuum is pulled, a small amount of refrigerant allowed into the system, and the procedure duplicated. This is often done three times.

Evacuación múltiple.   Procedimiento para evacuar o vaciar un sistema. Se crea un vacío, se permite la entrada de una pequeña cantidad de refrigerante al sistema, y se repite el procedimiento. Con frecuencia esto se lleva a cabo tres veces.

National Electrical Code® (NEC®).   A publication that sets the standards for all electrical installations, including motor overload protection.

Código estadounidense de electridad.   Publicación que establece las normas para todas las instalaciones eléctricas, incluyendo la protección contra la sobrecarga de un motor.

National Fire Protection Association (NFPA).   An association organized to prevent fires through establishing standards, providing research, and providing public education.

Asociación nacional para la protección contra incendios (NFPA en inglés).   Asociación cuyo objetivo es prevenir incendios estableciendo normativas y facilitando la investigación y concienciación del público.

National pipe taper (NPT).   The standard designation for a standard tapered pipe thread.

Cono estadounidense para tubos (NPT en inglés). Designación estándar para una rosca cónica para tubos estándar.

Natural convection.   The natural movement of a gas or fluid caused by differences in temperature.

Convección natural. Movimiento natural de un gas o fluido ocasionado por diferencias en temperatura.

Natural-draft tower. A water cooling tower that does not have a fan to force air over the water. It relies on the natural breeze or airflow.

Torre de corriente de aire natural. Una torre de enfriamiento de agua que no tiene un ventilador para forzar el aire sobre el agua; depende de la brisa o flujo natural del aire.

Natural gas. A fossil fuel formed over millions of years from dead vegetation and animals that were deposited or washed deep into the earth.

Gas natural. Combustible fósil formado a través de millones de años de la vegetación y los animales muertos que fueron depositados o arrastrados a una gran profundidad dentro la tierra.

NC. Normally closed, a designation that states that a valve is closed or a switch is connected when deactivated.

NC. Normalmente cerrado. Designación que indica que una válvula está cerrada o que se conecta un interruptor cuando se desactiva.

Near-azeotropic blend. Two or more refrigerants mixed together that will have a small range of boiling and/or condensing points for each system pressure. Small fractionation and temperature glides will occur but are often negligible.

Mezcla casi-azeotrópica. Mezcla de dos o más refrigerantes, que tiene un bajo rango de punto de ebullición y/o condensación para cada presión del sistema. Puede producirse una cierta fraccionación y variación de temperatura, aunque suelen ser insignificantes.

Needlepoint valve. A device having a needle and a very small orifice for controlling the flow of a fluid.

Válvula de aguja. Dispositivo que tiene una aguja y un orificio bastante pequeño para regular el flujo de un fluido.

Negative electrical charge. An atom or component that has an excess of electrons.

Carga eléctrica negativa. Átomo o componente que tiene un exceso de electrones.

Negative pressure. A situation in which an enclosed area has a lower pressure than does the area around it.

Presión negativa. Situación en la que un área cerrada tiene una presión menor que el área que la rodea.

Neoprene. Synthetic flexible material used for gaskets and seals.

Neopreno. Material sintético flexible utilizado en guarniciones y juntas de estanqueidad.

Net oil pressure. Difference in the suction pressure and the compressor oil pump outlet pressure.

Presión neta del aceite. Diferencia en la presión de aspiración y la presión a la salida de la bomba de aceite del compresor.

Net refrigeration effect (NRE). The quantity of heat in BTU/ lb that the refrigerant absorbs from the refrigerated space to produce useful cooling.

Efecto neto de refrigeración (NRE en inglés). La cantidad de calor expresado en BTU/lb que el refrigerante absorbe del espacio refrigerado para producir refrigeración útil.

Net stack temperature. The temperature difference between the ambient temperature and the flue gas temperature, typically for oil- and gas-burning equipment.

Temperatura neta de chimenea. La diferencia en temperatura entre la temperatura ambiental y la del conducto de gas, normalmente para equipo que quema aceite y gas.

Neutralizer. A substance used to counteract acids.

Neutralizador. Sustancia utilizada para contrarrestar ácidos.

Neutron. Neutrons and protons are located at the center of the nucleus of an atom. Neutrons have no charge.

Neutrón. Los neutrones y protones están situados en le centro del núcleo del átomo. Los neutrones carecen de carga.

Nitrogen. An inert gas often used to "sweep" a refrigeration system to help ensure that all refrigerant and contaminants have been removed.

Nitrógeno. Gas inerte utilizado con frecuencia para purgar un sistema de refrigeración. Esta gas ayuda a asegurar la remoción de todo el refrigerante y los contaminantes del sistema.

NO. Normally open, a designation that states that a valve is open or a switch is disconnected when deactivated.

NA (NO, por su sigla en inglés). Normalmente abierto. Designación que indica que una válvula está abierta o que se desconecta un interruptor cuando se desactiva.

Nominal. A rounded-off stated size. The nominal size is the closest rounded-off size.

Nominal. Tamaño redondeado establecido. El tamaño nominal es el tamaño redondeado más cercano.

Noncondensable gas. A gas that does not change into a liquid under normal operating conditions.

Gas no condensable. Gas que no se convierte en líquido bajo condiciones de funcionamiento normales.

Nonferrous. Metals containing no iron.

No férreos. Metales que no contienen hierro.

North Pole, magnetic. One end of a magnet or the magnetic north pole of the earth.

Polo norte magnético. El extremo de un imán o el polo norte magnético del mundo.

Nozzle. A drilled opening that measures liquid flow, such as an oil burner nozzle.

Tobera. Una apertura taladrada que mide el flujo de líquido, tal como la tobera de un quemador de aceite.

Nut driver. These tools have a socket head used primarily to turn hex head screws on air conditioning, heating, and refrigeration cabinets.

Extractor de tuercas. Estas herramientas tienen una cabeza hueca hexagonal usadas principalmente para darle vuelta a tuercas de cabeza hexagonal en gabinetes de acondicionamiento de aire, de calefacción y de refrigeración.

Occupied zone. That area of a building that is between the floor and the head level of the occupants and not within two feet of any wall.

Zona ocupada. Área de un edificio que está entre el suelo y el nivel de la cabeza de los ocupantes y a no menos de dos pies de una pared.

Off cycle. A period when a system is not operating.

Ciclo de apagado. Período de tiempo cuando un sistema no está en funcionamiento.

Off-cycle defrost. Used for medium-temperature refrigeration where the evaporator coil operates below freezing but the air in the cooler is above freezing. The coil is defrosted by the air inside the cooler while the compressor is off cycle.

Descongelación de período de reposo. Se usa para refrigeración de temperatura media en la cual el serpentín del evaporador funciona por debajo del punto de congelación, pero el aire en el enfriador está por encima del punto de congelación. El serpentín se descongela por el aire dentro del enfriador mientras el compresor está en reposo.

Offset. The absolute (not signed + or −) difference between the set point and the control point of a control process.

Compensación. La diferencia absoluta (sin signo + o −) entre el punto de ajuste y el punto de control de un proceso de control.

Offset. The position of ductwork that must be rerouted around an obstacle.

Desviación. El posición de un conducto que tiene que desviarse alrededor de un obstáculo.

Ohm. A unit of measurement of electrical resistance.

Ohmio. Unidad de medida de la resistencia eléctrica.

Ohmmeter. A meter that measures electrical resistance.

Ohmiómetro. Instrumento que mide la resistencia eléctrica.

Ohm's Law. A law involving electrical relationships discovered by Georg Ohm: $E = I \times R$.

Ley de Ohm. Ley que define las relaciones eléctricas, descubierta por Georg Ohm: $E = I \times R$.

Oil level regulator. A needle valve and float system located on each compressor of a parallel compressor system. It senses the oil level in the compressor's crankcase and adds oil if necessary. It receives its oil from the oil reservoir.

Regulador del nivel de aceite. Sistema de válvula de aguja y flotador que se encuentra en cada compresor de un sistema de compresores en paralelo. Detecta el nivel del aceite en el cárter del compresor y permite la entrada de más aceite procedente de un depósito, en caso necesario.

Oil-pressure safety control (switch). A control used to ensure that a compressor has adequate oil lubricating pressure.

Regulador de seguridad para la presión de aceite (conmutador). Regulador utilizado para asegurar que un compresor tenga la presión de lubrificación de aceita adecuada.

Oil, refrigeration. Oil used in refrigeration systems.

Aceite de refrigeración. Aceite utilizado en sistemas de refrigeración.

Oil reservoir. A storage cylinder for oil usually used on parallel compressor systems. It is located between the oil separator and the oil level regulators. It receives its oil from the oil separator.

Depósito de aceite. Cilindro en el que se almacena el aceite utilizado en los sistemas de compresores en paralelo. Está ubicado entre el separador de aceite y el regulador del nivel. Recibe el aceite del separador.

Oil separator. Apparatus that removes oil from a gaseous refrigerant.

Separador de aceite. Aparato que remueve el aceite de un refrigerante gaseoso.

One-time relief valve. A pressure relief valve that has a diaphragm that blows out due to excess pressure. It is set at a higher pressure than the spring-loaded relief valve in case it fails.

Válvula de alivio de una vez. Una válvula de alivio de presión que tiene un diafragma que revienta debido a la presión excesiva. La misma se fija a una presión más alta que la válvula de alivio de resorte para en caso de que ésta falle.

Open compressor. A compressor with an external drive.

Compresor abierto. Compresor con un motor externo.

Open-loop heat pump. Heat pump system that uses the water in the earth as the heat transfer medium and then expels the water back to the earth in some manner.

Bomba de calor de circuito abierto. Sistema de bomba de calor que utiliza el agua de la tierra como medio de transferencia del calor y luego devuelve el agua a la tierra de cierta manera.

Open winding. The condition that exists when there is a break and no continuity in an electric motor winding.

Devanado abierto. Condición que se presenta cuando hay una interrupción en la continuidad del devanado de un motor.

Operating pressure. The actual pressure under operating conditions.

Presión de funcionamiento. La presión real bajo las condiciones de funcionamiento.

Organic. Materials formed from living organisms.

Orgánico. Materiales formados de organismos vivos.

Orifice. A small opening through which fluid flows.

Orificio. Pequeña abertura a través de la cual fluye un fluido.

Outward clinch tacker. A stapler or tacker that will anchor staples outward and can be used with soft materials.

Grapadora de agarre hacia fuera. Grapadora o tachueladora que ancla las grapas hacia fuera y que puede usarse con materiales suaves.

Overload protection. A system or device that will shut down a system if an overcurrent condition exists.

Protección contra sobrecarga. Sistema o dispositivo que detendrá la marcha de un sistema si existe una condición de sobreintensidad.

Oxidation. The combining of a material with oxygen to form a different substance. This results in the deterioration of the original substance. Rust is oxidation.

Oxidación. La combinación de un material con oxígeno para formar una sustancia diferente, lo que ocasiona el deterioro de la sustancia original. Herrumbre es oxidación.

Ozone. A form of oxygen (O_3). A layer of ozone is in the stratosphere that protects the earth from certain of the sun's ultraviolet wavelengths.

Ozono. Forma de oxígeno (O_3). Una capa de ozono en la estratosfera protege la tierra de ciertos rayos ultravioletas del sol.

Ozone depletion. The breaking up of the ozone molecule by the chlorine atom in the stratosphere. Stratosphere ozone protects us from ultraviolet radiation emitted by the sun.

Reducción del ozono. Descomposición de la molécula de ozono por el átomo de cloro en la estratosfera. El ozono presente en la estratosfera nos protege de las radiaciones ultravioletas del sol.

Ozone depletion potential (ODP). A scale used to measure how much a substance will deplete stratospheric ozone.

Potencial de depleción de ozono (ODP en inglés). Una escala que se usa para medir cuánta depleción del ozono de la estratosfera una sustancia va a causar.

Package unit.   A refrigerating system where all major components are located in one cabinet.

Unidad completa.   Sistema de refrigeración donde todos los componentes principales se encuentran en un solo gabinete.

Packing.   A soft material that can be shaped and compressed to provide a seal. It is commonly applied around valve stems.

Empaquetadura.   Material blando que puede formarse y comprimirse para proveer una junta de estanqueidad. Comúnmente se aplica alrededor de los vástagos de válvulas.

Paraffinic oil.   A refrigeration mineral oil containing some paraffin wax, which is refined from eastern U.S. crude oil.

Aceite de parafina.   Aceite mineral que contiene parafina y que se utiliza en sistemas de refrigeración. El aceite se obtiene mediante el refinado de petróleo extraído en los EE.UU. del este.

Parallel circuit.   An electrical or fluid circuit where the current or fluid takes more than one path at a junction.

Circuito paralelo.   Corriente eléctrica o fluida donde la corriente o el fluido siguen más de una trayectoria en un empalme.

Parallel compressor.   Many compressors piped in parallel and mounted on a steel rack. The compressors are usually cycled by a microprocessor.

Compresor en paralelo.   Varios compresores conectados en paralelo y montados en un bastidor de acero. Normalmente, se sirve de un microprocesador para activarlos y desactivarlos.

Parallel flow.   A flow path in which many paths exist for the fluid to flow.

Flujo paralelo.   Vía de flujo que consta de varias vías que permiten el flujo de un fluido.

Part-winding start.   A large motor that is actually two motors in one housing. It starts on one and then the other is energized. This is to reduce inrush current at start-up. For example, a 100-hp motor may have two 50-hp motors built into the same winding. It will start using one motor followed by the start of the other one. They will both run under the load.

Arranque de bobina parcial.   Un motor grande que es, en efecto, dos motores bajo un mismo cárter. El mismo arranca con uno y luego se activa el segundo. El propósito de esto es reducir la corriente interna al arrancar. Por ejemplo, un motor de 100 caballos de fuerza puede tener dos motores de 50 caballos de fuerza construidos con la misma bobina. El motor arrancará usando un motor, seguido por el arranque del otro motor. Los dos motores correrán bajo carga.

Pascal.   A metric unit of measurement of pressure.

Pascal.   Unidad métrica de medida de presión.

Passive recovery.   Recovering refrigerant with the use of the refrigeration system's compressor or internal vapor pressure.

Recuperación pasiva.   Recuperación de un refrigerante utilizando el compresor del sistema de refrigeración o la presión del vapor interno.

Passive solar design.   The use of nonmoving parts of a building to provide heat or cooling, or to eliminate certain parts of a building that cause inefficient heating or cooling.

Diseño solar pasivo.   La utilización de piezas fijas de un edificio para proveer calefacción o enfriamiento, o para eliminar ciertas piezas de un edificio que causan calefacción o enfriamiento ineficientes.

PE (polyethylene).   Plastic pipe used for water, gas, and irrigation systems.

Polietileno.   Tubo plástico utilizado en sistemas de agua, de gas y de irrigación.

Percent refrigerant quality.   Percent vapor.

Calidad porcentual de refrigerante.   Porcentaje de vapor.

Permanent magnet.   An object that has its own permanent magnetic field.

Imán permanente.   Objeto que tiene su propio campo magnético permanente.

Permanent-split capacitor motor (PSC).   A split-phase motor with a run capacitor only. It has a very low starting torque.

Motor permanente de capacitador separado (PSC en inglés).   Motor de fase separada que sólo tiene un capacitador de funcionamiento. Su par de arranque es sumamente bajo.

Phase.   One distinct part of a cycle.

Fase.   Una parte específica de un ciclo.

Phase-change loop.   The loop of piping, usually in a geothermal heat pump system, where there is a change of phase of the heat transfer fluid from liquid to vapor or vapor to liquid.

Circuito de cambio de fase.   Circuito de tubería, generalmente en un sistema de bombeo de calor geotermal, en el cual hay un cambio de fase del fluido de transferencia de calor de líquido a vapor o de vapor a líquido.

Phase failure protection.   Used on three-phase equipment to interrupt the power source when one phase becomes deenergized. The motors cannot be allowed to run on the two remaining phases or damage will occur.

Protección de fallo de fase.   Se usa en equipo trifásico para interrumpir la fuente de potencia cuando se energiza una fase. No se puede permitir que los motores corran con las dos fases restantes o podría ocurrir una avería.

Phase reversal.   Phase reversal can occur if someone switches any two wires on a three-phase system. Any system with a three-phase motor will reverse, and this cannot be allowed on some equipment.

Inversión de fase.   La inversión de fase puede ocurrir si por cualquier razón alguien intercambia cualquier par de cables en un sistema trifásico. Cualquier sistema con un motor trifásico irá en dirección contraria y esto no puede permitirse en algunos sistemas.

Pictorial wiring diagram.   This type of diagram shows the location of each component as it appears to the person installing or servicing the equipment.

Diagrama representativo del cableado.   Este tipo de diagrama indica la ubicación de cada componente tal y como lo verá el personal técnico.

Piercing valve.   A device that is used to pierce a pipe or tube to obtain a pressure reading without interrupting the flow of fluid. Also called a line tap valve.

Válvula punzante.   Aparato que se usa para punzar un tubo para obtener una lectura de la presión sin interrumpir el flujo del fluido. También se llama una válvula de toma.

Pilot duty relay.   A small relay that is used in control circuits for switching purposes. It is small and cannot take a lot of current flow, such as to start a motor.

Relé de función piloto.   Un pequeño relé que se usa en los circuitos de control con propósitos de interrupción. Es pequeño y no puede tolerar un flujo alto de corriente, como para arrancar un motor.

Pilot light.   The flame that ignites the main burner on a gas furnace.

Llama piloto.   Llama que enciende el quemador principal en un horno de gas.

Piston.   The part that moves up and down in a cylinder.

Pistón.   La pieza que asciende y desciende dentro de un cilindro.

Piston displacement.   The volume within the cylinder that is displaced with the movement of the piston from top to bottom.

Desplazamiento del pistón.   Volumen dentro del cilindro que se desplaza de arriba a abajo con el movimiento del pistón.

Pitot tube.   A pressure anemometer used for measuring the relative speed of a fluid.

Tubo pitot.   Anemómetro de presión utilizado para medir la velocidad relativa de un fluido.

Plenum.   A sealed chamber at the inlet or outlet of an air handler. The duct attaches to the plenum.

Plenum.   Cámara sellada a la entrada o a la salida de un tratante de aire. El conducto se fija al plenum.

Pneumatic controls.   Controls operated by low-pressure air, typically 20 psig.

Controles neumáticos.   Controles que se operan por aire de baja presión, generalmente 20 libras por pulgada cuadrada de presión de manómetro (psig).

Polyalkylene glycol.   A popular synthetic glycol-based lubricant used with HFC refrigerants, mainly in automotive systems. This was the first generation of oil used with HFC refrigerants.

Glicol polialkilénico.   Lubricante sintético de uso común basado en glicol, usado con refrigerantes HFC, principalmente en sistemas de automóviles. Ésta fue la primera generación de aceites usados con refrigerantes HFC.

Polybutylene.   A material used for the buried piping in geothermal heat pumps.

Polibutileno.   Material utilizado para la fabricación de tubos enterrados, en sistemas de bombas de calor geotérmicas.

Polyethylene.   A material used for the buried piping in geothermal heat pumps.

Polietileno.   Material utilizado para la fabricación de tubos enterrados, en sistemas de bombas de calor geotérmicas.

Polyol ester.   A very popular ester-based lubricant often used in HFC refrigerant systems.

Poliol éster.   Lubricante muy popular basado en éster usado frecuentemente en sistemas con refrigerantes HFC.

Polyphase.   Three or more phases.

Polifase.   Tres o más fases.

Polyphosphate.   A scale inhibitor with many phosphate molecules.

Polifosfato.   Un inhibidor de escama que contiene muchas moléculas de fosfato.

Porcelain.   A ceramic material.

Porcelana.   Material cerámico.

Positive displacement.   A term used with a pumping device such as a compressor that is designed to move all matter from a volume such as a cylinder or it will stall, possibly causing failure of a part.

Desplazamiento positivo.   Término utilizado con un dispositivo de bombeo, como por ejemplo un compresor, diseñado para mover toda la materia de un volumen, como un cilindro, o se bloqueará, posiblemente causándole fallas a una pieza.

Positive electrical charge.   An atom or component that has a shortage of electrons.

Carga eléctrica positiva.   Átomo o componente que tiene una insuficiencia de electrones.

Positive pressure.   A situation in which an enclosed area has a higher pressure than does the area around it.

Presión positiva.   Situación en la que un área cerrada tiene una presión mayor que el área que la rodea.

Positive temperature coefficient start device.   A thermistor used to provide start assistance to a permanent-split capacitor motor.

Dispositivo de arranque de coeficiente de temperatura positiva.   Termistor utilizado para ayudar a arrancar un motor permanente de capacitador separado.

Potential relay.   A switching device used with hermetic motors that breaks the circuit to the start capacitor and/or start windings after the motor has reached approximately 75% of its running speed.

Relé de potencial.   Dispositivo de conmutación utilizado con motores herméticos que interrumpe el circuito del capacitador y/o de los devandos de arranque antes de que el motor haya alcanzado aproximadamente un 75% de su velocidad de marcha.

Potential voltage.   The voltage measured across the start winding in a single-phase motor while it is turning at full speed. This voltage is much greater than the applied voltage to the run winding. For example, the run winding may have 230 V applied to it, and a measured voltage across the start winding may be 300 V. This is created by the motor stator turning in the magnetic field of the run winding. Voltage potential is the difference in voltage between any two parts of a circuit.

Potencial de voltaje.   El voltaje medido a través de la bobina de arranque en un motor de fase sencilla mientras está girando a velocidad completa. Este voltaje es mucho mayor que el voltaje aplicado a la bobina de marcha. Por ejemplo, a la bobina de marcha se le puede aplicar 230 V, y el voltaje medido a través de la bobina de arranque puede ser 300 V. Este voltaje lo crea el estator del motor al girar en el campo magnético de la bobina de marcha. Potencial de voltaje es la diferencia en voltaje entre cualesquiera dos partes del circuito.

Potentiometer.   An instrument that controls electrical current.

Potenciómetro.   Instrumento que regula corriente eléctrica.

Powder actuated tool (PAT).   A tool with a powder load that forces a pin, threaded stud, or other fastener into masonry.

Herramienta activada por pólvora (PAT en inglés).   Una herramienta con una carga de pólvora que inserta una clavija, una clavija con rosca u otro sujetador en la albañilería.

Power.   The rate at which work is done.

Potencia.   Velocidad a la que se realiza un trabajo.

Power-consuming devices.   A power-consuming device is considered the electrical load. For example, in a lightbulb circuit, the switch is a power-passing device that passes power to the lightbulb that consumes the power and produces light.

Aparatos consumidores de potencia.   Un aparato consumidor de potencia se considera la carga eléctrica. Por ejemplo, en un circuito de bombilla de luz, el interruptor es un dispositivo que pasa corriente que pasa la corriente a la bombilla, la cual consume electricidad y produce luz.

PPE.   Personal protective equipment; refers to gloves, helmets, goggles, or other garments designed to protect the wearer's body from injury for occupational health and safety.

EPP (PPE, por su sigla en inglés).   Equipo de protección personal; incluye guantes, cascos, gafas protectoras u otras prendas diseñadas para proteger de lesiones a quien las usa por razones de salud y de seguridad ocupacional.

Pressure.   Force per unit of area.

Presión.   Fuerza por unidad de área.

Pressure access ports.   Places in a system where pressure can be taken or registered.

Puerto de acceso a presión.   Lugares en un sistema donde se puede tomar o registrar la presión.

Pressure differential valve.   A valve that senses a pressure differential and opens when a specific pressure differential is reached.

Válvula de presión diferencial.   Válvula que detecta diferencia de presiones y se abre cuando se alcanza una diferencia específica.

Pressure drop.   The difference in pressure between two points.

Caída de presión.   Diferencia en presión entre dos puntos.

Pressure/enthalpy diagram.   A chart indicating the pressure and heat content of a refrigerant and the extent to which the refrigerant is a liquid and vapor.

Diagrama de presión y entalpía.   Esquema que indica la presión y el contenido de calor de un refrigerante y el punto en que el refrigerante es líquido y vapor.

Pressure limiter.   A device that opens when a certain pressure is reached.

Dispositivo limitador de presión.   Dispositivo que se abre cuando se alcanza una presión específica.

Pressure-limiting TXV.   A valve designed to allow the evaporator to build only to a predetermined pressure when the valve will shut off the flow of refrigerant.

Válvula electrónica de expansión limitadora de presión.   Válvula diseñada para permitir que la temperatura del evaporador alcance una presión predeterminada cuando la válvula detenga el flujo de refrigerante.

Pressure regulator.   A valve capable of maintaining a constant outlet pressure when a variable inlet pressure occurs.

Used for regulating fluid flow such as natural gas, refrigerant, and water.

Regulador de presión.   Válvula capaz de mantener una presión constante a la salida cuando ocurre una presión variable a la entrada. Utilizado para regular el flujo de fluidos, como por ejemplo el gas natural, el refrigerante y el agua.

Pressure switch.   A switch operated by a change in pressure.

Conmutador accionado por presión.   Conmutador accionado por un cambio en presión.

Pressure tank.   A pressurized tank for water storage located in the water piping of an open-loop geothermal heat pump system. It prevents short cycling of the well pump.

Depósito de presión.   Depósito presurizado que sirve para almacenar el agua contenida en la tubería de un sistema de bomba de calor geotérmica de circuito abierto. Impide el funcionamiento de la bomba del pozo en ciclos cortos.

Pressure-temperature relationship.   This refers to the pressure-temperature relationship of a liquid and vapor in a closed container. If the temperature increases, the pressure will also increase. If the temperature is lowered, the pressure will decrease.

Relación entre presión y temperatura.   Se refiere a la relación entre la presión y la temperatura de un líquido y un vapor en un recipiente cerrado. Si la temperatura aumenta, la presión también aumentará. Si la temperatura baja, habrá una caída de presión.

Pressure transducer.   A pressure-sensitive device located in the piping of a refrigeration system that will transform a pressure signal to an electronic signal. The electronic signal will then feed a microprocessor.

Transductor de presión.   Dispositivo sensible a la presión, situado en la tubería del sistema de refrigeración y que convierte una señal de presión en señal eléctrica. Seguidamente, la señal eléctrica se envía a un microprocesador.

Primary air.   Air that is introduced to a furnace's burner before the combustion process has taken place.

Aire primario.   Aire que se introduce en el quemador de un calefactor antes de que ocurra el proceso de combustión.

Primary control.   Controlling device for an oil burner to ensure ignition within a specific time span, usually 90 seconds.

Regulador principal.   Dispositivo de regulación para un quemador de aceite pesado. El regulador principal asegura el encendido dentro de un período de tiempo específico, normalmente 90 segundos.

Programmable thermostat.   An electronic thermostat that can be set up to provide desired conditions at desired times.

Termostato programable.   Un termostato electrónico que se puede programar para proveer las condiciones deseadas en tiempos deseados.

Propane.   An LP (liquefied petroleum) gas used for heat.

Propano.   Gas de petróleo licuado que se utiliza para producir calor.

Propeller fan.   This fan is used in exhaust fan and condenser fan applications. It will handle large volumes of air at low-pressure differentials.

Ventilador helicoidal.   Se utiliza en ventiladores de evacuación y de condensador. Es capaz de mover grandes volúmenes de aire a bajas diferenciales de presión.

Proportional controller. A modulating control mode where the controller changes or modifies its output signal in proportion to the size of the change in the error.

Controlador proporcional. Un modo de control modulante donde el controlador cambia o modifica su señal de salida en proporción al tamaño del cambio en el error.

Propylene glycol. An antifreeze fluid cooled by a primary (phase-change) refrigerant, which then is circulated by pumps throughout the refrigeration system to absorb heat.

Glicol propílico. Líquido anticongelante enfriado por un refrigerante primario (cambio de fase). Seguidamente, una bomba lo hace circular por el sistema de refrigeración para que absorba el calor.

Proton. That part of an atom having a positive charge.

Protón. Parte de un átomo que tiene carga positiva.

Protozoa. A microscopic organism with a complex life cycle.

Protozoario. Un organismo microscópico con un ciclo de vida complejo.

PSC motor. See Permanent-split capacitor motor.

Motor PSC. Véase Motor permanente de capacitador separado.

psi. Abbreviation for pounds per square inch.

psi. Abreviatura de libras por pulgada cuadrada.

psia. Abbreviation for pounds per square inch absolute.

psia. Abreviatura de libras por pulgada cuadrada absoluta.

psig. Abbreviation for pounds per square inch gauge.

psig. Abreviatura de indicador de libras por pulgada cuadrada.

Psychrometer. An instrument for determining relative humidity.

Sicrómetro. Instrumento para medir la humedad relativa.

Psychrometric chart. A chart that shows the relationship of temperature, pressure, and humidity in the air.

Esquema sicrométrico. Esquema que indica la relación entre la temperatura, la presión y la humedad en el aire.

Psychrometrics. The study of air and its properties, particularly the moisture content.

Sicrometría. El estudio del aire y sus propiedades, particularmente el contenido de humedad.

P-type material. Semiconductor material with a positive charge.

Material tipo P. Material con carga positiva utilizado en semiconductores.

Pump. A device that forces fluids through a system.

Bomba. Dispositivo que introduce fluidos por fuerza a través de un sistema.

Pumpdown. To use a compressor to pump the refrigerant charge into the condenser and/or receiver.

Extraer con bomba. Utilizar un compresor para bombear la carga del refrigerante dentro del condensador y/o receptor.

Pure compound. A substance formed in definite proportions by weight with only one molecule present.

Componente puro. Una sustancia formada en proporciones por peso definidas con sólo una molécula presente.

Purge. To remove or release fluid from a system.

Purga. Remover o liberar el fluido de un sistema.

PVC (polyvinyl chloride). Plastic pipe used in pressure applications for water and gas as well as for sewage and certain industrial applications.

Cloruro de polivinilo (PVC en inglés). Tubo plástico utilizado tanto en aplicaciones de presión para agua y gas, como en ciertas aplicaciones industriales y de aguas negras.

Quench. To submerge a hot object in a fluid for cooling.

Entriamiento por inmersión. Sumersión de un objeto caliente en un fluido para enfriarlo.

Quick-connect coupling. A device designed for easy connecting or disconnecting of fluid lines.

Acoplamiento de conexión rápida. Dispositivo diseñado para facilitar la conexión o desconexión de conductos de fluido.

R-12. Dichlorodifluoromethane, once a popular refrigerant for refrigeration systems. It can no longer be manufactured in the United States and many other countries.

R-12. Diclorodiflorometano, que fue una vez un refrigerante muy utilizado en sistemas de refrigeración. Ya no se puede fabricar ni en los Estados Unidos ni en muchos otros países.

R-22. Monochlorodifluoromethane, a popular HCFC refrigerant for air conditioning systems.

R-22. Monoclorodiflorometano, refrigerante HCFC muy utilizado en sistemas de acondicionamiento de aire.

R-123. Dichlorotrifluoroethane, an HCFC refrigerant developed for low-pressure application.

R-123. Diclorotrifloroetano, refrigerante HCFC elaborado para aplicaciones de baja presión.

R-134a. Tetrafluoroethane, an HFC refrigerant developed for refrigeration systems and as a replacement for R-12.

R-134a. Tetrafloroetano, refrigerante HFC elaborado para sistemas de refrigeración y como sustituto del R-12.

R-410A. A mixture of difluoromethane and pentafluoroethane, a refrigerant developed to replace R-22 for air conditioning systems.

R-410A. Una mezcla de difluorometano y pentafluoroetano, refrigerante desarrollado para remplazar el R-22 para los sistemas de aire acondicionado.

Rack system. Many compressors piped in parallel and mounted on a steel rack. The compressors are usually cycled by a microprocessor.

Sistema de bastidor. Varios compresores conectados en paralelo y montados en un bastidor de acero. Normalmente, un microprocesador se encarga de activar y desactivar los compresores.

Radiant heat. Heat that passes through air, heating solid objects that in turn heat the surrounding area.

Calor radiante. Calor que pasa a través del aire y calienta objetos sólidos que a su vez calientan el ambiente.

Radiation. Heat transfer. See Radiant heat.

Radiación. Transferencia de calor. Véase Calor radiante.

Radon. A colorless, odorless, and radioactive gas. Radon can enter buildings through cracks in concrete floors and walls, floor drains, and sumps.

Radón. Gas incoloro, inodoro y radioactivo. El radón puede penetrar en los edificios a través de las grietas en el hormigón y suelos, desagües y sumideros.

Random or off-cycle defrost. Defrost provided by the space temperature during the normal off cycle.

Descongelación variable o de ciclo apagado. Descongelación llevada a cabo por la temperatura del espacio durante el ciclo normal de apagado.

Range. The pressure or temperature settings of a control defining certain boundaries of temperature or pressure.

Rango. Los valores de presión o temperatura para un control que definen los límites de la temperatura o presión.

Rankine. The absolute Fahrenheit scale with 0 at the point where all molecular motion stops.

Rankine. Escala absoluta de Fahrenheit con el 0 al punto donde se detiene todo movimiento molecular.

Rapid oxidation. A reaction between the fuel, oxygen, and heat that is known as rapid oxidation or the process of burning.

Oxidación rápida. Reacción producida entre el combustible y el oxígeno. El calor producido se conoce como oxidación rápida o proceso de quemado.

Reactance. A type of resistance in an alternating current circuit.

Reactancia. Tipo de resistencia en un circuito de corriente alterna.

Reamer. Tool to remove burrs from inside a pipe after it has been cut.

Escariador. Herramienta utilizada para remover las rebabas de un tubo después de haber sido cortado.

Receiver-drier. A component in a refrigeration system for storing and drying refrigerant.

Receptor-secador. Componente en un sistema de refrigeración que almacena y seca el refrigerante.

Reciprocating. Back-and-forth motion.

Movimiento alternativa. Movimiento de atrás para adelante.

Reciprocating compressor. A compressor that uses a piston in a cylinder and a back-and-forth motion to compress vapor.

Compresor alternativo. Compresor que utiliza un pistón en un cilindro y un movimiento de atrás para adelante a fin de comprir el vapor.

Recirculated water system. A system where water is used over and over, such as a chilled water or cooling tower system.

Sistema de agua recirculada. Sistema donde el agua se usa una y otra vez, tal como en un sistema de agua enfriada o de torre enfriamiento.

Recovery cylinder. A cylinder into which refrigerant is transferred; should be approved by the Department of Transportation as a recovery cylinder. The color code for these cylinders is a yellow top with a gray body.

Cilindro de recuperación. Cilindro al que se transfiere el refrigerante y que debe ser homologado por el departamento de transporte como cilindro de recuperación. Este tipo de cilindro se identifica pintando su parte superior en amarillo y el resto del cuerpo en gris.

Rectifier. A device for changing alternating current to direct current.

Rectificador. Dispositivo utilizado para convertir corriente alterna en corriente continua.

Reed valve. A thin steel plate used as a valve in a compressor.

Válvula de lámina. Placa delgada de acero utilizada como una válvula en un compresor.

Refrigerant. The fluid in a refrigeration system that changes from a liquid to a vapor and back to a liquid at practical pressures.

Refrigerante. Fluido en un sistema de refrigeración que se convierte de líquido en vapor y nuevamente en líquido a presiones prácticas.

Refrigerant blend. Two or more refrigerants blended or mixed together to make another refrigerant. Blends can combine as either azeotropic or zeotropic blends.

Mezcla de refrigerante. Dos o más refrigerantes mezclados para crear otro. Las mezclas pueden ser de tipo azeotrópico o zeotrópico.

Refrigerant loop. The heat pump's refrigeration system, which exchanges energy with the fluid in the ground loop and the air side of the system.

Circuito de refrigeración. Sistema de refrigeración de la bomba de calor que sirve para intercambiar energía entre el fluido en el circuito de tierra y la parte del sistema que contiene el aire.

Refrigerant receiver. A storage tank in a refrigeration system where the excess refrigerant is stored. Since many systems use different amounts of refrigerant during the season, the excess is stored in the receiver tank when not needed. The refrigerant can also be pumped to the receiver when repairs on the low-pressure side of the system are made.

Recibidor de refrigerante. Tanque de almacenamiento en un sistema de refrigeración donde se almacena el exceso de refrigerante. Como muchos sistemas usan diferentes cantidades de refrigerantes durante la temporada, el exceso se almacena en el tanque cuando no se necesita. El refrigerante también puede bombearse al recibidor cuando se hacen reparaciones al extremo de baja presión del sistema.

Refrigerant reclaim. To process refrigerant to new product specifications by means which may include distillation. It will require chemical analysis of the refrigerant to determine that appropriate product specifications are met. This term usually implies the use of processes or procedures available only at a reprocessing or manufacturing facility.

Recuperación del refrigerante. Procesar refrigerante según nuevas especificaciones para productos a través de métodos que pueden incluir la destilación. Se requiere un análisis químico del refrigerante para asegurar el cumplimiento de las especificaciones para productos a través de métodos que pueden incluir la destilación. Se requiere un análisis químico del refrigerante para asegurar el cumplimiento de las especificaciones para productos adecuadas. Por lo general este término supone la utilización de procesos o de procedimientos disponibles solamente en fábricas de reprocesamiento o manufactura.

Refrigerant recovery. To remove refrigerant in any condition from a system and store it in an external container without necessarily testing or processing it in any way.

Recobrar refrigerante líquido. Remover refrigerante en cualquier estado de un sistema y almacenarlo en un recipiente externo sin ponerlo a prueba o elaborarlo de ninguna manera.

Refrigerant recycling. To clean the refrigerant by oil separation and single or multiple passes through devices, such as replaceable core filter driers, which reduce moisture, acidity, and particulate matter. This term usually applies to procedures implemented at the job site or at a local service shop.

Recirculación de refrigerante. Limpieza del refrigerante por medio de la separación del aceite y pasadas sencillas o múltiples a traves de dispositivos, como por ejemplo secadores filtros con núcleos reemplazables que disminuyen la humedad, la acidez y las partículas. Por lo general este término se aplica a los procedimientos utilizados en el lugar del trabajo o en un taller de servicio local.

Refrigerated air driers. A device that removes the excess moisture from compressed air.

Secadores de aire refrigerados. Aparato que remueve el exceso de humedad del aire comprimido.

Refrigeration. The process of removing heat from a place where it is not wanted and transferring that heat to a place where it makes little or no difference.

Refrigeración. Proceso de remover el calor de un lugar donde no es deseado y transferirlo a un lugar donde no afecte la temperatura.

Register. A terminal device on an air distribution system that directs air but also has a damper to adjust airflow.

Registro. Dispositivo de terminal en un sistema de distribución de aire que dirige el aire y además tiene un desviador para ajustar su flujo.

Regulator. A valve used to control the pressure in liquid systems to some value. Many households have a water pressure regulator to reduce the pressure from the main to a more usable pressure in the house. Gas systems all have pressure regulators to stabilize the pressure to the burners.

Regulador. Una válvula que se usa para controlar y fijar la presión en los sistemas líquidos a algún valor. Muchas casas tienen un regulador de presión de agua para reducir la presión de la tubería principal a una presión más útil en la casa. Todos los sistemas de gas tienen reguladores de presión para estabilizar la presión en el quemador.

Relative humidity. The amount of moisture contained in the air as compared to the amount the air could hold at that temperature.

Humedad relativa. Cantidad de humedad presente en el aire, comparada con la cantidad de humedad que el aire pueda contener a dicha temperatura.

Relay. A small electromagnetic device to control a switch, motor, or valve.

Relé. Pequeño dispositivo electromagnético utilizado para regular un conmutador, un motor o una válvula.

Relief valve. A valve designed to open and release vapors at a certain pressure.

Válvula para alivio. Válvula diseñada para abrir y liberar vapores a una presión específica.

Remote system. Often called a split system where the condenser is located away from the evaporator and/or other parts of the system.

Sistema remoto. Llamado muchas veces sistema separado donde el condensador se coloca lejos del evaporador y/o otras piezas del sistema.

Resilient-mount motor. Electric motor that uses various materials to isolate the motor noise from metal framework. This type of motor requires a ground strap.

Motor con montaje antivibratorio. Motor eléctrico que utiliza varios materiales para aislar el ruido del bastidor metálico. Este tipo de motor requiere conexión a tierra.

Resistance. The opposition to the flow of an electrical current or a fluid.

Resistencia. Oposición al flujo de una corriente eléctrica o de un fluido.

Resistive load. An electrical load that does not have significant inrush current but rises to a steady-state value without first rising to a higher value.

Carga resistiva. Carga eléctrica que no posee una corriente de ingreso significativa, pero que aumenta a un valor estable sin elevarse antes a un valor mayor.

Resistor. An electrical or electronic component with a specific opposition to electron flow. It is used to create voltage drop or heat.

Resistor. Componente eléctrico o electrónico con una oposición específica al flujo de electrones; se utiliza para producir una caída de tensión o calor.

Restrictor. A device used to create a planned resistance to fluid flow.

Limitador. Dispositivo utilizado para producir una resistencia proyectada al flujo de fluido.

Retrofit guidelines. Guidelines intended to make the transition from a CFC/mineral oil system to a system containing an alternative refrigerant and its appropriate oil.

Directrices de reconversión. Directrices destinadas a facilitar la transición entre un sistema de aceite CFC/mineral y otro con refrigerante alternativo y su aceite correspondiente.

Return well. A well for return water after it has experienced the heat exchanger of the geothermal heat pump.

Pozo de retorno. Pozo donde se acumula el agua de retorno una vez ha pasado por el intercambiador térmico de la bomba de calor geotérmica.

Reverse cycle. The ability to direct the hot gas flow into the indoor or the outdoor coil in a heat pump to control the system for heating or cooling purposes.

Ciclo invertido. Capacidad de dirigir el flujo de gas caliente dentro de la bobina interior o exterior en una bomba de calor a fin de regular el sistema para propósitos de calentamiento o enfriamiento.

Rigid-mount motor. Electric motor that is bolted metal-to-metal to a frame. This type of motor will transmit noise.

Motor con montaje rígido. Motor eléctrico que se encuentra sujeto directamente a un bastidor mediante pernos. Este tipo de motor genera ruido.

Rod and tube. The rod and tube are each made of a different metal. The tube has a high expansion rate and the rod a low expansion rate.

Varilla y tubo. La varilla y el tubo se fabrican de un metal diferente. El tubo tiene una tasa de expansión alta y la varilla una tasa de expansión baja.

Room air change.   A measure of the rate at which air in a room is being exchanged with fresh or conditioned air.
Cambio del aire ambiente.   Medida de la velocidad a la que se intercambia el aire ambiente por aire fresco o acondicionado.

Room heater.   A gas stove or appliance considered by ANSI to be a heating appliance. This heater will have an efficiency rating.
Calentador de sala.   Estufa de gas u otro dispositivo que, según ANSI, es un dispositivo de calefacción. Este tipo de calentador cuenta con una clasificación de eficacia.

Root mean square (RMS) voltage.   The alternating current voltage effective value. This is the value measured by most voltmeters. The RMS voltage is 0.707 the peak voltage.
Voltaje de la raíz del valor medio cuadrado (RMS en inglés).   El valor efectivo del voltaje de corriente alterna. Este valor es el que miden la mayoría de los voltímetros. El voltaje RMS es 0.707 por el voltaje pico.

Rotary compressor.   A compressor that uses rotary motion to pump fluids. It is a positive displacement pump.
Compresor giratorio.   Compresor que utiliza un movimiento giratorio para bombear fluidos. Es una bomba de desplazamiento positivo.

Rotor.   The rotating or moving component of a motor, including the shaft.
Rotor.   Componente giratorio o en movimiento de un motor, incluyendo el arbol.

Run-load amperage (RLA).   The amperage at which a motor can safely operate while under full load, unless it has a service (reserve) factor allowing more amperage.
Amperaje de operación con carga (RLA en inglés).   Amperaje bajo el cual el motor puede operar seguramente bajo carga completa, a menos que tenga un factor de servicio (reserva) que permite más amperaje.

Running time.   The time a unit operates. Also called the on time.
Período de funcionamiento.   El período de tiempo en que funciona una unidad. Conocido también como período de conexión.

Run winding.   The electrical winding in a motor that draws current during the entire running cycle.
Devanado de funcionamiento.   Devanado eléctrico en un motor que consume corriente durante todo el ciclo de funcionamiento.

Rupture disk.   Pressure safety device for a centrifugal low-pressure chiller.
Disco de ruptura.   Dispositivo de seguridad para un enfriador centrífugo de baja presión.

Saddle valve.   A valve that straddles a fluid line and is fastened by solder or screws. It normally contains a device to puncture the line for pressure readings.
Válvula de silleta.   Válvula que está sentada a horcajadas en un conducto de fluido y se fija por medio de la soldadura o tornillos. Por lo general contiene un dispositivo para agujerear el conducto a fin de que se puedan tomar lecturas de presión.

Safety control.   An electrical, electronic, mechanical, or electromechanical control to protect the equipment or public from harm.
Regulador de seguridad.   Regulador eléctrico, electrónico, mecánico o electromecánico para proteger al equipo de posibles averías o al público de sufrir alguna lesión.

Safety plug.   A fusible plug that blows out when high temperature occurs.
Tapón de seguridad.   Tapón fusible que se sale cuando se presentan temperaturas altas.

Sail switch.   A safety switch with a lightweight, sensitive sail that operates by sensing an airflow.
Conmutador con vela.   Conmutador de seguridad con una vela liviana sensible que funciona al advertir el flujo de aire.

Salt solution.   Antifreeze solution used in a closed water loop of geothermal heat pumps.
Solución de sal.   Líquido anticongelante utilizado en el circuito cerrado de agua de una bomba de calor geotérmica.

Satellite compressor.   The compressor on a parallel compressor system that is dedicated to the coldest evaporators.
Compresor auxiliar.   Compresor montado en un sistema paralelo y que está dedicado a los evaporadores más fríos.

Saturated vapor.   The refrigerant when all of the liquid has just changed to a vapor.
Vapor saturada.   El refrigerante cuando todo el líquido acaba de convertirse en vapor.

Saturation.   A term used to describe a substance when it contains all of another substance it can hold.
Saturación.   Término utilizado para describir una sustancia cuando contiene lo más que puede de otra sustancia.

Scale inhibitor.   Chemical or surface treatments that inhibit the depositing of solids that form flakes or scale.
Inhibidor antiincrustante.   Tratamientos químicos o de superficie que inhiben el depósito de sólidos que forman cascarillas o costras.

Scavenger pump.   A pump used to remove the fluid from a sump.
Bomba de barrido.   Bomba utilizada para remover el fluido de un sumidero.

Scheduled maintenance.   The action of performing regularly scheduled maintenance on a unit, including inspection, cleaning, and servicing.
Mantenimiento programado.   La acción de dar mantenimiento regularmente programado a una unidad incluyendo inspección, limpieza y servicio.

Schematic wiring diagram.   Sometimes called a line or ladder diagram, this type of diagram shows the electrical current path to the various components.
Diagrama de cableado esquematizado.   También conocido como de línea o de escalera, este tipo de diagrama muestra la ruta actual que sigue la electricidad para llegar a los diferentes componentes.

Schraeder valve.   A valve similar to the valve on an auto tire that allows refrigerant to be charged or discharged from the system.

Válvula Schraeder.   Válvula similar a la válvula del neumático de un automóvil que permite la entrada o la salida de refrigerante del sistema.

Scotch yoke.   A mechanism used to create reciprocating motion from the electric motor drive in very small compressors.

Yugo escocés.   Mecanismo utilizado para producir movimiento alternativo del accionador del motor eléctrico en compresores bastante pequeños.

Screw compressor.   A form of positive displacement compressor that squeezes fluid from a low-pressure area to a high-pressure area, using screw-type mechanisms.

Compresor de tornillo.   Forma de compresor de desplazamiento positivo que introduce por fuerza el fluido de un área de baja presión a un área de alta presión, a través de mecanismos de tipo de tornillo.

Scroll compressor.   A compressor that uses two scroll-type components, one stationary and one orbiting, to compress vapor.

Compresor espiral.   Compresor que utiliza dos componentes de tipo espiral para comprimir el vapor.

Sealed unit.   The term used to describe a refrigeration system, including the compressor, that is completely welded closed. The pressures can be accessed by saddle valves.

Unidad sellada.   Término utilizado para describir un sistema de refrigeración, incluyendo el compresor, que es soldado completamente cerrado. Las presiones son accesibles por medio de válvulas de dilleta.

Seasonal energy efficiency ratio (SEER).   An equipment efficiency rating that takes into account the start-up and shutdown for each cycle.

Relación del rendimiento de energía temporal (SEER en inglés).   Clasificación del rendimiento de un equipo que toma en cuenta la puesta en marcha y la parada de cada ciclo.

Seat.   The stationary part of a valve that the moving part of the valve presses against for shutoff.

Asiento.   Pieza fija de una válvula contra la que la pieza en movimiento de la válvula presiona para cerrarla.

Secondary air.   Air that is introduced to a furnace after combustion takes place and that supports combustion.

Aire secundario.   Aire que se introduce en un calefactor después que occure la combustión y que ayuda la combustión.

Secondary fluid.   An antifreeze fluid cooled by a primary (phase-change) refrigerant, which is then circulated by pumps throughout the refrigeration system to absorb heat.

Fluido secundario.   Líquido anticongelante refrigerado por otro primario (cambio de fase), que luego es propulsado por las bombas a través del sistema de refrigeración para absorber el calor.

Semiconductor.   A component in an electronic system that is considered neither an insulator nor a conductor but a partial conductor. It conducts current in a controlled and predictable manner.

Semiconductor.   Componente en un sistema eléctrico que no se considera ni aislante ni conductor, sino conductor parcial. Conduce la corriente de una manera controlada y predecible.

Semihermetic compressor.   A motor compressor that can be opened or disassembled by removing bolts and flanges. Also known as a serviceable hermetic.

Compresor semihermético.   Compresor de un motor que puede abrirse o desmontarse al removerle los pernos y bridas. Conocido también como compresor hermético utilizable.

Sensible heat.   Heat that causes a change in temperature.

Calor sensible.   Calor que produce un cambio en la temperatura.

Sensor.   A component for detection that changes shape, form, or resistance when a condition changes.

Sensor.   Componente para la detección que cambia de forma o de resistencia cuando cambia una condición.

Sequencer.   A control that causes a staging of events, such as a sequencer between stages of electrical heat.

Regulador de secuencia.   Regulador que produce una sucesión de acontecimientos, como por ejemplo etapas sucesivas de calor eléctrico.

Series circuit.   An electrical or piping circuit where all of the current or fluid flows through the entire circuit.

Circuito en serie.   Circuito eléctrico o de tubería donde toda la corriente o todo el fluido fluye a través de todo el circuito.

Series flow.   A flow path in which only one path exists for fluid to flow.

Flujo en serie.   Ruta de flujo única para el líquido.

Serviceable hermetic.   See Semihermetic compressor.

Compresor hermético utilizable.   Véase Compresor semihermético.

Service valve.   A manually operated valve in a refrigeration system used for various service procedures.

Válvula servicio.   Válvula de un sistema de refrigeración accionada manualmente que se utiliza en varios procedimientos de servicio.

Servo pressure regulator.   A sensitive pressure regulator located inside a combination gas valve that senses the outlet or working pressure of the gas valve.

Regulador de presión por servomotor.   Un regulador de presión sensible que está ubicado dentro de una válvula de gas de combinación que detecta la presión de salida o de trabajo de la válvula de presión.

Set point.   The desired control point's magnitude in a control process.

Punto de ajuste.   La magnitud deseada de un punto de control en un proceso de control.

Shaded-pole motor.   An alternating current motor used for very light loads.

Motor polar en sombra.   Motor de corriente alterna utilizado en cargas sumamente livianas.

Shell and coil.   A vessel with a coil of tubing inside that is used as a heat exchanger.

Coraza y bobina.   Depósito con una bobina de tubería en su interior que se utiliza como intercambiador de calor.

Shell and tube.   A heat exchanger with straight tubes in a shell that can normally be mechanically cleaned.

Coraza y tubo.   Intercambiador de calor con tubos rectos en una coraza que por lo general puede limpiarse mecánicamente.

Short circuit.   A circuit that does not have the correct mea-surable resistance: too much current flows and will overload the conductors.

Cortocircuito.   Corriente que no tiene la resistencia medible correcta: un exceso de corriente fluye a través del circuito provocando una sobrecarga de los conductores.

Short cycle.   The term used to describe the running time (on time) of a unit when it is not running long enough.

Ciclo corto.   Término utilizado para describir el período de funcionamiento (de encendido) de una unidad cuando no funciona por un período de tiempo suficiente.

Shorted motor winding.   Part of an electric motor winding is shorted out because one part of the winding touches another part, where the insulation is worn or in some way defective.

Devanado de motor en cortocircuito.   Debido a un aisla-miento deficiente u otro defecto, una parte de los elementos del devanado en un motor eléctrico entran en contacto con otra, causando un cortocircuito.

Shroud.   A fan housing that ensures maximum airflow through the coil.

Boveda.   Alojamiento del abanico que asegura un flujo máximo de aire a través de la bobina.

Sight glass.   A clear window in a fluid line.

Mirilla para observación.   Ventana clara en un conducto de fluido.

Silica gel.   A chemical compound often used in refrigerant driers to remove moisture from the refrigerant.

Gel silíceo.   Compuesto químico utilizado a menudo en secadores de refrigerantes para remover la humedad del refrigerante.

Silver brazing.   A high-temperature (above 800°F) brazing process for bonding metals.

Soldadura con plata.   Soldadura a temperatura alta (sobre los 800°F o 430°C) para unir metales.

Sine wave.   The graph or curve used to describe the charac-teristics of alternating current and voltage.

Onda sinusoidal.   Gráfica o curva utilizada para describir las características de tensión y de corriente alterna.

Single phase.   The electrical power supplied to equipment or small motors, normally under $7\frac{1}{2}$ hp.

Monofásico.   Potencia eléctrica suministrada a equipos o motores pequeños, por lo general menor de $7\frac{1}{2}$ hp.

Single-phase hermetic motor.   A sealed motor, such as with a small compressor, that operates off single-phase power.

Motor de fase sencilla hermético.   Un motor sellado, tal como un compresor pequeño, que opera con electricidad de fase sencilla.

Single phasing.   The condition in a three-phase motor when one phase of the power supply is open.

Fasaje sencillo.   Condición en un motor trifásico cuando una fase de la fuente de alimentación está abierta.

Slab on grade.   Any concrete slab poured over excavated soil. The term generally refers to a floor made in this way.

Losa sobre pendiente.   Cualquier losa de hormigón colocada sobre suelos excavados. En general, el término se refiere a un piso fabricado de este modo.

Slinger ring.   A ring attached to the blade tips of a con-denser fan. This ring throws condensate onto the condenser coil, where it is evaporated.

Anillo tubular.   Anillo instalado en los extremos de las palas de un ventilador de condensación. Sirve para lanzar la condensación sobre el serpentín de refrigeración donde es evaporada.

Sling psychrometer.   A device with two thermometers, one a wet bulb and one a dry bulb, used for checking air condi-tions, wet-bulb and dry-bulb.

Sicrómetro con eslinga.   Dispositivo con dos termómetros, uno con una bombilla húmeda y otro con una bombilla seca, utilizados para revisar las condiciones del aire, de la tempera-tura y de la humedad.

Slip.   The difference in the rated rpm of a motor and the actual operating rpm when under a load.

Deslizamiento.   Diferencia entre las rpm nominales de un motor y las rpm de funcionamiento reales.

Slugging.   A term used to describe the condition when large amounts of liquid enter a pumping compressor cylinder.

Relleno.   Término utilizado para describir la condición donde grandes cantidades de líquido entran en el cilindro de un compresor de bombeo.

Snap-disc.   An application of the bimetal. Two different metals fastened together in the form of a disc that provides a warping condition when heated. This also provides a snap action that is beneficial in controls that start and stop cur-rent flow in electrical circuits.

Disco de acción rápida.   Aplicación del bimetal. Dos metales diferentes fijados entre sí en forma de un disco que provee un deformación al ser calentado. Esto provee también una acción rápida, ventajosa para reguladores que ponen en marcha y detienen el flujo de corriente en circuitos eléctricos.

Software.   Computer programs written to give specific instructions to computers.

Software.   Programas de computadoras escritos para darles instrucciones específicas a las computadoras.

Solar collectors.   Components of a solar system designed to collect the heat from the sun, using air, a liquid, or refriger-ant as the medium.

Colectores solares.   Componentes de un sistema solar dise-ñados para acumular el calor emitido por el sol, utilizando el aire, un líquido o un refrigerante como el medio.

Solar heat.   Heat from the sun's rays.

Calor solar.   Calor emitido por los rayos del sol.

Solar influence.   The heat that the sun imposes on a structure.

Influencia solar.   El calor que el sol impone en una estructura.

Solar radiant heat.   Solar-heated water or an antifreeze solution is piped through heating coils embedded in concrete in the floor or in plaster in ceilings or walls.

Calor de radiación solar.   El agua o líquido anticongelante calentado por energía solar se canaliza a través de serpentines de calefacción instalados en el hormigón del suelo o en el yeso de los techos o paredes.

Soldering. Fastening two base metals together by using a third, filler metal that melts at a temperature below 800°F.

Soldadura. La fijación entre sí de dos metales bases utilizando un tercer metal de relleno que se funde a una temperatura menor de 800°F (430°C).

Solderless terminals. Used to fasten stranded wire to various terminals or to connect two lengths of stranded wire together.

Terminales sin soldadura. Se usan para fijar cable trenzado a varios terminales o para unir dos pedazos de cable.

Solenoid. A coil of wire designed to carry an electrical current producing a magnetic field.

Solenoide. Bobina de alambre diseñada para conducir una corriente eléctrica generando un campo magnético.

Solid. Molecules of a solid are highly attracted to each other, forming a mass that exerts all of its weight downward.

Sólido. Las moléculas de un sólido se atraen entre sí y forman una masa que ejerce todo su peso hacia abajo.

Solid state. Pertaining to circuits where electricity passes through solid semiconductor material.

Estado sólido. Relativo a circuitos en los cuales la electricidad pasa a través de un material semiconductor sólido.

Space cooling and heating thermostat. The device used to control the temperature of a space, such as a home thermostat that controls the temperature in a home or office.

Termostato de enfriamiento o calefacción de espacio. Aparato usado para controlar la temperatura de un espacio, tal como un termostato de hogar que controla la temperatura en un hogar u oficina.

Specific gravity. The weight of a substance compared to the weight of an equal volume of water.

Gravedad específica. El peso de una sustancia comparada con el peso de un volumen igual de agua.

Specific heat. The amount of heat required to raise the temperature of 1 lb of a substance 1°F.

Calor específico. La cantidad de calor requerida para elevar la temperatura de una libra de una sustancia 1°F (–17°C).

Specific volume. The volume occupied by 1 lb of a fluid.

Volumen específico. Volumen que ocupa una libra de fluido.

Splash lubrication system. A system of furnishing lubrication to a compressor by agitating the oil.

Sistema de lubrificación por salpicadura. Método de proveerle lubrificación a un compresor agitando el aceite.

Splash method. A method of water dropping from a higher level in a cooling tower and splashing on slats with air passing through for more efficient evaporation.

Método de salpicaduras. Método dé dejar caer agua desde un nivel más alto en una torre de refrigeración y salpicándola en listones, mientras el aire pasa a través de los mismos con el propósito de lograr una evaporación más eficaz.

Split-phase motor. A motor with run and start windings.

Motor de fase separada. Motor con devandos de funcionamiento y de arranque.

Split suction. When the common suction line of a parallel compressor system has been valved in such a way as to provide for multiple temperature applications in one refrigeration package.

Succión dividida. Cuando la línea común de succión de un sistema de compresor paralelo ha sido dividida de tal manera que provee para aplicaciones de múltiples temperaturas en un empaque de refrigeración.

Split system. A refrigeration or air conditioning system that has the condensing unit remote from the indoor (evaporator) coil.

Sistema separado. Sistema de refrigeración o de acondicionamiento de aire cuya unidad de condensación se encuentra en un sitio alejado de la bobina interior del evaporador.

Spray pond. A pond with spray heads used for cooling water in water-cooled air conditioning or refrigeration systems.

Tanque de rociado. Tanque con una cabeza rociadora utilizada para enfriar el agua en sistemas de acondicionamiento de aire o de refrigeración enfriados por agua.

Spring-loaded relief valve. A fluid (refrigerant, air, water, or steam) relief valve that can function more than one time because a spring returns the valve to a seat.

Válvula de alivio de resorte. Una válvula de alivio de fluido (refrigerante, aire, agua o vapor de agua) que puede funcionar más de una vez porque el resorte regresa la válvula a su asiento.

Squirrel cage fan. A cylindrically shaped fan assembly used to move air.

Abanico con jaula de ardilla. Conjunto cilíndrico de abanico utilizado para mover el aire.

Squirrel cage rotor. Describes the construction of a motor rotor.

Rotor de jaula de ardilla. Describe la construcción del rotor de un motor.

Stamped evaporator. An evaporator that has stamped refrigerant passages in sheet steel or aluminum.

Evaporador estampado. Un evaporador que tiene pasajes para el refrigerante estampados en lata o aluminio.

Standard atmosphere or standard conditions. Air at sea level at 70°F when the atmosphere's pressure is 14.696 psia (29.92 in. Hg). Air at this condition has a volume of 13.33 ft³/lb.

Atmósfera estándar o condiciones estándares. El aire al nivel del mar a una temperatura de 70°F (15°C) cuando la presión de la atmósfera es 14.696 psia (29.92 pulgadas Hg). Bajo esta condición, el aire tiene un volumen de 13.33 ft³/lb (pies³/libras).

Standing pilot. Pilot flame that remains burning continuously.

Piloto constante. Llama piloto que se quema de manera continua.

Start capacitor. A capacitor used to help an electric motor start.

Capacitador de arranque. Capacitador utilizado para ayudar en el arranque de un motor eléctrico.

Starting relay.   An electrical relay used to disconnect the start capacitor and/or start winding in a hermetic compressor.

Relé de arranque.   Relé eléctrico utilizado para desconectar el capacitador y/o el devanado de arranque en un compresor hermético.

Starting winding.   The winding in a motor used primarily to give the motor extra starting torque.

Devanado de arranque.   Devanado en un motor utilizado principalmente para proveerle al motor mayor para el arranque.

Starved coil.   The condition in an evaporator when the metering device is not feeding enough refrigerant to the evaporator.

Bobina estrangulada.   Condición que ocurre en un evaporador cuando el dispositivo de medida no le suministra suficiente refrigerante al evaporador.

Static pressure.   The bursting pressure or outward force in a duct system.

Presión estática.   La presión de estallido o la fuerza hacia fuera en un sistema de conductos.

Stator.   The component in a motor that contains the windings: it does not turn.

Estátor.   Componente en un motor que contiene los devanados y que no gira.

Steady-state condition.   A stabilized condition of a piece of heating or cooling equipment where not much change is taking place.

Condición de régimen estable.   Condición estabilizada de un dispositivo de calefacción o de refrigeración en el cual no hay muchos cambios.

Steam.   The vapor state of water.

Vapor.   Estado de vapor del agua.

Step motor.   An electric motor that moves with very small increments or "steps," usually in either direction, and is usually controlled by a microprocessor with input and output controlling devices.

Motor a pasos.   Un motor eléctrico que se mueve en incrementos muy pequeños o "pasos", generalmente en cualquier dirección, y generalmente son controlados por un microprocesador de aparatos de control con entradas y salidas.

Stoichiometric.   The science of stoichiometry studies the relationship of oxygen and hydrogen combustion and their ability to combine to create water. The combination can only occur in set amounts under certain conditions. By measuring the amount of water formed, a technician can determine the amount of oxygen and hydrogen.

Estequiométrico.   La ciencia de la estequiometría estudia la relación entre la combustión del oxígeno y el hidrógeno, y su capacidad de combinarse para crear agua. La combinación sólo puede ocurrir en cantidades fijas y en ciertas circunstancias. Al medir la cantidad de agua que se forma, un técnico puede determinar la cantidad de oxígeno y de hidrógeno.

Strainer.   A fine-mesh device that allows fluid flow and holds back solid particles.

Colador.   Dispositivo de malla fina que permite el flujo de fluido a través de él y atrapa partículas sólidas.

Stratification.   The condition where a fluid appears in layers.

Estratificación.   Condición que ocurre cuando un fluido aparece en capas.

Stratosphere.   An atmospheric level that is located from 7 to 30 miles above the earth. Good ozone is found in the stratosphere.

Estratosfera.   Capa del atmósfera que se encuentra a una altura entre 11 a 48 kilómetros encima de la tierra. Contiene una buena capa de ozono.

Stress crack.   A crack in piping or other component caused by age or abnormal conditions such as vibration.

Grieta por tensión.   Grieta que aparece en una tubería u otro componente ocasionada por envejecimiento o condiciones anormales, como por ejemplo vibración.

Subbase.   The part of a space temperature thermostat that is mounted on the wall and to which the interconnecting wiring is attached.

Subbase.   Pieza de un termóstato que mide la temperatura de un espacio que se monta sobre la pared y a la que se fijan los conductores eléctricos interconectados.

Subcooled.   The temperature of a liquid when it is cooled below its condensing temperature.

Subenfriado.   La temperatura de un líquido cuando se enfría a una temperatura menor que su temperatura de condensación.

Sublimation.   When a substance changes from the solid state to the vapor state without going through the liquid state.

Sublimación.   Cuando una sustancia cambia de sólido a vapor sin covertirse primero en líquido.

Suction gas.   The refrigerant vapor in an operating refrigeration system found in the tubing from the evaporator to the compressor and in the compressor shell.

Gas de aspiración.   El vapor del refrigerante en un sistema de refrigeración en funcionamiento presente en la tubería que va del evaporador al compresor y en la coraza del compresor.

Suction line.   The pipe that carries the heat-laden refrigerant gas from the evaporator to the compressor.

Conducto de aspiración.   Tubo que conduce el gas de refrigerante lleno de calor del evaporador al compresor.

Suction-line accumulator.   A reservoir in a refrigeration system suction line that protects the compressor from liquid floodback.

Acumulador de la línea de succión.   Un estanque en la línea de succión de un sistema de refrigeración que protege al compresor de una inundación de líquido.

Suction pressure.   The pressure created by the boiling refrigerant on the evaporator or low-pressure side of the system.

Presión de succión.   La presión creada por el refrigerante hirviendo en el evaporador o en el lado de baja presión del sistema.

Suction service valve.   A manually operated valve with front and back seats located at the compressor.

Válvula de aspiración para servicio.   Válvula accionada manualmente que tiene asientos delanteros y traseros ubicados en el compresor.

Suction valve.   The valve at the compressor cylinder that allows refrigerant from the evaporator to enter the compressor cylinder and prevents it from being pumped back out to the suction line.

Válvula de succión.   La válvula en el cilindro de un compresor que permite que el refrigerante del evaporador entre al cilindro del compresor y evita que se bombee nuevamente a la línea de succión.

Suction valve lift unloading.   The suction valve in a reciprocating compressor cylinder is lifted, causing that cylinder to stop pumping.

Descarga por levantamiento de la válvula de aspiración.   La válvula de aspiración en el cilindro de un compresor alternativo se levanta, provocando que el cilindro deje de bombear.

Sulfur dioxide.   A combustion pollutant that causes eye, nose, and respiratory tract irritation and possibly breathing problems.

Dióxido de azufre.   Un contaminante por combustión que causa irritación en los ojos, la nariz y las vías respiratorias, y posiblemente problemas respiratorios.

Sump.   A reservoir at the bottom of a cooling tower to collect the water that has passed through the tower.

Sumidero.   Tanque que se encuentra en el fondo de una torre de refrigeración para acumular el agua que ha pasado a través de la torre.

Superheat.   The temperature of vapor refrigerant above its saturation change-of-state temperature.

Sobrecalor.   Temperatura del refrigerante de vapor mayor que su temperatura de cambio de estado de saturación.

Surge.   When the head pressure becomes too great or the evaporator pressure too low, refrigerant will flow from the high- to the low-pressure side of a centrifugal compressor system, making a loud sound.

Movimiento repentino.   Cuando la presión en la cabeza aumenta demasiado o la presión en el evaporador es demasiado baja, el refrigerante fluye del lado de alta presión al lado de baja presión de un sistema de compresor centrífugo. Este movimiento produce un sonido fuerte.

Swaged joint.   The joining of two pieces of copper tubing by expanding or stretching the end of one piece of tubing to fit over the other piece.

Junta estampada.   La conexión de dos piezas de tubería de cobre dilatando o alargando el extremo de una pieza de tubería para ajustarla sobre otra.

Swaging tool.   A tool used to enlarge a piece of tubing for a solder or braze connection.

Herramienta de estampado.   Herramienta utilizada para agrandar una pieza de tubería a utilizarse en una conexión soldada o broncesoldada.

Swamp cooler.   A slang term used to describe an evaporative cooler.

Nevera pantanosa.   Término del argot utilizado para describir una nevera de evaporación.

Sweating.   A word used to describe moisture collection on a line or coil that is operating below the dew point temperature of the air.

Exudación.   Término utilizado para describir la acumulación de humedad en un conducto o una bobina que está funcionando a una temperatura menor que la del punto de rocío de aire.

System charge.   The refrigerant in a system, both liquid and vapor. The correct charge is a balance where the system will give the most efficiency.

Carga del sistema.   El refrigerante en un sistema, tanto líquido y vapor. La carga correcta es un balance donde el sistema dará la mayor eficiencia.

System lag.   The temperature drop of the controlled space below the set point of the thermostat.

Retardo del sistema.   Caída de temperatura de un espacio controlado, por debajo del nivel programado en el termostato.

Tank.   A closed vessel used to contain a fluid.

Tanque.   Depósito cerrado utilizado para contener un fluido.

Tap.   A tool used to cut internal threads in a fastener or fitting.

Macho de roscar.   Herramienta utilizada para cortar roscas internas en un aparto fijador o en un accesorio.

Technician.   A person who performs maintenance, service, testing, or repair to air conditioning or refrigeration equipment. Note: The EPA defines this person as someone who could reasonably be expected to release CFCs or HCFCs into the atmosphere.

Técnicos.   Una persona que lleva a cabo mantenimiento, servicio o reparaciones a equipos de aire acondicionado o refrigeración. Nota: Esta persona, según defunido por la EPA, es una persona del cual razonablemente se estaría esparado que libere CFC (clorofluorocarbonos) a la atmósfera.

Temperature.   A word used to describe the level of heat or molecular activity, expressed in Fahrenheit, Rankine, Celsius, or Kelvin units.

Temperatura.   Término utilizado para describir el nivel de calor o actividad molecular, expresado en unidades Fahrenheit, Rankine, Celsio o Kelvin.

Temperature-measuring instruments.   Devices that accurately measure the level of temperature.

Instrumentos que miden temperatura.   Aparatos que miden el nivel de la temperatura con precisión.

Temperature reference points.   Various points that may be used to calibrate a temperature-measuring device, such as boiling or freezing water.

Puntos de referencia de temperatura.   Varios puntos que pueden usarse para calibrar un aparato que mide temperatura, tales como agua congelada o hirvienda.

Temperature-sensing elements.   Various devices in a system that are used to detect temperature.

Elementos que detectan temperatura.   Varios aparatos en un sistema que se usan para detectar temperatura.

Temperature swing.   The temperature difference between the low and high temperatures of the controlled space.

Oscilación de temperatura. Diferencia existente entre las temperaturas altas y bajas de un espacio controlado.

Testing, Adjusting and Balancing Bureau (TABB). A certification bureau for individuals involved in working with ventilation.

Agencia de Prueba, Ajuste y Balanceo (TABB en inglés). Agencia de certificación para los individuos involucrados en el trabajo de ventilación.

Test light. A lightbulb arrangement used to prove the presence of electrical power in a circuit.

Luz de prueba. Arreglo de bombillas utilizado para probar la presencia de fuerza eléctrica en un circuito.

Therm. Quantity of heat, 100,000 BTU.

Therm. Cantidad de calor, mil unidades térmicas inglesas.

Thermistor. A semiconductor electronic device that changes resistance with a change in temperature.

Termistor. Dispositivo eléctrico semiconductor que cambia su resistencia cuando se produce un cambio en temperatura.

Thermocouple. A device made of two unlike metals that generates electricity when there is a difference in temperature from one end to the other. Thermocouples have a hot and cold junction.

Termopar. Dispositivo hecho de dos metales distintos que genera electricidad cuando hay una diferencia en temperatura de un extremo al otro. Los termopares tienen un empalme caliente y uno frío.

Thermometer. An instrument used to detect differences in the level of heat.

Termómetro. Instrumento utilizado para detectar diferencias en el nivel de calor.

Thermopile. A group of thermocouples connected in series to increase voltage output.

Pila termoeléctrica. Grupo de termopares conectados en serie para aumentar la salida de tensión.

Thermostat. A device that senses temperature change and changes some dimension or condition within to control an operating device.

Termostato. Dispositivo que advierte un cambio en temperatura y cambia alguna dimensión o condición dentro de sí para regular un dispositivo en funcionamiento.

Thermostatic expansion valve (TXV). A valve used in refrigeration systems to control the superheat in an evaporator by metering the correct refrigerant flow to the evaporator.

Válvula de gobierno termostático para expansión. Válvula utilizada en sistemas de refrigeración para regular el sobrecalor en un evaporador midiendo el flujo correcto de refrigerante al evaporador.

Three-phase power. A type of power supply usually used for operating heavy loads. It consists of three sine waves that are out of phase by 120° with each other.

Potencia trifásica. Tipo de fuente de alimentación normalmente utilizada en el funcionamiento de cargas pesadas. Consiste de tres ondas sinusoidales que no están en fase la una con la otra por 120°.

Throttling. Creating a planned or regulated restriction in a fluid line for the purpose of controlling fluid flow.

Estrangulamiento. Que ocasiona una restricción intencional o programada en un conducto de fluido, a fin de controlar el flujo del fluido.

Thrust surface. A term that usually applies to bearings that have a pushing pressure to the side and that therefore need an additional surface to absorb the push. Most motor shifts cradle in their bearings because they operate in a horizontal mode, like holding a stick in the palm of your hand. When a shaft is turned to the vertical mode, a thrust surface must support the weight of the shaft along with the load the shaft may impose on the thrust surface. The action of a vertical fan shaft that pushes air up is actually pushing the shaft downward.

Superficie de empuje. Un término que generalmente se aplica a cojinetes que sostienen una presión de empuje a un lado y que necesitan una superficie adicional para absorber este empuje. La mayoría de los ejes de motor están al abrigo en sus cojinetes, porque funcionen en una modalidad horizontal, como cuando uno sostiene una vara en la mano. Cuando un eje está sintonizado a la modalidad vertical, una superficie de empuje debe sostener el peso del eje junto con la carga que el eje puede imponer sobre la superficie de empuje. La moción de un eje vertical de un ventilador que empuja aire hacia arriba está en realidad empujando el eje hacia abajo.

Time delay. A device that prevents a component from starting for a prescribed time. For example, many systems start the fans and use a time-delay relay to start the compressor at a later time to prevent too much in-rush current.

Retraso de tiempo. Un aparato que evita que un componente se encienda por un período prescrito de tiempo. Por ejemplo, muchos sistemas encienden los ventiladores y usando un relé de retraso, encienden el compresor un tiempo después para evitar mucha corriente interna.

Timers. Clock-operated devices used to time various sequences of events in circuits.

Temporizadores. Dispositivos accionados por un reloj utilizados para medir el tiempo de varias secuencias de eventos en circuitos.

Toggle bolt. Provides a secure anchoring in hollow tiles, building block, plaster over lath, and gypsum board. The toggle folds and can be inserted through a hole, where it opens.

Tornillo de fiador. Proveen un anclaje seguro en losetas huecas, bloques de construcción, yeso sobre listón y tablón de yeso. El fiador se dobla y puede insertarse a través de un roto y después se abre.

Ton of refrigeration. The amount of heat required to melt a ton (2,000 lb) of ice at 32°F in 24 hours, 288,000 BTU/24 h, 12,000 BTU/h, or 200 BTU/min.

Tonelada de refrigeración. Cantidad de calor necesario para fundir una tonelada (2000 libras) de hielo a 32°F (0°C) en 24 horas, 288,000 BTU/24 h, 12,000 BTU/h o 200 BTU/min.

Torque. The twisting force often applied to the starting power of a motor.

Par de torsión. Fuerza de torsión aplicada con frecuencia a la fuerza de arranque de un motor.

Torque wrench.   A wrench used to apply a prescribed amount of torque or tightening to a connector.

Llave de torsión.   Llave utilizada para aplicar una cantidad específica de torsión o de apriete a un conector.

Total equivalent warming impact (TEWI).   A global warming index that takes into account both the direct effects of chemicals emitted into the atmosphere and the indirect effects caused by system inefficiencies.

Impacto de calentamiento equivalente total (TEWI en inglés).   Índice de calentamiento de la tierra que tiene en cuenta los efectos directos de los productos químicos emitidos en la atmósfera, y los efectos indirectos causados por la ineficacia de un sistema.

Total heat.   The total amount of sensible heat and latent heat contained in a substance from a reference point.

Calor total.   Cantidad total de calor sensible o de calor latente presente en una sustancia desde un punto de referencia.

Total pressure.   The sum of the velocity and the static pressure in an air duct system.

Presión total.   La suma de la velocidad y la presión estática en un sistema de conducto de aire.

Transformer.   A coil of wire wrapped around an iron core that induces a current to another coil of wire wrapped around the same iron core. Note: A transformer can have an air core.

Transformador.   Bobina de alambre devanado alrededor de un núcleo de hierro que induce una corriente a otra bobina de alambre devanado alrededor del mismo núcleo de hierro. Nota: un transformador puede tener un núcleo de aire.

Transistor.   A semiconductor often used as a switch or amplifier.

Transistor.   Semiconductor que suele utilizarse como conmutador o amplificador.

Traversing.   A method of measuring velocity with a pitot tube in which many readings are taken and then averaged.

Desplazamiento transversal.   Método utilizado para medir la velocidad con un tubo Pitot en el que se toman muchas lecturas que luego se promedian.

TRIAC.   A semiconductor switching device.

TRIAC.   Dispositivo de conmutación para semiconductores.

Tube-in-tube coil.   A coil used for heat transfer that has a pipe in a pipe and is fastened together so that the outer tube becomes one circuit and the inner tube another.

Bobina de tubo en tubo.   Bobina utilizada en la transferencia de calor que tiene un tubo dentro de otro y se sujeta de manera que el tubo exterior se convierte en un circuito y el tubo interior en otro circuito.

Tubing.   Pipe with a thin wall used to carry fluids.

Tubería.   Tubo que tiene una pared delgada utilizada para conducir fluidos.

Two-speed compressor motor.   Can be a four-pole motor that can be connected as a two-pole motor for high speed (3,450 rpm) and connected as a four-pole motor for running at 1,725 rpm for low speed. This is accomplished with relays outside the compressor.

Motor de compresor de dos velocidades.   Puede ser un motor de 4 polos que puede conectarse como un motor de 2 polos para velocidades altas (3,450 revoluciones por minuto) y conectarse como un motor de 4 polos para correr a 1,725 revoluciones por minuto para velocidades bajas. Esto se hace con relés fuera del compresor.

Two-temperature valve.   A valve used in systems with multiple evaporators to control the evaporator pressures and maintain different temperatures in each evaporator. Sometimes called a hold-back valve.

Válvula de dos temperaturas.   Válvula utilizada en sistemas con evaporadores múltiples para regular las presiones de los evaporadores y mantener temperaturas diferentes en cada uno de ellos. Conocida también como válvula de retención.

Ultrasound leak detector.   Detectors that use sound from escaping refrigerant to detect leaks.

Detector de escapes ultrasónico.   Detectores que usan sonido del refrigerante que está escapando para detectar escapes.

Ultraviolet.   Light waves that can only be seen under a special lamp.

Ultravioleta.   Ondas de luz que pueden observarse solamente utilizando una lámpara especial.

Ultraviolet light.   Light frequency between 200 and 400 nanometers.

Luz ultravioleta.   Luz con frecuencia entre 200 y 400 nanómetros.

Upflow furnace.   This furnace takes in air from the bottom or from sides near the bottom and discharges hot air out the top.

Horno de flujo ascendente.   La entrada del aire en este tipo de horno se hace desde abajo o en los laterales, cerca del suelo. La evacuación se realiza en la parte superior.

Urethane foam.   A foam that can be applied between two walls for insulation.

Espuma de uretano.   Espuma que puede aplicarse entre dos paredes para crear un aislamiento.

U-tube mercury manometer.   A U-tube containing mercury, which indicates the level of vacuum while evacuating a refrigeration system.

Manómetro de mercurio de tubo en U.   Tubo en U que contiene mercurio y que indica el nivel del vacío mientras vacía un sistema de refrigeración.

U-tube water manometer.   Indicates natural gas and propane gas pressures. It is usually calibrated in inches of water.

Manómetro de agua de tubo en U.   Indica las presiones del gas natural y del propano. Se calibra normalmente en pulgadas de agua.

Vacuum.   The pressure range between the earth's atmospheric pressure and no pressure, normally expressed in inches of mercury (in. Hg) vacuum.

Vacío.   Margen de presión entre la presión de la atmósfera de la tierra y cero presión, por lo general expresado en pulgadas de mercurio (pulgadas Hg) en vacío.

Vacuum gauge.   An instrument that measures the vacuum when evacuating a refrigeration, air conditioning, or heat pump system.

Medidor del vacío.   Instrumento que se utiliza para medir el vacío al vaciar un sistema de refrigeración, de aire acondicionado, o de bomba de calor.

Vacuum pump.   A pump used to remove some fluids such as air and moisture from a system at a pressure below the earth's atmosphere.

Bomba de vacío.   Bomba utilizada para remover algunos fluidos, como por ejemplo aire y humedad de un sistema a una presión menor que la de la atmósfera de la tierra.

Valve.   A device used to control fluid flow.

Válvula.   Dispositivo utilizado para regular el flujo de fluido.

Valve plate.   A plate of steel bolted between the head and the body of a compressor that contains the suction and discharge reed or flapper valves.

Placa de válvula.   Placa de acero empernado entre la cabeza y el cuerpo de un compresor que contiene la lámina de aspiración y de descarga o las chapaletas.

Valve seat.   That part of a valve that is usually stationary. The movable part comes in contact with the valve seat to stop the flow of fluids.

Asiento de la válvula.   Pieza de una válvula que es normalmente fija. La pieza móvil entra en contacto con el asiento de la válvula para detener el flujo de fluidos.

Valve stem depressor.   A service tool used to access pressure at a Schrader valve connection.

Depresor de vástago de válvula.   Una herramienta de servicio que se usa para acceder la presión en una conexión de válvula Schrader.

Vapor.   The gaseous state of a substance.

Vapor.   Estado gaseoso de una sustancia.

Vapor barrier.   A thin film used in construction to keep moisture from migrating through building materials.

Película impermeable.   Película delgada utilizada en construcciones para evitar que la humedad penetre a través de los materiales de construcción.

Vapor charge bulb.   A charge in a thermostatic expansion valve bulb that boils to a complete vapor. When this point is reached, an increase in temperature will not produce an increase in pressure.

Válvula para la carga de vapor.   Carga en la bombilla de una válvula de expansión termostática que hierve a un vapor completo. Al llegar a este punto, un aumento en temperatura no produce un aumento en presión.

Vaporization.   The changing of a liquid to a gas or vapor.

Vaporización.   Cuando un líquido se convierte en gas o vapor.

Vapor lock.   A condition where vapor is trapped in a liquid line and impedes liquid flow.

Bolsa de vapor.   Condición que ocurre cuando el vapor queda atrapado en el conducto de líquido e impide el flujo de líquido.

Vapor pressure.   The pressure exerted on top of a saturated liquid.

Presión del vapor.   Presión que se ejerce en la superficie de un líquido saturado.

Vapor pump.   Another term for compressor.

Bomba de vapor.   Otro término para compresor.

Vapor refrigerant charging.   Adding refrigerant to a system by allowing vapor to move out of the vapor space of a refrigerant cylinder and into the low-pressure side of the refrigeration system.

Carga del refrigerante de vapor.   Agregarle refrigerante a un sistema permitiendo que el vapor salga del espacio de vapor de un cilindro de refrigerante y que entre en el lado de baja presión del sistema de refrigeración.

Variable-frequency drive (VFD).   An electrical device that varies the frequency (hertz) for the purpose of providing a variable speed.

Propulsión de frecuencia variable (VFD en inglés).   Un aparato eléctrico que varía la frecuencia (hertz) con el propósito de proveer velocidad variable.

Variable pitch pulley.   A pulley whose diameter can be adjusted.

Polea de paso variable.   Polea cuyo diámetro puede ajustarse.

Variable resistor.   A type of resistor where the resistance can be varied.

Resistor variable.   Tipo de resistor donde la resistencia puede variarse.

Variable-speed motor.   A motor that can be controlled, with an electronic system, to operate at more than one speed.

Motor de velocidad variable.   Un motor que puede controlarse, con un sistema electrónico, para operar a más de una velocidad.

V belt.   A belt that has a V-shaped contact surface and is used to drive compressors, fans, or pumps.

Correa en V.   Correa que tiene una superficie de contacto en forma de V y se utiliza para accionar compresores, abanicos o bombas.

Velocity.   The speed at which a substance passes a point.

Velocidad.   Rapidez a la que una sustancia sobrepasa un punto.

Velocity meter.   A meter used to detect the velocity of fluids, air, or water.

Velocimetro.   Instrumento utilizado para medir la velocidad de fluidos, aire o agua.

Velometer.   An instrument used to measure the air velocity in a duct system.

Velómetro.   Instrumento utilizado para medir la velocidad del aire en un conducto.

Vent-free.   Certain gas stoves or gas fireplaces are not required to be vented; therefore, they are called "vent-free."

Sin ventilación.   Ciertas estufas de gas y chimeneas de gas no requieren una ventilación, por lo que se conocen como estufas "sin ventilación".

Ventilation.   The process of supplying and removing air by natural or mechanical means to and from a particular space.

Ventilación.   Proceso de suministrar y evacuar el aire de un espacio determinado, utilizando procesos naturales o mecánicos.

Venting products of combustion.   Venting flue gases that are generated from the burning process of fossil fuels.

Descargar productos de combustión.   Descargar los gases de la chimenea que se generan del proceso de combustión de los combustibles fósiles.

Vertical feet of head.   A unit of measure that specifies the total pumping head in feet of liquid.

Pies verticales de presión.   Unidad de medida que indica la presión de bombeo total en pies de líquido.

Voltage.   The potential electrical difference for electron flow from one line to another in an electrical circuit.

Voltaje.   Diferencia de potencial eléctrico del flujo de electrones de un conducto a otro en un circuito eléctrico.

Voltage feedback.   Voltage potential that travels through a power-consuming device when it is not energized.

Retroalimentación de voltaje.   El voltaje que viaja a través de un aparato de consumo de electricidad cuando no está energizado.

Volt-ohm-milliammeter (VOM).   A multimeter that measures voltage, resistance, and current in milliamperes.

Voltio-ohmio-miliamperimetro (VOM en inglés).   Multímetro que mide tensión, resistencia y corriente en miliamperios.

Volumetric efficiency.   The pumping efficiency of a compressor or vacuum pump that describes the pumping capacity in relationship to the actual volume of the pump.

Rendimiento volumétrico.   Rendimiento de bombeo de un compresor o de una bomba de vacío que describe la capacidad de bombeo con relación al volumen real de la bomba.

Vortexing.   A whirlpool action in the sump of a cooling tower.

Acción de vórtice.   Torbellino en el sumidero de una torre de refrigeración.

Walk-in cooler.   A large refrigerated space used for storage of refrigerated products.

Nevera con acceso al interior.   Espacio refrigerado grande utilizado para almacenar productos refrigerados.

Water column (w.c.).   The pressure it takes to push a column of water up vertically. One inch of water column is the amount of pressure it would take to push a column of water in a tube up one inch.

Columna de agua (w.c. en inglés).   Presión necesaria para levantar una columna de agua verticalmente. Una pulgada de columna de agua es la cantidad de presión necesaria para levantar una columna de agua a una distancia de una pulgada en un tubo.

Water-cooled condenser.   A condenser used to reject heat from a refrigeration system into water.

Condensador enfriado por agua.   Condensador utilizado para dirigir el calor de un sistema de refrigeración al agua.

Water-regulating valve.   An operating control regulating the flow of water.

Válvula reguladora de agua.   Regulador de mando que controla el flujo de agua.

Watt.   A unit of power applied to electron flow. One watt equals 3.414 BTU.

Watio.   Unidad de potencia eléctrica aplicada al flujo de electrones. Un watio equivale a 3.414 BTU.

Watt-hour.   The unit of power that takes into consideration the time of consumption. It is the equivalent of a 1-watt bulb burning for 1 hour.

Watio hora.   Unidad de potencia eléctrica que toma en cuenta la duración de consumo. Es el equivalente de una bombilla de 1 watio encendida por espacio de una hora.

WB.   See Wet-bulb temperature.

TH (WB, por su sigla en inglés).   Véase Temperatura de una bombilla húmeda.

Weep holes.   Holes that connect each cell in an ice machine's cell-type evaporator that allow air entering from the edges of the ice to travel along the entire ice slab to relieve the suction force and allow the ice to fall off of the evaporator.

Aberturas de exudación.   Aberturas que conectan cada célula en el evaporador de tipo celular de una hielera, que permiten que el aire que entra de los bordes del hielo viaja a lo largo de todo el pedazo de hielo para aliviar la fuerza de succión y permitir que el hielo se caiga del evaporador.

Welded hermetic compressor.   A compressor that is completely sealed by welding, versus a semihermetic compressor that is sealed by bolts and flanges.

Compresor hermético soldado.   Un compresor que está completamente sellado por soldadura, contrario a un compresor semihermético que está sellado con tornillos y bridas.

Wet-bulb temperature.   A wet-bulb temperature of air is used to evaluate the humidity in the air. It is obtained with a wet thermometer bulb to record the evaporation rate with an airstream passing over the bulb to help in evaporation.

Temperatura de una bombilla húmeda.   La temperatura de una bombilla húmeda se utiliza para evaluar la humedad presente en el aire. Se obtiene con la bombilla húmeda de un termómetro para registrar el margen de evaporación con un flujo de aire circulando sobre la bombilla para ayudar en evaporar el agua.

Wet heat.   A heating system using steam or hot water as the heating medium.

Calor húmedo.   Sistema de calentamiento que utiliza vapor o agua caliente como medio de calentamiento.

Winding thermostat.   A safety device used in electric motor windings to detect over-temperature conditions.

Termostato de bobina.   Un aparato de seguridad usado en un motor eléctrico para detectar condiciones de exceso de temperatura.

Window unit.   An air conditioner installed in a window that rejects the heat outside the structure.

Acondicionador de aire para la ventana.   Acondicionador de aire instalado en una ventana que desvía el calor proveniente del exterior de la estructura.

Wire connectors (screw-on).   Used to connect two or more wires together.

Conectores de cables (de rosca).   Se usan para conectar dos o más cables.

Work.   A force moving an object in the direction of the force. Work = Force × Distance.

Trabajo. Fuerza que mueve un objeto en la dirección de la fuerza. Trabajo = Fuerza × Distancia.

Wye-Delta. A configuration of motor windings that form the shape of a triangle (called a Delta) and are electrically changed to form the letter Y (called a Wye or "Y"). The Wye-Delta starter starts the motor as a Delta-configured three-phase motor and then operates the motor as a Wye-configured three-phase motor.

Estrella-triángulo. Configuración de bobinados de motor que forman un triángulo (llamado delta) y se cambian eléctricamente para formar la letra Y (llamada *wye* en inglés o "Y"). Un reóstato de arranque estrella-triángulo arranca el motor como un motor de tres fases con configuración triángulo y, luego, hace funcionar el motor como un motor de tres fases con configuración de estrella.

Wye transformer connection. Typically furnishes 208 and 115 V to a customer.

Conexión de transformador Wye. Típicamente provee 208 y 115 voltios a los consumidores.

Zeotropic blend. Two or more refrigerants mixed together that will have a range of boiling and/or condensing points for each system pressure. Noticeable fractionation and temperature glide will occur.

Mezcla zeotrópica. Mezcla de dos o más refrigerantes que tiene un rango de ebullición y/o punto de condensación para cada presión en el sistema. Se produce una fraccionación y una variación de temperatura notables.

Zone damper. A zone control system for air delivery systems that controls air volume to individual rooms or groups of rooms within a zone.

Regulador de tiro de zona. Sistema de control de zona para sistemas de circulación de aire que controla el volumen de aire que va a las habitaciones individuales o los grupos de habitaciones en una zona.

Zone valve. Zone control valves are thermostatically controlled valves that control water flow in various zones in a hydronic heating system.

Válvula de sector. Se trata de válvulas controladas termostáticamente, que controlan el flujo del agua en varios sectores en un sistema de calefacción hidrónico.

Index

Italic page numbers indicate material in tables or figures.